Tracking the Transition from Fossil Fuels to Clean Energy

RIVER PUBLISHERS SERIES IN ENERGY MANAGEMENT

Series Editor

MICHELE ALBANO
Aalborg University,
Denmark

The "River Publishers Series in Energy Management" is a series of comprehensive academic and professional books focussing on management theory and applications for energy related industries and facilities. Books published in the series serve to provide discussion and exchange information on management strategies, techniques, methodologies and applications, with a focus on the energy industry.

Topics include management systems, handbooks for facility management, safety, security, industrial strategies, maintenance and financing, impacting organizational communications, processes and work practices. Content is also featured for energy resilient and high-performance buildings.

The main aim of this series is to serve as a useful reference for academics, researchers, managers, engineers, and other professionals in related matters with energy management practices.

Books published in the series include research monographs, edited volumes, handbooks and textbooks. The books provide professionals, researchers, educators, and advanced students in the field with an invaluable insight into the latest research and developments.

Topics covered in the series include, but are not limited to:

- Facility management;
- Safety and security;
- Management systems and solutions;
- Industrial energy strategies;
- Financing and costs;
- Energy resilient buildings;
- Green buildings management.

For a list of other books in this series, visit www.riverpublishers.com

Tracking the Transition from Fossil Fuels to Clean Energy

Editor

Stephen A. Roosa

River Publishers

Routledge
Taylor & Francis Group
NEW YORK AND LONDON

Published 2026 by River Publishers
River Publishers
Broagervej 10, 9260 Gistrup, Denmark
www.riverpublishers.com

Distributed exclusively by Routledge
605 Third Avenue, New York, NY 10017, USA
4 Park Square, Milton Park, Abingdon, Oxon OX14 4RN

Tracking the Transition from Fossil Fuels to Clean Energy / Stephen A. Roosa.

Routledge is an imprint of the Taylor & Francis Group, an informa business

ISBN 978-87-438-1129-9 (hardback)
ISBN 978-87-438-1135-0 (paperback)
ISBN 978-87-438-1134-3 (online)
ISBN 978-8-743-81131-2 (ebook master)

DOI: 10.1201/9788743811312

Dedication

This book is dedicated to my three wonderful children, Sarah, Daniel, and Maryelle. They have grown into caring and responsible adults. I know that during their lifetimes, they and others like them will continue to work to help ensure the sustainability of our planet. After all, we must focus our efforts on creating a better world for future generations.

Contents

Foreword

The transition to clean energy is one of the most important and urgent challenges we face today. As climate change continues to impact communities worldwide, the need to reduce greenhouse gas emissions has become a shared responsibility. The implementation of more sustainable energy systems is no longer a future ambition—it is a present-day necessity that calls for innovation, cooperation, and strong leadership.

This book, *Tracking the Transition from Fossil Fuels to Clean Energy*, offers a clear and comprehensive guide to understanding how energy systems are evolving. It is written for professionals in the energy industry, policymakers, researchers, educators, and advocates who are working toward sustainable solutions. Whether you are shaping policy, building infrastructure, or exploring new technologies, this book provides the knowledge and tools to support your efforts.

Each chapter examines the systems and strategies that drive the energy transition, as well as the increasing electricity demands for air conditioning, electric vehicles, and data centers. It will allow readers to gain insight into emerging clean energy technologies, global policy frameworks, and the modernization of electric grids. It examines economic approaches that promote energy efficiency and empower consumers, as well as sustainability models that consider the entire lifecycle of energy production and use.

Through case studies, modeling techniques, and practical examples, this book bridges the gap between theory and real-world application. It encourages readers to think in systems methodologies and to understand the broader context of energy transformation, ultimately helping professionals make informed decisions and contribute meaningfully to the energy transition.

This book is more than just a reference; it serves as a foundation for action. Preparing readers to engage with the complex realities of energy change and equipping them to lead with clarity and purpose. The path toward decarbonization is more than a technical endeavor; it is also an ethical, economic, and deeply human one that requires a commitment to fairness, resilience, and creativity.

President Barack Obama once said, "We're the first generation to feel the impacts of climate change, and the last generation that can do something about it." This book embraces that challenge and invites readers to do the same.

Let this work inspire you to think with courage, collaborate with others, and act with conviction because the future of energy is being shaped right now. With the insights offered in these pages, you can be a part of a story of progress, possibility, and hope.

Bill Kent, CEM
Executive Director
Association of Energy Engineers
Atlanta, Georgia

Preface

"Decarbonization"… take a few quiet moments and consider its consequences. Rarely has one word had such far-reaching implications, fomented such controversy, and yet offered such promise. The energy transition from using mostly fossil fuels to clean energy resources is upon us. Energy transitions are decades-long events. Literally powerful, they are also revolutionary, disruptive, and costly. These are international trends, and the enormity and breadth of their worldwide impacts can be surprising.

Consider the implications of the pathways that advance sustainability and the energy transition we are experiencing. Think about how you can make changes that lead to reductions in greenhouse gas emissions. The decisions you make, and those that we make collectively, are impactful and provide hope for generations to come. This transition will not be easily accomplished. Solutions are needed. There will be trade-offs and sacrifices. The sooner we act, the less climate chaos we will endure. Despite the enormity of the challenges we face, I believe that our concerted actions will be effective and that our efforts to advance decarbonization will yield success.

Acknowledgments

During my career, I have been blessed with many opportunities to learn and share experiences with friends and associates. All have helped me understand how important it is to offer solutions to overarching problems, particularly those that are the most difficult to resolve. I include our efforts to reduce greenhouse gas emissions among them. In this endeavor, I have studied the works of those who support efforts toward successfully achieving the transition to clean energy. This book compiles some of the directions that these initiatives have taken.

It is difficult to acknowledge all those who have helped in this pursuit; there have been many, and some are found in this book's chapter references. I thank Al Thumann, Barney Capehart, Victor Ottaviano, Bill Payne, and Shirley Hansen, all who have passed on, for their insights, support, and friendship over a vital period of my growth as an energy professional. I miss them all and am in awe of the contributions they made. Others who have helped in the past with various publications and supporting initiatives include Wayne C. Turner, Jerry A. Taylor, Bill Kent, Samer Zaweydeh, Fred Hauber, Joy Maugans, Eric Woodroof, John Whitney, Laurie Wiegand-Jackson, Arun Jhaveri, John Gilderbloom, and Steven Parker. All have supported my career or assisted with my journal articles and books.

I also thank everyone at the Association of Energy Engineers (AEE), Atlanta, Georgia, and River Publishers, LLC, Aalborg, Denmark, who supported this project since its onset. Articles from issues of the *International Journal of Strategic Energy and Environmental Planning* and the *Alternate Energy and Distributed Generation Journal*, many peer-reviewed, have been updated and published herein with the permission of both the AEE and the authors as chapters in this book.

Now in the twilight of my professional endeavors, it is thrilling to make yet another contribution, that of sharing my thoughts and, more importantly, the perspectives of my colleagues. My sincere thanks to all the contributing authors. Their contributions helped define the policies, programs, and technological applications that define the relevance of the drive toward decarbonization, which underpins the present energy transition. Without their support, this book would lack its robust and varied perspectives. I also thank the students and professionals who have taken or supported my topical workshops and seminars in sustainable development, renewable energy, and microgrid development. It was in those seminars that many of the various perspectives about the energy transition evolved, were debated, and came to fruition.

List of Contributors

Kritika Bhandari is a graduate student in industrial engineering and management at Oklahoma State University (OSU). She received her Bachelor in Engineering degree in mechanical engineering from Tribhuvan University in Nepal in 2022. Since joining the OSU Industrial Assessment Center (IAC) in 2023, she has contributed to 13 industrial assessments and serves as a lead student for the OSU IAC. Kritika's interests include energy management and supply chain management. She is passionate about finding innovative solutions to optimize energy use and streamline processes, contributing to more sustainable and efficient operations.

Satish C. Boregowda received his Ph.D. in mechanical engineering from Old Dominion University. He is a senior lecturer at the School of Mechanical Engineering, Purdue University. His research and teaching interests include energy and thermal-fluid sciences, bio-inspired design, and energy conversion. His research involves the integration of thermodynamics and optimization techniques to assess the complexity of design at the interface of energy, environment, and economics. Prior to this current position, Dr. Boregowda worked at the NASA Langley Research Center, Sandia National Laboratories, the University of Chicago, Purdue Krannert School of Management, and the University of Illinois.

Suresh Chand is the Louis A. Weil Jr., Chair of Management and Professor of Operations and Supply Chain Management at the Krannert School of Management, Purdue University. His research and teaching interests include the analysis and improvement of production processes on dimensions of cost, quality, time, and flexibility. His research focuses on capacity and production planning to improve volume flexibility and develop supply chain models that match supply with demand. Professor Chand served as the associate dean of graduate programs from 2013 to 2016.

Shannon Dancler has worked in the power systems industry for 3.5 years. She is affiliated with Ulteig. She has a background in physics and electrical engineering with an emphasis on power systems and renewable technologies. She has gained experience in energy market simulations, capacity expansion analysis, steady-state transmission studies for both the near-term and long-term planning horizon, and the interconnection processes of various ISOs for utility-scale generation projects.

Ahmed Ghazal, CEM, has more than 10 years of experience in industrial utilities and facilities management, working in the pharmaceutical industry. He has also worked with Veolia and CBRE companies at different facilities for various energy management projects. This has given him the desire and initiative to support the great need for energy management in all aspects of industrial applications.

Adam Gribb is a chemical engineer with over 30 years of experience spanning operations, consulting, and compliance. His career has included leadership roles, including plant superintendent and lead engineer on international projects. He holds a Bachelor of Science degree in chemical engineering from Michigan State University, a Certificate of Graduate Studies in Hazardous Waste Control from Wayne State University, and a Master of Science degree in fire protection engineering from Worcester Polytechnic Institute. Mr. Gribb currently leads Cicada Consulting, a nonprofit organization dedicated to environmental and manufacturing issues, including the prevention of greenwashing in environmental reporting.

Samantha Gribb recently graduated with her Bachelor of Science degree in economics from the University of Oregon. She worked as a contractor developing language learning models, mainly in output correction and development. During her time at the University of Oregon, she completed over 100 hours of volunteer work for domestic violence awareness.

Abinash Kumar Gupta earned his Master of Science degree in industrial engineering and management from Oklahoma State University in 2024. He joined the OSU IAC in 2022 and has participated in 30 energy assessments for industrial manufacturing companies. From January 2023 to May 2024, he served as the lead student for the OSU IAC. Abinash is passionate about supply chain and energy management, with a strong dedication to supporting a clean and equitable energy transition.

Andrew R. Kramer has worked in the electric industry for 13 years. He is affiliated with Ulteig. He has a research background in energy efficiency and renewable energy. He has gained experience in energy engineering through work in nodal market simulations, integrated resource planning, demand side management, renewable energy, energy efficiency, generation operations, and natural gas pipeline expansion projects.

Robert A. Kramer is a professor of physics at Purdue University Northwest and director of the energy efficiency center. In this role, he develops programs for and conducts research in energy production, transmission, utilization, and efficiency. He also leads ongoing research for biological hydrogen production.

Nancy L. Najarian is a clean energy and climate expert working for the Government of Finland and the circular economy sectors. She began her career in international development, focused on the private sector and economic development, and then built expertise in clean energy and climate, committing her career to addressing climate change. Ms. Najarian has a B.A. from New York University and a M.S. from Georgetown University's School of Foreign Service, and has been active in AEE's National Capital Chapter and Council on Women in Energy and Environmental Leadership.

Marianne Ramos has a career that spans 40 years of utility engineering work in Colorado, with 30 years as a licensed professional engineer. She started in plant engineering in large coal-fired generators and eventually moved to mechanical/electrical positions in equipment design and system planning engineering departments for various IPO electric and natural gas transmission utilities, as well as a non-profit electric transmission cooperative. Of late, she has worked for the Colorado Public Utilities Commission as a Pipeline and Hazardous Materials Safety Administration gas safety inspector and a staff engineer assigned to litigate steam, natural gas, and electric utility rate cases.

Nergui Rentsen works in the energy market modeling industry and is affiliated with Ulteig. She has a background in applied and engineering physics, with a specialization in mechanical and thermodynamic physics. She has gained experience in energy market simulations and capacity expansion analysis related to integrated resource planning.

Shivani Saikar is an energy manager at Fermi National Accelerator Laboratory where she leads strategic initiatives to enhance sustainability, improve energy efficiency, and effectively manage utilities across the laboratory's facilities. She specializes in energy markets, cost optimization, renewable energy integration, and reducing carbon footprints. Shivani's experience includes energy procurement, audits, analytics, and project development, as well as energy market modeling for strategic advisory services. Shivani has a Bachelor's degree in electrical engineering and a Master's degree in energy systems.

Nicolette Sowa is the CEO and Founder of the Founder Frequency application, which helps startups become investor-ready with customized education pathways. She is a leadership coach at Premier Consulting Group. Nicolette's work in the energy industry includes managing demand response programs at Pacific Gas and Electric Company (PG&E), where she drove program growth, improved cost-effectiveness, and distributed energy resource (DER) integration. Her expertise spans renewable energy policy, market design, distributed grid technologies, and customer engagement. Nicolette is a speaker, author, former national mentorship chair for the Council on Women in Energy and Environmental Leadership (CWEEL), former mentorship director for PG&E's Women in Science, Technology, Engineering, and Mathematics (W-STEM) employee resource group, and current mentor with Oregon Bio Women (OBW). Nicolette holds a Bachelor of Science degree in environmental studies from the University of Oregon and a Master's degree in Environmental Law and Policy with an Energy Law Certificate from Vermont Law School.

Omer Vejzovic has worked in the electric industry for 11 years. He is affiliated with Ulteig. He has gained experience in utility transmission planning, where he performed various steady-state transmission studies for both the near- and long-term planning horizons.

Hitesh D. Vora is an endowed associate professor of mechanical engineering technology, mechatronics, and robotics at Oklahoma State University. He received his Ph.D. in materials science and engineering (2013) and his Master of Science degree in mechanical engineering technology (2008) from the University of North Texas. He directs the IAC at OSU, funded by the U.S. Department of Energy. Dr. Vora conducts research in smart manufacturing and energy management, securing over $12.5 million in federal funding. Since 2019, he has been an executive board member and student chapter director of the Association of Energy Engineers, Oklahoma Chapter. He is the faculty advisor for the OSU student chapters of the AEE and the Society of Manufacturing Engineers.

Samer A. Zawaydeh works with owners, developers, and donor agencies on energy and environment projects. As a Jordan power expert, he studied options for Jordan's National E-Mobility Strategy, which included analyzing policy incentives for e-mobility and developing uptake strategies. Samer contributed to a study on legal and regulatory aspects of smart grids in Jordan, which provided technical advice to stakeholders. As a lecturer at Al Hussein Technical University, he has been teaching an energy policy, regulation, and economics course since 2019. As an owner's representative, he has developed over 250 solar PV projects since 2012. He worked on the Regional Initiative for Promoting Small-Scale Renewable Energy Applications in rural areas of the Arab Region with the UNESCWA, Carbon Markets and Carbon Financing with South Pole, and the Renewable Readiness Assessment, Jordan 2020–2025 Report issued by IRENA. Samer is a past president of the AEE, a carbon reduction manager course trainer, and has developed 38 AEE national and student chapters.

The Energy Transition: What Is It Really About?

Stephen A. Roosa

What is the energy transition all about? There have been several energy transitions in the past. What makes the current one important? Perhaps the most notable energy transition occurred in the mid-1800s as the world began transitioning away from using mostly renewable biomass feedstocks, such as wood and whale oil, to using coal as a primary energy resource. Wind power was widely used for transportation over our seas, and hydropower for mechanical applications. It was a time before automobiles, electricity, and air conditioning became widely available. Initially fueling economic growth on a massive scale, our dependence on fossil fuels gave way to economic, social, and environmental problems. Today's energy transition is about evolving away from an over-reliance on fossil fuels to a greater portion of renewables in the energy mix. This contemporary transition represents a fundamental redirection of how we produce, distribute, and consume energy. It has been called the *renewable energy transition*. It is a "back to the future" transition, one that is causing us to once again focus on regionally available renewable energy resources.

Among the drivers of this transformation is the need to decarbonize energy production and delivery systems. Decarbonization is now a driving force in the energy sector [1]. The concept is important and transformative because it challenges countries to move away from fossil fuels to environmentally cleaner fuels. Clean energy is a category of energy resources that normally refers to renewables plus nuclear power. The benefits of the energy transition include reducing air pollution and greenhouse gas emissions while improving ecosystem and human health. Other drivers include improving energy resource availability and reducing costs. In some countries, renewables

Table 0.1 Total energy production 2023 by leading countries [3].

Ranking	Country	quadrillion Btu
1	China	127
2	United States	103
3	Russia	60
4	Saudi Arabia	28
5	Canada	22
6	India	21
7	Indonesia	19
8	Iran	19
9	Australia	17
10	Brazil	11
11	United Arab Emirates	11
12	Qatar	10
13	Iraq	10
14	Norway	9
15	Kazakhstan	8

such as solar and wind power are now the least expensive means of adding electrical generation capacity.

For the energy transition to be effective, it requires the support of those countries that are the leaders in energy production. Where are the world's centers of energy production? Consider the leading energy-producing countries in 2023. China, the world's largest energy producer, produced 127 quadrillion Btu, and the U.S. was second with 103 quadrillion Btu (see Table 0.1). Between them, their production was more than the next 11 largest energy-producing countries combined.

The differences in energy production are profound across countries. Each has a unique set of circumstances regarding energy supplies. The Russian Federation, the world's third-largest energy producer, is reeling from economic sanctions and damage to energy infrastructure due to its ongoing war with Ukraine. It produces 60% of its energy from fossil fuels, 20% from nuclear energy, and about 20% from renewables. Saudi Arabia is a major oil producer. Canada produces more natural gas than it can use and generates much of its electricity with hydropower. Only three provinces use primarily coal for electricity generation. India, the world's most populous country, produces less than 1/6 of the energy that China produces, and coal is the dominant fuel source. There are also many countries that lack the resources to produce the energy needed to drive economic expansion and address energy poverty. Often unable to afford to import energy, they rely on regional resources that are readily available.

The energy consumption trends in the U.S. are informative (see Figure 0.2). After years of being heavily dependent on imported oil, the U.S. can now claim to be energy independent. Still the world's largest economy, it has become the world's leading producer of crude oil and exporter of liquefied natural gas. Energy consumption in 2023 totaled 94 quadrillion Btu, and excess production was available for export. In 2024, fossil fuels accounted for 82% of total energy consumption (petroleum, 38%, natural gas, 36%, and coal, 8%). Clean energy resources accounted for 18% of total consumption (renewables, 9%, and nuclear 9%). While China leads by a substantial margin, the U.S. is the world's second-largest producer of renewable electricity, followed by Brazil and Canada.

The share of renewable energy production in the U.S. energy consumption mix has been increasing. A substantial portion of the energy from renewables in the U.S. is from hydropower generation. In the states of Vermont, Washington, and Idaho, hydropower provides the largest amounts of power. In New Mexico, Oklahoma, Kansas, South Dakota, Minnesota, and Iowa, wind power is the leading power producer.

Nuclear generation is the leading way to produce electricity in New Hampshire, Maryland, Illinois, Tennessee, and South Dakota. With the addition of two costly utility-scale nuclear reactors in the southeastern state of Georgia, electricity production using this nuclear technology will likely increase in 2026 [2]. Regardless, the energy transition for the U.S. over the last couple of decades has not been solely about transitioning to clean energy resources. It has been a transition from less coal use to greater use of natural gas and, to a lesser extent, renewables (see Figure 0.1). Perhaps more importantly, U.S. domestic energy production has been increasing and diversifying toward a mix of fuels. The country was dependent on imported energy for 34% of its needs in 2006. Due to increased domestic energy production, this dependency has declined to 17% in 2024—the lowest share in almost 40 years (see Figure 0.2). Energy imports are offset by exports.

In certain situations, renewables have economic advantages over fossil fuels. Development times can be shortened. With years-long wait times for the delivery of natural gas turbines and development times up to seven years, solar projects can be operational in less than three years or less. Longer development times invariably increase project costs. It may be necessary to expand infrastructure, such as rail, pipelines, tank farms, and offloading facilities, just to deliver fossil fuels to the plant site. Not so if using hydropower, solar, or wind energy.

Such considerations are especially important for developed countries such as Norway and Sweden and developing countries such as Costa Rica, Paraguay, and Panama [4]. For these countries and others like them, the

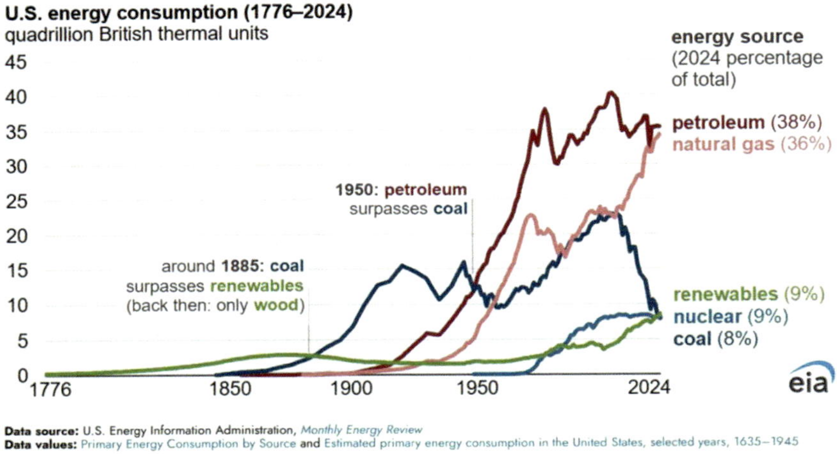

Figure 0.1 U.S. energy consumption from 1776 to 2024 [3].

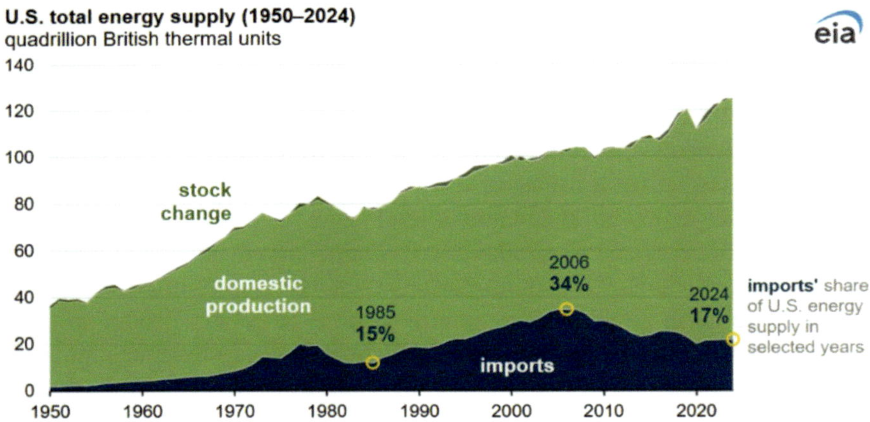

Figure 0.2 U.S. energy supplies from 1950 to 2024 (EIA, Monthly Energy Review) [3].

energy transition has been about advancing the use of renewables. Norway has vast hydropower resources from which it obtains 99% of its electricity. Sweden gets 68% of its total energy consumed from renewables that primarily include biofuels and hydropower [5]. In 2022, Costa Rica produced 98% of its electricity from renewable sources and often exports its excess electricity [4].

In many countries, there has been a decoupling of gross domestic product from energy use—meaning that their economies are expanding despite reductions in per capita energy use. In some remote regions of the world where water is scarce, solar and wind power are the first type of energy used for electrification. This is because fossil-fuel generation stations require large quantities of water, and the costs of extending high-voltage power lines from central plants to remote locations enable renewables to achieve cost parity.

Energy transitions are long-term trends that develop over decades. This energy transition involves substituting fossil fuels for renewables and nuclear energy resources. Differences in energy transitions are evident among countries. Energy infrastructure improvements tend to occur incrementally. Drivers supporting the transition to renewables include decarbonization, energy resource availability, and economics. Economics has an important role. As alternative technologies improve, become less costly, and are accepted by engineering professionals, there is a greater impetus to implement them. They become key components of sustainable development policy agendas [6].

This transition leads us to an alternative energy future. The various countries of the world view the current energy transition differently. However, country-scale energy transitions require massive investments in infrastructure that use public funds and subsidies. It is usually less costly in the short-term to maintain and improve existing infrastructure than to add new energy systems; this hinders the implementation of new technological solutions. As a result, renewable energy development tends to occur at the margin when capacity additions are needed or infrastructure is retired. For countries with smaller populations and limited energy consumption, expanding power systems and supplying electricity at the lowest cost are the principal issues, regardless of the energy resources used.

References

[1] Marr, B. (2022, February 11). The 3 biggest future trends (and challenges) in the energy sector. https://www.forbes.com/sites/bernardmarr/2022/02/11/the-3-biggest-future-trends-and-challenges-in-the-energy-sector/?sh=71f4a87127b7, accessed 25 January 2026.

[2] Roosa, S. (2024, September). Nuclear energy—lessons from Georgia's new power plant. *Alternative Energy and Distributed Generation Journal*, 6(3), 5–12.

[3] U.S. Energy Information Administration (2023). Rankings, total energy production 2023. https://www.climatecouncil.org.au/11-countries-leading-the-charge-on-renewable-energy/https://www.eia.gov/international/rankings/world?pa=12&u=0&f=A&v=none&y=01%2F01%2F2023, accessed 6 July 2025.

[4] Climate Council (2022, August 15). 11 countries leading the charge on renewable energy.

[5] Statista (2025, March 18). Renewable energy in Sweden—statistics and facts. https://www.statista.com/topics/10734/renewable-energy-in-sweden/#editorsPicks.

[6] Roosa, S. (2010) *The Sustainable Development Handbook*, 2nd edition. The Fairmont Press: Lilburn Georgia.

List of Figures

List of Tables

List of Abbreviations

3D	Three-dimensional
AC	Alternating current
ACOST	Annualized cost
ADB	Asian development bank
AEE	Association of Energy Engineers
AI	Artificial intelligence
AL	Average load
AR	Assessment recommendation
ARC	Assessment recommendation codes
ASTM	American Society for Testing and Materials
BEMS	Building energy management system
BESS	Battery energy storage system
BOE	Barrels of oil equivalent
BREEAM	Building Research Establishment Environmental Assessment Method
BTU	British thermal unit
BWRs	Boiling water reactors
CAGR	Compound annual growth rate
CCG	Combined-cycle gas
CCGT	Combined-cycle gas turbine
CCOST	Capital cost
CCS	Carbon capture and sequestration
CDD	Cooling degree day
CDM	Clean development mechanism
CdTe	Cadmium telluride
CEO	Chief executive officer
CES	Clean energy standards
CF	Capacity factor
CH	Clean hydrogen
CHP	Combined heat and power
CIGS	Copper indium gallium selenide
CLS	Canadian light sweet

CNTs	Carbon nanotubes
CO_2	Carbon dioxide
CO_2e	Carbon dioxide emissions
COP	Coefficient of performance
C_P	Power coefficient
CPI	Consumer proce index
CRE	Energy Regulatory Commission of Mexico
CRF	Capital recovery factor
c-Si	Monocrystalline silicon
CSP	Concentrating solar power
DC	Direct current
DDC	Direct digital control(s)
DEG	Daily electricity generation
DER	Distributed energy resources
DFI	Development Financial Institution
DG	Distributed generation
DHW	Domestic hot water
DLR	Dynamic line rating
DoE	Department of Energy (U.S.)
DPV	Distributed photovoltaic
DRI	Direct reduced iron
DSM	Demand-side management
DSO	Distribution system operator
EDCO	Electricity distribution company
EIA	Energy information administration
EJ	Exajoule (1 billion joules)
EMRC	Energy and mineral regulatory commission
EOL	End of life
EPA	Environmental Protection Agency (U.S.)
EPR	Extended producer responsibility
EPRI	Electric power research institute
ERCOT	Electric Reliability Council of Texas
EU	European Union
EV	Electric vehicle
EVA	Ethylene vinyl acetate
FCOST	Fuel cost
FEMP	Federal Energy Management Program
FERC	Federal Energy Regulatory Commission
FPV	Floating photovoltaics
FY	Fiscal year

GaAs	Gallium arsenide multijunction
GDP	Gross domestic product
GE	General Electric
GEF	Global Environment Facility
GETS	Grid enhancing technologies
GHG	Greenhouse gases
GOM	Government of Mexico
GW	Gigawatt
GWh	Gigawatt hour
HAWTs	Horizontal axis wind turbines
HDD	Heating degree day
HENG	Hydrogen enriched natural gas
HF	Hydrogen fluoride
HFC	Hydrofluorocarbon
HOMER	Hybrid optimization of multiple energy resources (software tool)
HVAC	Heating, ventilation, and air conditioning
IAC	Industrial assessment center
IDECO	Irbid Distribution Company
IEA	International Energy Agency
IEEE	Institute of electrical and electronics engineers
IMSR	Integral molten salt reactor
IoT	Internet of things
IRA	Inflation reduction act
IRENA	International Renewable Energy Agency
IRR	Internal rate of return
IRP	Integrated resource planning
ISO	Independent system operator
IT	Information technology
ITC	Investment tax credits
JEPCO	Jordan electric power company
kW	Kilowatt
kWh	Kilowatt hour
kWp	Kilowatt peak
LCOE	Levelized/Levelized cost of energy
LCOH	Levelized cost of hydrogen
LED	Light-emitting diode
LEDS GP	Low emission development strategies global partnership
LF	Load factor
LFP	Lithium iron phosphate

L-I	Lithium-ion
LLNL	Lawrence Livermore National Laboratory
LMP	Locational marginal price
LNG	Liquefied natural gas
LSE	Load serving entity
M&V	Measurement and verification
MCC	Marginal cost of congestion
mc-Si	Multi-crystalline silicon
MEC	Marginal energy cost
MEMR	Ministry of Energy and Mineral Resources
MINAE	Ministry of Environment and Energy
MISO	Midcontinent Independent System Operator
MIT	Massachusetts Institute Of Technology
MLC	Marginal loss component
MMT	Million metric tonnes
MSW	Municipal solid waste
MW	Megawatt
MWh	Megawatt hour
MySE	Mingyang Smart Energy
NaS, Na-S	Sodium, sulfur
NDCs	Nationally determined contributions
NDP	National decarbonization plan
NEPCO	National Electric Power Company
NERC	North American Electric Reliability Corporation
NIB	Sodium-ion battery
NG	Natural gas
NPPs	Nuclear power plants
NPV	Net present value
NREL	National Renewable Energy Laboratory
NYSERDA	New York State Energy Research and Development Authority
O&M	Operation and maintenance
OMCOST	Operation and maintenance cost
OPEC	Organization of the petroleum exporting countries
PECAN	Polyester covalently adaptable network
PJM	Pennsylvania-New Jersey-Maryland
PPA	Power purchase agreement
PPP	Public private partnerships
PSH	Pumped storage hydropower
PV	Photovoltaic

PWRs	Pressurized water reactors
R&D	Research and development
RCM	Reliability centered maintenance
RE	Renewable energy
REC	Renewable energy certificate
RECs	Renewable energy credits
RF	Reserve factor
ROI	Return on investment
RPS	Refined petroleum products
RPS	Renewable portfolio standard
SAM	System advisor model
SCADA	Supervisory control and data acquisition
SCED	Security constrained economic dispatch
SDGs	Sustainable development goals
SGIP	Smart grid interoperability panel
SJEC	Saint Joseph Energy Center
SMR	Steam methane reforming
SMRs	Small modular reactors
SNRs	Small nuclear reactors
SPP	Simple payback period
SPR	Strategic petroleum reserve
TCECs	Tidal current energy converters
TOU	Time-of-use
TW	Terawatt
TWh	Terawatt hours
UK	United Kingdom
UN	United Nations
UNFCCC	United Nations framework convention on climate change
UNT	Uniform National tariff
US	United States
USAID	United States Agency for International Development
US-CERT	United States - Computer Emergency Readiness Team
USD	United States dollar
USDN	Urban sustainability directors network
USMCA	United States-Mexico-Canada Agreement
UV	Ultra-violet
VAWT	Vertical axis wind turbines
VPP	Virtual power plant
VRB	Vanadium-redox battery
VRE	Variable renewable energy/electricity

W	Watt
Wac	Watts alternating current
WCS	Western Canadian select
Wdc	Watts direct current
WTG	Wind turbine generators
WTI	West Texas intermediate
ZnBr	Zinc-bromide

Introduction

The energy transition is upon us. Like most past transitions from dependency on one or more forms of energy to others, this transition is evolutionary, distinct, and innovative. There are risks and costs. This incremental process is one that requires decades yet will ultimately transform our energy systems and infrastructure. It will redefine energy policies, change programs that support energy development, and require innovative energy technologies. In fact, this process is already underway. Research is being redirected. Resources are being reallocated. Energy systems are in flux and being reinvented. New technologies are being tested and deployed. This book highlights the drivers, policies, and technologies that provide impetus to the transformation of our energy systems.

Chapter 1, Tough Decisions on the Path to Decarbonization by Stephen A. Roosa

Reducing greenhouse gas emissions is a central problem that involves the interplay of policies, programs, and technologies. The term decarbonization literally means the reduction of carbon. It is often included in lists that identify future trends. We are seeing it becoming policy today. The term refers to the transition to an economic system that sustainably reduces and compensates for atmospheric emissions of carbon dioxide (CO_2), a potent greenhouse gas (GHG). Decarbonization has been identified as one of the top three trends in the energy sector. The concept is indeed transformative because it challenges us to transition from using fossil fuels to clean energy resources. There are tough decisions to be made as countries take actions to reduce greenhouse gas emissions. This chapter explores the issue of decarbonization. The extent of the dependency on fossil fuels is addressed by using North America and Europe as case examples.

Chapter 2, Decarbonization for a Sustainable Energy Future by Kritika Bhandari, Abinash Kumar Gupta, and Hitesh D. Vora

This chapter presents a comprehensive analysis of strategies for reducing industrial sector greenhouse gas (GHG) emissions to help combat climate change. It emphasizes the sector's significant contribution to global GHG emissions, particularly CO_2, and outlines the technical and economic challenges in achieving decarbonization. This chapter discusses various approaches, such as hydrogen-enriched natural gas (HENG), carbon capture and storage (CCS), policy frameworks, and innovative technologies for iron, steel, and glass production. It advocates for collaborative efforts among governments, industries, and communities to overcome barriers, align with global climate goals, and highlight the urgent need for transition to low-carbon energy resources. It considers the correlation between decarbonization and energy savings. This study serves as a guide for policymakers, business executives, and industry stakeholders in effectively pursuing sustainable decarbonization strategies.

Chapter 3, Software Analysis of Solar PV System Efficacy by Stephen A. Roosa

There are many times when an energy engineer is tasked with developing a solar project using virtual tools. This chapter provides an overview of how an energy engineer can perform preliminary site investigations of solar projects by virtually assessing their generation and economic potential without the need to physically visit the site. It discusses ways to perform a solar photovoltaic analysis of a sample site using virtual surveying and the National Renewable Energy Laboratory's PVWatts, a web-based tool used to estimate the electrical generation from a solar photovoltaic system.

Two case studies for hypothetical sites are provided using PVWatts to assess the solar potentials of orientation options and selected module racking systems. One example considers a new solar array mounted on the roof of an existing commercial building; the other assesses a canopy-mounted system for an electric vehicle charging station. The case studies are designed to compare the electrical production of the solar PV systems based on the orientation and type of racking system. The use of solar thermal systems for water heating is considered.

Chapter 4, Wind Turbine Generators—They Just Keep Getting Bigger by Stephen A. Roosa

The wind turbine industry is evolving. One notable trend is to design, develop, and install ever-larger wind turbine generators. Is bigger really better? When using machines of larger capacities, their efficiencies increase. Fewer are required to generate the electricity needed. This chapter discusses the shift in the wind power industry to producing larger capacity machines for use in the development of both land-based and offshore wind farms.

In the U.S., renewable energy resources—biomass, geothermal, hydropower, solar, and wind—provide nearly 30% of total electrical generating capacity. Together, they comprise the largest portion of electrical generation using clean energy. One notable trend is to use wind turbine generators with larger capacities and higher efficiencies in new and existing wind farms. This means that fewer machines are required to generate the electricity needed. This chapter discusses the status of wind power in the U.S. and the primary ways wind energy is harnessed to generate electricity. The growth of U.S. wind farm developments is limited by technical, economic, maintenance, and developmental challenges.

This chapter further considers the reasons why manufacturers are racing to build larger wind turbine generators, discusses development limitations, and identifies the electrical power capacities of the largest ones that have been installed. The basic mechanics of wind turbines are detailed. The importance of making on-site operational performance assessments after prototypes have been placed into operation is emphasized. The author concludes that there are opportunities for efficiency improvements and growth in the wind power industry that will drive future innovation and create the ability to produce machines with larger capacities.

Chapter 5, Clean Energy Technologies, by Stephen A. Roosa

This chapter discusses the status of selected advanced clean energy technologies. The chapter considers hydropower, tidal energy, floating solar installation, and the situation with nuclear energy. Hydropower is a proven, mature, predictable, price-competitive, and scalable technology. Hydropower has a key role in providing renewable and emissions-free generation of electricity. There is growing interest in harnessing tidal power to generate electricity. One of the key benefits of tidal power is that tides are highly predictable. Floating solar farms offer the potential to take advantage of unusable or underused lake surface areas. By using solar PV systems over stable areas of water, they provide opportunities to generate revenue. Nuclear energy technologies promise reliable and resilient energy supplies. They are not dependent on the transportation of large volumes of fossil fuels to the plant site or on variable weather conditions. Nuclear power plants can produce large amounts of energy in a limited land area due to their high energy efficiency and power density.

Chapter 6, Nuclear Energy—Lessons from Georgia's New Power Plant by Stephen A. Roosa,

Electricity is a key enabler of contemporary life. The opportunities for nuclear power as a clean energy resource seem limitless. Nuclear energy has the potential to solve many of the world's most pressing energy problems. It can provide nearly unlimited supplies of energy and, more importantly, offer base-load electrical power generation. Some see nuclear energy as key to meeting decarbonization goals.

Nuclear energy accounts for almost 10% of the world's electrical energy production with about 440 operating reactors. The need to achieve ambitious decarbonization objectives has driven increases in nuclear power investments. During the 28th Conference of the Parties held in Dubai in December 2023, twenty of the world's countries, including the U.S. committed to increasing the output of nuclear energy worldwide. The financial risks associated with utility-scale energy projects are mitigated when construction and financing costs are predictable. This chapter considers the costs and benefits of the first new nuclear reactors in the U.S. in over 30 years, which are now operating in Waynesboro, Georgia.

Chapter 7, Energy Storage Capacities Are Increasing by Stephen A. Roosa

One of the more exciting developments in electrical generation is the recent focus on energy storage technologies. The deployment of storage technologies for utility-scale electrical energy applications is fast becoming mainstream. For utility-scale energy storage, pumped hydropower systems have been the mainstay as a means of leveling intermittent electrical loads. Energy storage systems are being reinvented, and the technologies they use are improving. This chapter explores the applications of energy storage systems, the various uses for energy storage, and discusses its applications for renewable energy systems. Several types of battery energy storage systems are discussed, and those used for utility-scale applications are identified.

The chapter concludes by suggesting that energy storage systems offer a wide range of benefits for distributed generation and will become more specialized to meet the needs of future applications. Energy storage technologies will continue to be incorporated with renewable generation systems and will likely utilize hybrid generation configurations.

Chapter 8, Matching Electricity Supply with Demand: Integrating Renewable Energy Systems into the U.S. Power Grid by Satish C. Boregowda, and Suresh Chand

This chapter presents a conceptual approach to assess the impacts of integrating renewable energy systems into the conventional utility grid. It considers performance, economic, and reliability. There is a push for reliance on renewable energy systems due to concerns about climate change. Renewable portfolio standards expect that U.S. states will produce a certain percentage of electricity using renewable energy sources (e.g., solar, wind, and geothermal) by a certain future date. The U.S. Environmental Protection Agency (EPA) has mandated strict emission controls on greenhouse gases (GHGs) and harmful pollutants such as sulfur dioxide, nitric oxide, lead, and mercury. This poses major challenges for utility companies and their suppliers to meet increasing demand for electricity as renewable energy systems are integrated with conventional utility systems. The methodology begins with a theoretical example of a daily load (demand) curve of a utility firm. Three scenarios are considered: a) conventional power systems providing 100% electricity; b) hybrid power system with renewable energy systems providing 25% of the electricity without battery storage systems; and c) hybrid power system with renewable energy systems providing 25% of the electricity supplemented with battery storage systems. The impact of these scenarios on three techno-economic assessments, performance, economics, and reliability, is examined at a system level.

Conventional power systems include coal, natural gas, and nuclear generation units while the renewable systems include solar photovoltaic (PV), solar thermal, wind, geothermal, and battery energy storage units. In this study, all the conventional power units were retained as part of the hybrid system while the renewables were integrated to meet 25% of the electricity demand. The results demonstrate that the integration of renewable energy systems leads to an increase in the levelized cost of electricity and a modest increase in reliability. This conceptual study provides a futuristic framework to investigate the techno-economic complexity of the integration of renewable energy systems to match electricity supply with demand.

Chapter 9, Electricity Generation Capacity Planning with Nuclear, Renewables, and Batteries: Defining Pathways to Energy Security and Decarbonization by Satish C. Boregowda, Suresh Chand, and Stephen A. Roosa.

This chapter presents a capacity planning approach to assess the impact of integrating nuclear energy systems on techno-economics and CO_2 emissions.

While the U.S. Environmental Protection Agency (EPA) has imposed strict rules and standards to reduce pollution from fossil fuel-fired power plants, energy security and reliability are pervasive issues in light of the climate crisis. Issues related to sustainability need to be addressed. The complexity of the problem at the interface of climate change and energy security poses major challenges to utility companies and electric grid operators who provide cost-effective and reliable electricity to their customers.

The main goal of our study is to better understand the integration of nuclear, renewables, and battery storage with the help of a model utility system. To this end, this study utilizes a daily load curve from a model utility system to demonstrate electricity generation capacity planning and portfolio selection. It investigates three possible capacity planning scenarios: Scenario I comprised of coal, nuclear, and natural gas units; Scenario II comprised of nuclear and natural gas units; and Scenario III comprised of nuclear, natural gas, renewable, and battery energy storage units. Using U.S. Energy Information Administration data, the impact of capacity planning scenarios on techno-economic elements such as plant performance factors, economics, reliability, and CO_2 emissions is examined. The results indicate a significant variation in performance factors, the levelized cost of electricity (LCOE), and CO_2 emissions, while the reliability remains the same across all three scenarios. With nuclear energy, renewables, and battery energy storage replacing coal, a reduction in CO_2 emissions of 99% is observed. In summary, this study provides a framework to assess the techno-economic complexity of electricity generation capacity planning with nuclear energy, renewables, and battery storage units.

Chapter 10, Distributed Energy Resources and Demand Response: The Tension Between Reliability and Flexibility by Nicolette Sowa

Electric grids are trending toward the use of more renewable energy resources, which creates more intermittent electrical loads. Between 2010 and 2020, the growth of renewable energy resources outpaced all other forms of energy generation in the U.S., increasing by 42%. Distributed energy resources (DERs) are on track to grow exponentially nationwide, with capacity expected to increase by 217 GW from 2024 to 2028. Renewable portfolio standards (RPS) and clean energy standards (CES) established in 31 states plus the District of Columbia are critical drivers of DER growth, with half of renewable energy generation in the U.S. attributed to states' renewable energy requirements.

Demand response is a valuable tool to mitigate unpredictable loads from renewables and utilize DERs to effectively shift loads. Demand response programs incentivize customers to reduce their energy usage during peak times of load and stress on the grid. However, there is an inherent tension that exists in asking DER owners to participate in demand response programs because they include customers with variable energy use needs that are often determined by external factors (heat, drought, unexpected business trends, etc.). DER, such as solar panels, add the most value by responding in real-time to periods of peak energy consumption, while traditional demand response programs require load reduction commitments months in advance.

In the evolving energy landscape, demand response will integrate aggregated DERs at a rapid pace, fueled by Federal Energy Regulatory Commission (FERC) Order 2222, initially issued in September 2020. Demand response programs must become more responsive to real-time signals while providing more accurate load reduction forecasts. Energy providers, such as utilities, need to be able to rely on demand response resources with certainty to address electrical system peak loads while also responding in real-time to usage spikes. This requires the flexibility to achieve as much load reduction as possible at any given moment. This chapter explores how demand response can expand and stack programs to resolve the tension between reliability and flexibility with an approach that includes: 1) program options; 2) policy that appropriately values DERs; 3) implementation and utilization of technology solutions; and 4) a focus on customer experiences.

Chapter 11, Initial Systemwide Capital Cost Estimate for Clean Hydrogen Electricity Produced by a 500 MWh Baseload Plant by Adam Gribb, Marianne Ramos, and Samantha Gribb

In 2023, the U.S. federal government awarded $7 billion to develop regional clean hydrogen (CH) hubs; several U.S. states began allowing investor-owned utilities to operate CH projects. For example, the U.S. state of Colorado has laws that define CH projects as those ventures using specific types of hydrogen-producing equipment. These include electrolyzers, dedicated renewable energy resources for electrolysis, upgraded turbine-generators at an electric generating station that use hydrogen as an energy source, plus purchasing clean hydrogen from third parties. This chapter develops an initial (systemwide) capital cost estimate for a 500 MWh CH-combusting facility that continually supplies electricity to the grid (i.e., baseload generation plant).

The method used in the estimate is based on an initial energy balance of a new 500 MW electric generation system. The energy source for electrolyzers comes from renewable energy generation comprised of 50% wind and 50% solar generation, adequately sized transmission, and storage components to provide continuous generation. An appropriately sized storage system is assumed to have 10 units—one unit in operation, six units charged and ready, two units charging, and one spare unit. Given these assumptions, a preliminary capital cost estimate was developed based on recently published wind, solar, transmission, and energy storage costs.

Chapter 12, Reinventing the Electric Grid by Stephen A. Roosa.

It is time to reinvent the electric grids, transforming them into infrastructure that is more usable, durable, and efficient. Most of today's electric grids are based on early 20th-century models. They are configured to meet the requirements of an antiquated and centralized electricity distribution system. They are designed to deliver electricity in a one-way direction, which provides generation, transmission, and distribution to customers. The power grids of the future need to be more responsive, resilient, and capable of providing uninterrupted and expanded services. Improving transmission and distribution system efficiencies requires upgrading technologies and modifying system configurations. Tomorrow's electric grids need to be able to quickly incorporate the added generation from renewable and nuclear energy resources. How can this be accomplished? Selectively upgrading electrical distribution systems, adding energy storage systems, and using smart microgrids are ways to improve the operation of electric grids. Other possibilities include reconfiguring electrical systems into peer-to-peer networks and virtual power plants. This chapter explores opportunities to structure electrical energy delivery systems to better meet the needs of tomorrow's electrical consumers.

Chapter 13, Electrical Transmission System Modeling within Utility Integrated Resource Planning to Accommodate Increasing Renewable Penetration by Andrew R. Kramer and Robert A. Kramer

Across the electric utility industry, traditional thermal electrical generation is being retired and replaced with renewable energy alternatives that are often supplemented with energy storage and transmission enhancement technologies. Utility long-range planning, also known as integrated resource planning (IRP), seeks to model system costs over a 20+ year time horizon. Marginal energy cost (MEC), a component of the locational marginal price (LMP), is forecast in this modeling for various scenarios and sensitivities. Within this analysis, the marginal cost of congestion (MCC), another key component of the LMP, is typically based on historic averages rather than forecasts. This can result in model costs deviating from true system costs when considering electrical generation retirements and replacement scenarios.

This research shows that the averages for the hourly standard deviations and maximum hourly peaks for the MCC have generally been increasing for the electric power hubs in the U.S. states of Illinois, Indiana, and Michigan, during recent winter seasons. An industrial power node near Chicago, Illinois, is used as an example to show significant swings in both the average of the hourly standard deviations and the average of maximum hourly peaks of the MCC during the spring season. Large electric power nodes of this type are particularly vulnerable to transmission system constraints. The results of this research indicate rising electric transmission costs due to the increasing penetration of intermittent electrical generation resources. Consequently, it is advantageous for utilities to work with independent system operators (ISO) and other organizations, such as the North American Electric Reliability Corporation (NERC), to forecast the MCC for IRPs when electrical generation retirement scenarios are being considered. This can lead to a more representative IRP process. It also improves forecasting accuracy in cases where utilities are unaware of potential neighboring generator retirements. Including MCC forecasts within IRPs may lead to reduced purchased electric power costs, expanded power sales opportunities, and improved system reliability.

Chapter 14, Using Dynamic Line Ratings to Improve Electric Generation and Load Interconnection by Andrew R. Kramer, Robert A. Kramer, Omer Vejzovic, Nergui Rentsen, and Shannon Dancler

Much emphasis has been placed upon the use of grid-enhancing technologies (GETS) to assist with the ongoing evolution of electric grids in the United States. Dynamic line rating (DLR) is often associated with GETS because new technologies and equipment are needed to implement and validate dynamic transmission line ratings during system operations or within long-term planning models. Given the expected rapid load growth associated with data centers and electrification initiatives across the U.S., the ability to provide additional transmission capacity may help reduce or avoid reliability issues and restrictions on economic development. DLR can contribute to this by allowing system planners and operators to assume additional transmission capacity subject to site ambient conditions and electrical system configuration. DLR does not modify the physical structure of the transmission system but rather incorporates improved data into transmission rating values to allow for greater generation dispatch than would be captured with static, conservative rating values. A measurement of this increase in transmission capacity is reflected in the changes in transmission congestion and generation curtailment.

This research incorporates DLR transmission ratings into the nodal market simulation tool (PLEXOS®) to study the impact of these rating changes in terms of transmission congestion and generation curtailment. The power node studied was in the Pennsylvania-New Jersey-Maryland (PJM) independent system operator (ISO) territory near the Indiana, Michigan, and Ohio borders. This region has potential for future data center load additions due to a robust transmission network and readily available supplies of natural gas. A theoretical data center load of 1.5 GW was considered in the model near the Saint Joseph Energy Center (SJEC) in New Carlisle, Indiana. In addition, a speculative, non-dispatchable 1.5 GW solar resource was also simulated near the energy center to provide power and offset carbon emissions. Upon analysis, two primary transmission binding constraints were identified. Using the DLR for transmission increases rated capacity from 10% (low case) to 40% (high case), which reduces transmission congestion. The curtailment at the solar resource node was reduced by 53% and 99% respectively, for the low and high cases relative to the modelled baseline. The results of this research illustrate how DLR may be used to increase transmission line capacities while helping to defer reconductoring or other costly initiatives.

Chapter 15, Cutting Electricity Costs: A Guide for Organizations in Deregulated Markets by Shivani Saikar

In deregulated U.S. electricity markets, electricity costs make up a substantial share of an organization's operating expenses. While the flexibility to choose providers and negotiate pricing structures presents cost-saving opportunities, it also brings challenges such as price volatility, contract complexities, and electrical demand charges. Without a strategic approach, businesses may face unpredictable electricity bills and financial uncertainty.

This chapter discusses key electricity pricing models, such as fixed-rate, variable-rate, and time-of-use plans, and outlines how organizations can manage costs through effective procurement strategies, energy efficiency improvements, and demand-side management. It also highlights the increasing importance of data-driven decision-making and renewable energy integration in reducing long-term electricity expenditures.

As energy markets evolve, adopting proactive strategies—like optimized procurement, real-time energy monitoring, and effective contract management—can help organizations reduce expenses, strengthen energy resilience, and improve their financial stability in competitive deregulated markets.

Chapter 16, How Can Eco-friendly Power Be Suppled to Data Centers? by Stephen A. Roosa, Ph.D.

To support the growth of crypto-mining and artificial intelligence in the United States, companies have filed for permits to build new data processing centers. These data centers require substantial amounts of electricity mainly from fossil fuel-fired generation. The larger ones use as much electricity and a medium-size U.S. city. The inefficient use of fossil fuels is inherently counter to the idea of reducing greenhouse gas emissions.

Decarbonization has been identified as one of the top three trends the energy sector [1]. This transformative concept challenges us to transition from our present reliance on fossil fuels to using more sustainable energy resources. As the energy transition gains momentum, data centers present a perplexing conundrum. How can the goals of the energy transition be achieved in the face of data center growth? Adding capacity to electric grids to meet the increased demand created by regional data centers is difficult. What are the alternative ways to provide electricity? Among the ways to provide this additional capacity is to use either renewables or nuclear energy. Regardless, natural gas is often chosen as a primary fuel to expand electrical generation in the U.S. This chapter explores the issues surrounding the use of natural gas to provide electricity for data centers and discusses their implications.

Chapter 17, Policy Design for Prosumer Viability: Evaluating Solar PV Grid Connection Mechanisms in Jordan's Energy Transition by Samer Adnan Zawaydeh, MSC, CRM, REP

This chapter evaluates the economic and environmental impacts of three grid connection mechanisms—net billing, zero-to-grid, and buy-all/sell-all—on residential solar photovoltaic (PV) prosumers in Jordan. Drawing on empirical data from Jordan's Ministry of Energy and Mineral Resources, transmission system operators, distribution system operators, and international energy reports, a techno-economic analysis was conducted to assess the viability and performance of distributed solar PV systems under evolving regulatory frameworks.

Findings indicate that residential solar PV systems contributed approximately 9% of Jordan's total electricity generation in 2024, avoiding an estimated 1.026 million metric tons of carbon dioxide emissions annually. An economic evaluation reveals variation across connection mechanisms: net billing yields the highest internal rate of return (IRR) at 27.7% for high-consumption households, while buy-all/sell-all offers up to 15.6% IRR, particularly under declining PV system costs. The zero-to-grid mechanism emerges as advantageous for consumers with substantial daytime usage due to Jordan's tiered tariff structure.

This research provides a comprehensive assessment of prosumer economics and policy design in an emerging market context, offering actionable insights for accelerating the transition to sustainability. Policy recommendations are presented that aim to harmonize regulatory frameworks, promote standardized data collection, incentivize energy storage integration, and enable equitable access to solar PV adoption.

Chapter 18, An Asset Manager's Conundrum: Sunsetting Solar Farms by Stephen A. Roosa,

This chapter considers the process of sunsetting distributive generation infrastructure from the perspective of an asset manager. While there are ways to extend their lifecycles, when renewable energy developments such as solar farms reach the end of their efficient operations, they must at some point be decommissioned. When projects are being developed, the costs of decommissioning and equipment disposal at the end of the development's useful lifecycle are often ignored. What is involved in this decommissioning process, and what sort of costs must be incurred? The example of decommissioning solar farms is presented to highlight the challenges involved. Planning for sunsetting assets considers decommissioning costs, types of regulation, and the available financial mechanisms. The decommissioning process for distributed generation technologies, such as those used for solar arrays, is evolving. In some jurisdictions in the United States, this process is regulated. Asset managers are encouraged to survey the changing landscape and include equipment decommissioning in their lifecycle planning. There are opportunities for related industries to establish standards, bolster recycling capacity, and advocate for common regulations across the United States.

Chapter 19, Sustainable Adaptive Reuse of Renewable Energy Components and Materials by Ahmed Gazel and Stephen Roosa

Renewable energy technologies have seen exponential growth over recent decades and are poised to dominate new power installations through 2050. This transition is crucial for mitigating climate change, yet it brings new sustainability challenges. Chief among these is the management of end-of-life waste from solar panels, wind turbines, batteries, geothermal, and hydroelectric systems. Large-scale deployments imply that by mid-century, tens of millions of tonnes of solar photovoltaic modules, composite turbine blades, spent chemical batteries, and other system components will retire. These components will require reuse, recycling, or disposal. If not properly handled, this waste could offset environmental gains and strain waste management infrastructure. To avoid such unintended consequences, it is essential to understand the types of wastes and their volumes under realistic growth scenarios. Strategies for extending the lifespan of renewable infrastructure and closing the material and supply loops are needed.

This chapter presents a comprehensive analysis of global waste generation from key renewables for 2025 to 2050 using linear growth models and current policy scenarios. It proposes multidisciplinary engineering solutions—in design, maintenance, materials, chemistry, and economics—to enhance sustainability. We show how integrative approaches can accelerate progress toward reducing emissions. The objective of this chapter is to chart a path for renewables to sustainably support the world's circular economy.

Chapter 20, Tariffs and the U.S.-Canada Energy Interdependence by Stephen A. Roosa

In the world's developed countries, access to low-cost energy is a prime mover that enables productivity and economic expansion. This chapter considers the energy interdependencies of two major North American countries, namely the United States and Canada. Together, these countries jointly produce more oil than anywhere else on Earth. They are complementary and interdependent regarding their primary supplies of energy. This chapter details the resources and trade relationships of these two allies regarding key sources of energy, particularly oil, natural gas, uranium, and electricity. The infrastructure of these two counties is designed to support bilateral energy distribution and trading. Mapping the energy relationships between countries is a complex process. Canada exports almost all its oil and natural gas production to the U.S. Regions of the U.S., such as the upper midwestern states and the Northeast, are highly dependent on Canadian crude oil, natural gas, and electricity. The U.S. also exports energy to Canada, albeit in lesser volumes. This chapter delves into the issue of tariffs applied to the importation and exportation of energy resources such as petroleum and electricity. The application of tariffs across international borders changes the cost structures of fuels, materials, components, and energy systems, which impacts the energy transition. The author concludes that the new tariffs being applied to energy resources by the U.S. on Canadian energy imports are counterproductive and create uncertainties for businesses and industries on both sides of the border.

Chapter 21, Sharing Best Renewable Energy Practices in the Developing World by Nancy L. Najarian

Increasing the use of renewable energy and energy efficiency in developing countries is critical to combating climate change and improving the lives of residents. An array of tools, online and in print, is available to these countries to assist with project preparation and management. The ultimate objective is to help countries and their citizens achieve carbon reduction goals set on a national or community level. To this end, this chapter examines the successful development and use of renewable energy resources in the countries of Costa Rica, Uruguay, India and Haiti. Are these tools and technical assistance being accessed as fully as intended? This chapter offers tools to help determine if and how these tools are being applied by countries, leadership, businesses, and residents to share lessons learned and best practices.

1

Tough Decisions on the Path to Decarbonization

Stephen A. Roosa

Reducing greenhouse gas emissions is a central problem that involves the interplay of policies, programs, and technologies. The term *decarbonization* literally means the reduction of carbon. It is often included in lists that identify future trends, but we are seeing it becoming policy today. The term refers to the transition to an economic system that sustainably reduces and compensates for atmospheric emissions of carbon dioxide (CO_2), a potent greenhouse gas (GHG). Decarbonization has been identified as one of the top trends in the energy sector [1]. The concept is indeed transformative because it challenges us to transition from using fossil fuels to clean energy resources. There are tough decisions to be made as countries take actions to reduce greenhouse gas emissions. This chapter explores the issue of decarbonization. The extent of the dependency on fossil fuels is addressed by using North America and Europe as case examples.

1.1 Introduction

Gaining traction in the 1950s, the world's historical CO_2 trend shows a trajectory of emissions that increases unabated since the 1950s (see Figure 1.1). Though global GHG emissions declined in 2020 due to the economic impact of COVID-19, they increased to 51.8 billion tonnes in 2023 as the world's economic situation improved. Energy-related emissions of CO_2 hit an all-time high of 53.2 Gt in 2024.

The goal of decarbonization is to create a CO_2 emissions-free global economy [2]. This is a daunting, if not seemingly impossible, objective to

DOI: 10.1201/9788743811312-1

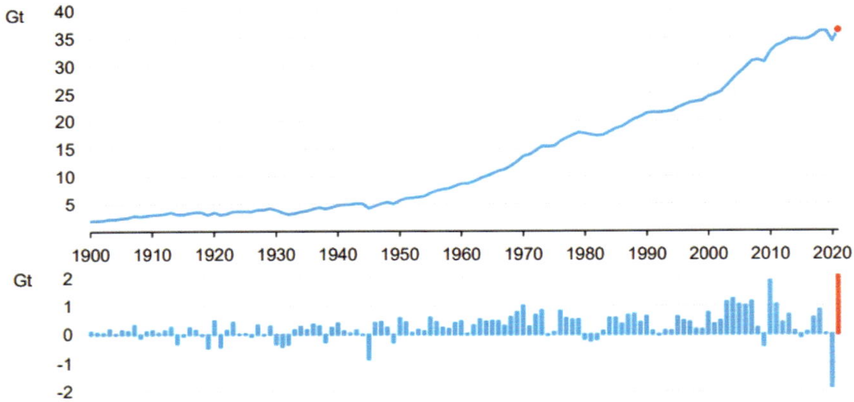

Figure 1.1 Total CO_2 emissions from energy combustion and industrial processes (top graph) and their annual change (bottom graph) [3].

achieve in the short term. Long-term reductions in GHG emissions will require substantial increases in infrastructure expenditures by industrial and governmental entities. Many advanced economies have emphasized decarbonization measures in their recent economic improvement plans [3]. Given the past trajectory of greenhouse gas emissions, getting to zero emissions is a questionable goal. Regardless, we are not without tools that aid in this effort.

How can GHG emissions be reduced? Numerous ways have been proposed to achieve decarbonization. One key strategy is to lower or eliminate the use of fossil fuels in combustion processes, ultimately lowering total GHG emissions. As the world's population increases, per capita fossil fuel consumption will show incremental declines. Some argue that this is beginning to happen. Per capita fossil fuel consumption began to level off in 2011, peaked in 2018, and has since declined. This has been attributed to energy conservation and efficiency improvements, fuel-switching (e.g., substituting renewables and natural gas for coal), and economic conditions. An increase in this measure was recorded in 2021 and attributed primarily to economic growth [4].

Another solution is to expedite the transition to using energy that does not emit GHGs during combustion processes or otherwise lacks a combustion process. Realistic approaches include creating pathways to incrementally transition to clean energy. *Clean energy* usually refers to renewables plus nuclear energy since these resources produce negligible quantities of GHGs [5]. The vision of decarbonization is to evolve to clean energy electric grids using local generation from renewable energy sources (e.g., wind, geothermal, hydropower, and solar power). Excess power can be stored for later use by using batteries (e.g., lithium-ion, sodium-sulfur, vanadium-redox, etc.), compressed air, pumped hydropower storage, or directed to a process storage solution, such as using water for electrolysis to create hydrogen and oxygen [5]. Another alternative is to generate electricity with packaged nuclear power systems [5]. In 2021, renewable energy sources and nuclear power combined provided a larger share of global electricity generation than coal [3]. Renewable energy, led by wind and solar power, continued to grow and now accounts for 13% of total power generation [4]. Renewable electrical generation increased by about 17% in 2021 [4]. It has accounted for over half of the increase in global power generation over the past few years [4].

Pathways to decarbonization are often sector-specific [6]. Instituting decarbonization measures for buildings is a workable solution, and there are many recent trend-setting examples [3,5]. Technological approaches include developing net-zero buildings and processes using green hydrogen for mobility applications [8]. Another approach is to convert all appliances and heating systems to operate using electricity and generate all the electricity required using renewable resources. To make this happen, industries would need to make major infrastructure changes. Key elements of an industrial decarbonization strategy would include gathering energy data, establishing emissions reporting methodologies, developing decarbonization drivers and projects, creating a marginal abatement cost curve, expanding electrification initiatives, and enforcing decarbonization checklists [9]. For industrial concerns, making changes to functional processes often faces formidable internal resistance.

Most years since the 1950s have yielded increased CO_2 emissions (see Figure 1.1). This trend must be reversed. As energy engineers and managers, the world looks to us to identify the pathways that lead to successful decarbonization. Not all efforts toward decarbonization will be effective. There are unforeseen risks. Implementing our solutions will require rigorous analysis, compromise, and ultimately trade-offs. Thankfully, central governments (i.e., China and the European Union) are starting to provide more funding for decarbonization efforts that will enable more of our projects to go forward. Regardless of the approaches we choose, now is the time to advance the policies, programs, and technologies that will successfully enable us to achieve decarbonization.

1.2 Electricity Generation in North America

The types of primary energy we use can be broadly classified as either fossil fuels or clean energy resources. Fossil fuels are the predominant category of fuels that drive the world's economies. They include coal, oil, natural gas, and others. These fuels are used for heating, electrical generation, transportation, and other applications. China leads the world in fossil fuel use, consuming 171 equivalent exajoules (EJ) in 2023.[1,2] Its CO_2 emissions have been increasing when measured by total and per capita emissions since 2000. The U.S. is ranked second in energy consumption. Despite becoming the world's largest producer of oil, the country has been gradually lowering its total and per capita emissions since 2000.

1.2.1 Dependency on fossil fuels

North America provides an example of how the dependence on fossil fuels makes efforts that advance decarbonization policies challenging. The U.S. and Canada are significant users of fossil fuels. The U.S. is the world's second-largest user and consumed 140 exajoules in 2023; Canada is the ninth-largest user, consuming over 9 exajoules.[3]

There have been notable shifts in the types of fuels consumed over time. The principal fossil fuels used worldwide for electrical generation include coal and natural gas. North America has transitioned from being coal-dependent to greater use of natural gas and renewables, historically harnessing hydropower and more recently expanding the use of solar and wind power. Increased investments in renewable energy align with world trends. Beginning in 2013, the world's annual total investments in renewables (including hydropower) have exceeded investments in non-renewables in all subsequent years.

1.2.2 Fuels used for generating electricity

In North America, the U.S. uses primarily natural gas, coal, and nuclear power for electrical generation, while Canada relies on both hydraulic and nuclear power [10]. Currently, 22% of U.S. and 7% of Canada's electricity is generated using coal [10]. U.S. states such as West Virginia (89% of electricity is coal-based), Missouri (70%), Kentucky (68%), and Canadian provinces like Nova Scotia (51%) and Saskatchewan (40%) rely mainly on coal as their primary fuel for the generation of electricity [10]. The states that

---

[1] One exajoule = 174 million barrels of oil equivalent.

[2] Data for China for 2023 includes Hong Kong.

[3] India ranks 3rd using 35 exajoules, Russia is 4th consuming 27 exajoules, and Japan ranks 5th using 15 exajoules.

rely heavily on natural gas include Rhode Island (90.9%), Delaware (85.8%), Massachusetts (76.9%), and Florida (72.1%) [10].

Coal is a critical energy source for producing electricity and cannot be entirely replaced in the foreseeable future [11]. However, coal is combusted using Rankine cycle processes that produce greenhouse gases, nitrogen (N_2), and sulfur oxides (SO_x), which cause smog, acid rain, and introduce particulate matter into Earth's atmosphere [11]. Petroleum is used as the primary fuel for electrical generation in the U.S. state of Hawaii and the Canadian territory of Nunavut (established in 1999). Using fossil fuels to produce electricity consumes large amounts of water. In several western U.S. states where water is a scarce resource, this is a major disadvantage.

For a total of 19 U.S. states, natural gas is now the primary energy source used to produce electricity. These include California, Texas, most of the remaining Sunbelt states, plus New York, New Jersey, Ohio, and Pennsylvania.

There are other sustainable ways to supply electricity to North America's electric grids. Alternatives to fossil fuels that substantially reduce pollution and greenhouse gases include numerous clean energy resources. Clean energy is usually defined as renewable energy (hydropower, solar, wind, tidal, geothermal, etc.) plus nuclear power.

About 19% of U.S. electricity and 15% of Canadian electricity comes from nuclear power [10]. In the U.S., Illinois, Pennsylvania, and South Carolina are the leading producers of nuclear power [10]. In Canada, the province of Ontario, with a 2024 population of 16.0 million, gets about 50% of its electricity from nuclear generation [12].

While greenhouse gas emissions are low, nuclear power has its drawbacks. These include substantial water consumption and long periods of time for permitting, development, and construction. Once operational, radioactive wastes must be managed; in the U.S., they are typically stored on site. High-level nuclear waste consists largely of spent fuel rods from nuclear reactors [13]. Existing and planned nuclear waste sites operate under the assumption that the management of waste storage sites will require time frames from 10,000 to 100,000 years [13]. Perhaps of more immediate concern for development in the U.S. is that constructing new utility-scale nuclear plants has proven to be a very expensive way to add electrical generation capacity [14]. Ratepayers in a utility's service territory are ultimately burdened with the substantial costs associated with amortizing the investments in nuclear plants through increased electrical tariffs [14]. There is also the lingering problem of decommissioning the plants when they need to be retired.

1.2.3 Renewable energy

Some renewables provide intermittent electrical generation—a major concern for electric utilities. For example, the electrical generation potential for solar power is only available when the sun is shining. Solar panel efficiencies vary based on geographic location, time of day, cloud cover, shading, outdoor air temperatures, and other factors [15]. Connecting renewable generation to the electric grids can cause delays, particularly if the sites for solar and wind power are distant from utility substations. However, water use is minimal, which can be a regional advantage.

Hydropower is used to supply 60% of the electricity for Canada and 6% for the U.S. [10]. Without hydropower, it is difficult to provide utility-scale and reliable baseload electricity using a renewable energy resource. In the U.S., states that generate significant shares of their electricity from hydropower include Washington (65%), Idaho (51%), Vermont (50%), and Oregon (46%) [10]. The Canadian provinces that use primarily hydropower include Manitoba (97%), Newfoundland and Labrador (95%), Quebec (94%), and British Columbia (87%) [10]. Quebec (population of 9.0 million) and British Columbia (5.6 million) are the second and third most populated provinces, following Ontario, which highlights the importance of hydropower [12].

Wind power supplies notable shares of electricity for the U.S. states of Iowa (62%), South Dakota (55%), Kansas (46%), Oklahoma (42%), and South Dakota (55%). The state of Texas produces the most electricity annually using wind, yet it accounts for only about 16% of total generation.

1.3 Renewables in Europe's Energy Mix

Renewable energy has moved from being a set of emerging technologies to dominance in the energy mix of some countries. Worldwide, renewable energy made up 93% of all new power capacity in 2024; non-renewables were just 7% [16]. This lopsided contrast underscores a continuing structural transition from fossil fuels to renewables, especially for electrical generation. Renewable energy resources added 582 GW of capacity in 2024 [16]. Yet current growth in production remains below the levels required to meet the increase in global demand for energy and the need to further add capacity.

The use of renewable energy is expanding in many European countries. They have worked in concert to reduce their reliance on fossil fuels and, to a lesser extent, nuclear power. The use of coal to generate electricity fell in mid-2025 to a 6.1% share, dropping from a 25% share in 2016. This incremental transition has been supported by fuel substitution efforts and energy efficiency improvements. There is a growing focus on the use of renewable energy resources for energy production. In 2023, renewable energy represented a 24.5% share of energy consumed in the European Union (EU), up from 23.0% in 2022 [17].

What are the reasons for this change in focus? An important one is changes in energy policies that provide sustained support for the increased use of renewables despite changes in EU leadership. EU Directive 2023/2413 centers on the promotion of the use of renewable sources; it has revised upwards the EU's 2030 renewable energy target above the present share of 32% [17]. The directive aims to end EU dependency on Russian natural gas, oil, and uranium by 2027 while boosting renewable energy. It increases the EU's binding renewable energy target that needs to be collectively achieved by 2030 from 32% to 42.5%, with a notable aspiration of reaching 45% [18]. It also strengthens the sub-targets for renewables to be achieved across various economic sectors, including industry [18].

1.3.1 Successful policies in the European Union

This revised renewable energy directive is a cornerstone of REPowerEU and the European Green Deal [18]. REPowerEU is an initiative by the European Commission to respond to the energy crisis caused by Russia's invasion of Ukraine [19]. It is backed by almost €300 billion in additional funding [19]. The European Green Deal is a comprehensive policy initiative aimed at making the EU climate-neutral by 2050 to promote sustainability and economic transformation. A key driver was to reduce dependence on natural gas imports that have been used periodically by Russia as an economic weapon [18]. Though most of its liquified natural gas terminals are underutilized, the EU is building new ones to enable increased imports from other sources, such as the U.S. Other policy objectives include boosting industrial decarbonization, faster deployment of renewables, increasing investments in energy infrastructure and interconnections, and improving energy efficiency [19].

What successes have EU countries had in reaching these goals? Figure 1.2 indicates the shares of renewable sources in gross energy consumption. Each European country has a unique pathway to advance renewables, and some have been more successful than others. Sweden ranked first among the EU countries, with 66.4% of its gross final energy consumption coming from renewable sources in 2023 [20]. Sweden primarily relies on solid biofuels, hydropower, and wind turbine generation. Finland followed with 50.8%, also relying on solid biofuels, wind, and hydropower [20].

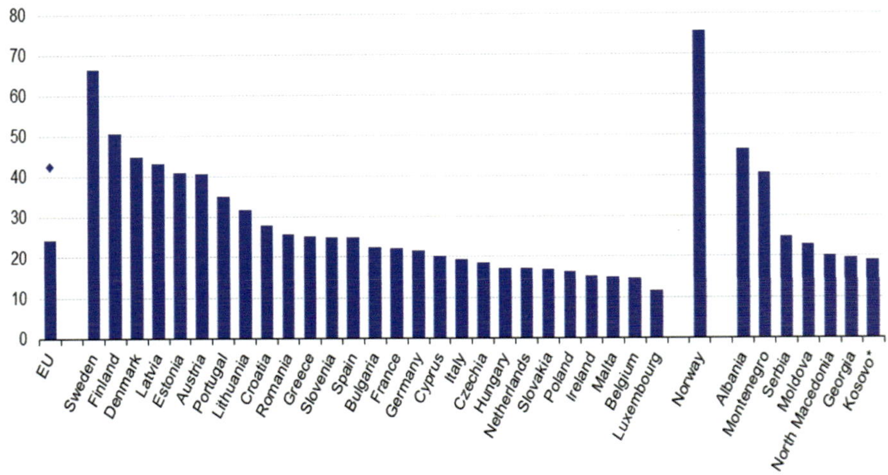

Figure 1.2 Share of energy from renewable sources in European countries, 2023 (%) [20].

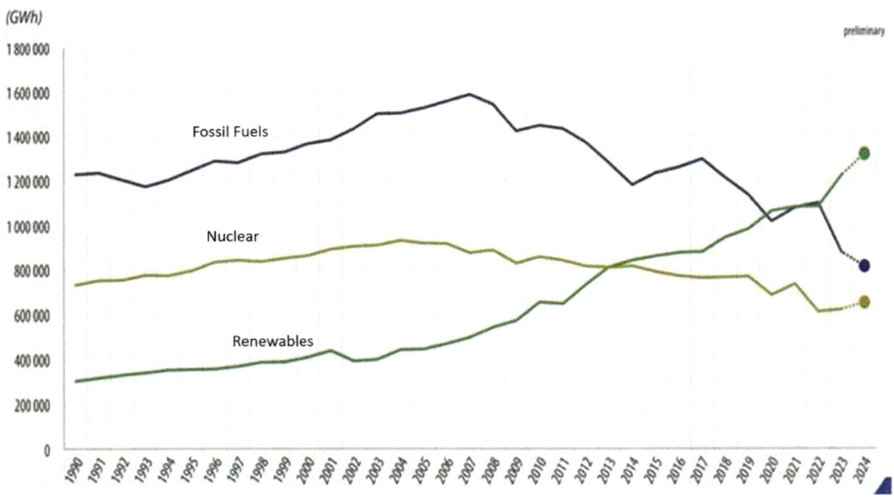

Figure 1.3 Electricity production in Europe, 1990–2024 (GWh) [source: Eurostat].

Denmark is ranked third with 44.9%; most of its renewable energy is sourced from solid biofuels and wind [20].

Germany leads in the development of onshore wind power and uses compressed air storage facilities for wind smoothing. There are currently 26 pumped storage hydropower stations in Germany with a total capacity of

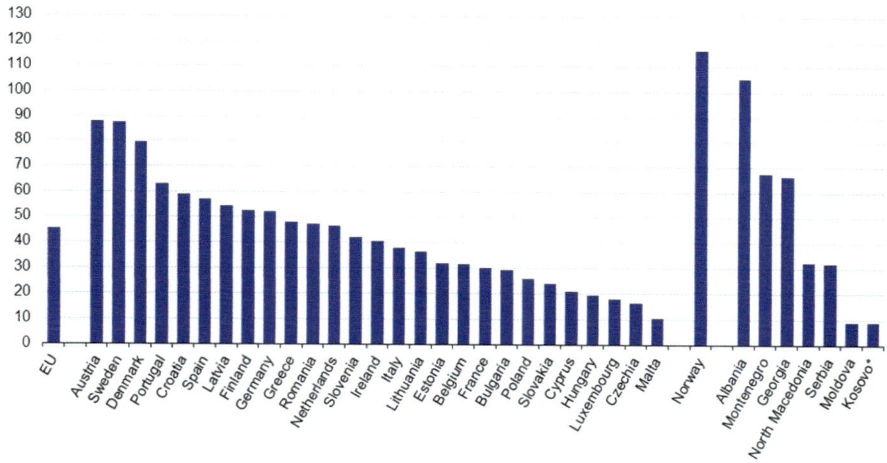

Figure 1.4 Share of energy from renewable sources in gross electricity production, 2023 (%) [20].

6.3 GW, and a further 1.4 GW are planned to be added by 2030 [22]. The Netherlands gets 100% of the electricity for its extensive railway system from wind power. Other European nations struggle to incorporate renewable energy despite lofty objectives. The lowest shares of renewables were recorded in Malta (15.1%), Belgium (14.7%), and Luxemburg (11.6%) [5]. With just a few years left, Belgium's goal of having half of its electricity sourced from green energy sources by 2030 seems unachievable.

Over the last two decades, electrical power generation using fossil fuels and nuclear energy has declined in the EU. In the last few years, renewable energy resources have displaced fossil fuels as the primary means of generating electricity in Europe (see Figure 1.4). Albania, Austria, Denmark, Norway, and Sweden generate over 70% of their electricity from renewables (see Figure 1.4). The advent of inexpensive solar power has driven the price of electricity downward during some mid-day periods, causing electricity prices to be negative, especially in Sweden, Germany, Spain, and the Netherlands [21]. So much electricity is generated at times that it has nowhere to go. This is being driven by the continued expansion of renewable capacity without a commensurate increase in underlying demand that matches the generation curve [21]. For these countries, when supply exceeds demand, electricity prices tend to drop rapidly [21]. When generation from solar collectors declines precipitously during evenings, it is necessary to ramp up the output from energy storage systems or natural gas plants. Renewable energy use has increased in France, but the country remains a leader in the use of nuclear energy, which accounts for about 65% of its electrical output.

Many EU countries will soon be capable of ridding themselves of dependence on Russian fossil fuels while continuing to pursue the transition to green energy infrastructure [19]. With the increased use of renewables for generation, energy storage systems, which reached 89 GW of capacity in 2024 with over 3,000 projects, will need to be further expanded [23]. Because of concerted efforts to align policies, programs, and technological advancements, the use of renewables will continue to increase and hold a larger share in Europe's energy mix in the future.

1.4 Conclusion

Providing reliable power supplies and resilient electrical generation can be challenging. In North America, regional differences in resource availability and cost are important factors in fuel selection for electrical generation. The U.S. remains heavily dependent on fossil fuels to generate electricity, while Canada has a high share of renewables in its generation mix. Particularly for the U.S., the dependency on fossil fuels makes a transition to clean energy challenging. Infrastructure supporting the use of fossil fuels is in place and has not been entirely amortized.

While some U.S. states and a few Canadian provinces rely heavily on coal, it is not just coal that keeps the lights on. In North America, nuclear energy is used for electric generation in two Canadian provinces and five U.S. states. It is surprising to many that hydropower is the primary resource used to generate electricity in seven Canadian provinces and territories and five U.S. states. With hydropower supplying 60% of Canada's electricity and nuclear power another 15%, the country has proved that it is dedicated to the use of clean energy fuels for electrical generation. While there are no U.S. states or Canadian provinces that use solar energy as the primary source of electrical generation, there are a few U.S. states in the Great Plains that use predominantly wind power. This transition to greater use of clean energy resources for electrical generation is already reducing water use in some areas of the U.S.

In the U.S., the dependence on coal is gradually diminishing. Coal-fired generation in the U.S. is being replaced by the increased use of natural gas and renewables such as solar and wind power. This trend could be easily reversed in the face of increased demand for electricity due to the development of data centers, greater use of electric vehicles, and the need to add air conditioning systems in buildings. We will likely see a push to extend the lifecycles of some coal-fired generating stations. This will be justified primarily by the heavy, continuous electricity demand required by new facilities designed for data processing to support the growing artificial intelligence industry. The need for additional generation will increase consumer costs because power from newly constructed stations will be more expensive to generate than power from existing ones.

The results for European Union countries are mixed, but they have made advancements toward decarbonization. Despite continuing challenges such as regulatory hurdles, financial constraints, and technical issues, the EU has successfully met many of the ambitious targets set out in its REPowerEU plan [19]. The EU has dedicated its policies and instituted programs to support its transition to clean energy. With an evolving dependency on solar and wind power, the EU will need contingency plans and infrastructure to generate electricity during days that are both cloudy and windless. Solutions may entail adding hydropower resources, energy storage, and programmed use of natural gas generation.

There are tough decisions to be made as countries take actions to reduce carbon emissions. Each of the world's countries will have a different pathway to achieving decarbonization. This energy transition from fossil fuels to clean energy resources will continue due to increased investments in renewables. Unless we experience another industry disaster, there will be growing support to revisit the benefits of nuclear energy. For the U.S., the energy transition involves using a greater mix of energy resources. To offset the decline in the use of coal, there will be greater reliance on natural gas and renewables. Electrical generation in the U.S. from the use of solar and wind power may eventually exceed that of coal-fired generation. For the EU, the energy transition is centered on freeing itself of dependence on energy from Russia and elsewhere. To accomplish this, there has been a focus on greater use of renewable energy.

References

[1] Marr, B. (2022, February 11). The 3 biggest future trends (and challenges) in the energy sector. https://www.forbes.com/sites/bernardmarr/2022/02/11/the-3-biggest-future-trends-and-challenges-in-the-energy-sector/?sh=71f4a87127b7, accessed 2 August 2022.

[2] Volkswagen AG (2022). Decarbonization–what is it? https://www.volkswagenag.com/en/news/stories/2019/03/decarbonization-what-is-it.html, accessed 2 August 2022.

[3] International Energy Agency. Global energy review: CO_2 emissions in 2021. https://iea.blob.core.windows.net/assets/c3086240-732b-4f6a-89d7-db01be018f5e/ GlobalEnergyReviewCO2Emissionsin2021.pdf, accessed 22 January 2026.

[4] bp (2022). Statistical review of world energy, 71st edition. https://www.bp.com/content/dam/bp/business-sites/en/global/corporate/pdfs/energy-economics/statistical-review/bp-stats-review-2022-full-report.pdf, accessed 29 August 2022.

[5] Roosa, S. (2022, May). What to do about climate change? *International Journal of Strategic Energy and Environmental Policy*, 4(3), 8–30.

[6] Roosa, S. (2022, September). Creating pathways to clean energy. *International Journal of Strategic Energy and Environmental Policy*, 4(5), 69–76.

[7] Annan, G. (2022, September). Building decarbonization transformation. *International Journal of Strategic Energy and Environmental Policy*, 4(5), 61–68.

[8] Wiegand-Jackson, L. (2022, September). On the path to net zero: green hydrogen as a decarbonization solution. *International Journal of Strategic Energy and Environmental Policy*, 4(5), 8–23.

[9] Miller, R. (2022, September). Decarbonization strategies for industrials. *International Journal of Strategic Energy and Environmental Policy*, 4(2), 26–37.

[10] Oguz, S., Dander, A. and Wadswoth, C. (2023, January 12). Mapped: biggest sources of electricity by state and province. https://elements.visualcapitalist.com/biggest-sources-of-electricity-by-state-and-province, accessed 27 August 2024.

[11] McIntyre, D. (2024, August 16). Coal is energy king, and nothing could be worse. https://www.msn.com/en-us/money/companies/coal-is-energy-king-and-nothing-could-be-worse/ar-AA1oUjeo?ocid=msedgdhp&pc=U531&cvid=c2d43926426240d88e07e9936773de28, accessed 30 August 2024.

[12] Wikipedia. Population of Canada by province and territory. https://en.wikipedia.org/wiki/Population_of_Canada_by_province_and_territory.

[13] Ro, C. (2019, November 26). The staggering timescales of nuclear waste disposal. https://www.forbes.com/sites/christinero/2019/11/26/the-staggering-timescales-of-nuclear-waste-disposal, accessed 30 August 2024.

[14] Roosa, S. (2024, September). Nuclear energy—lessons from Georgia's new power plant. *Alternative Energy and Distributed Generation Journal*, 6(3), 5–12.

[15] Roosa, S. (2024, June). Software analysis of solar PV system efficacy. *Alternative Energy and Distributed Generation Journal*, 6(2), 30–52.
[16] International Renewable Energy Agency (2025). Renewable energy statistics. https://www.irena.org/-/media/Files/IRENA/Agency/Publication/2025/Jul/IRENA_DAT_RE_Statistics_2025.pdf.
[17] Eurostat (2025). Statistics explained, renewable energy statistics. https://ec.europa.eu/eurostat/statistics-explained/index.php?title=Renewable_energy_statistics, accessed 14 July 2025.
[18] European Commission (2024, February 9). Guidance on the targets for the consumption of renewable fuels of non-biological origin in the industry and transport sectors laid down in Articles 22a, 22b, and 25 of Directive (EU) 2018/2001 on the promotion of energy from renewable sources, as amended by Directive (EU) 2023/2413.
[19] European Commission (2025, June 17). REPowerEU, affordable, secure and sustainable energy for Europe. https://commission.europa.eu/topics/energy/repowereu_en, accessed 17 July 2025.
[20] Eurostat (2025). Statistics explained, renewable energy statistics https://ec.europa.eu/eurostat/statistics-explained/index.php?title=Renewable_energy_statistics#Share_of_renewable_energy_almost_tripled_between_2004_and_2023, accessed 14 July 2025.
[21] Energy Live News (2025, July 15). Europe's clean power paradox: record solar sends prices below zero. https://www.energylivenews.com/2025/07/15/europes-clean-power-paradox-record-solar-sends-prices-below-zero, accessed 15 July 2025.
[22] Wehrmann, B. (2020, July 24). German government says pumped hydro power capacity to grow by 1.4 GW by 2030. https://www.cleanenergywire.org/news/german-government-says-pumped-hydro-power-capacity-grow-14-gw-2030, accessed 19 July 2025.
[23] European Association for Energy Storage (2025, March). EMMES 9.0. https://ease-storage.eu/publication/emmes-9-0-march-2025, accessed 19 July 2025.

2

Decarbonization for a Sustainable Energy Future

Kritika Bhandari, Abinash Kumar Gupta, and Hitesh D. Vora

This chapter presents a comprehensive analysis of strategies for reducing industrial sector greenhouse gas (GHG) emissions to help combat climate change. It emphasizes the sector's contribution to global GHG emissions, particularly CO_2, and outlines the technical and economic challenges in achieving decarbonization. This chapter discusses various approaches, such as hydrogen-enriched natural gas (HENG), carbon capture and storage (CCS), policy frameworks, and innovative technologies for iron, steel, and glass production. It advocates for collaborative efforts among governments, industries, and communities to overcome barriers, align with global climate goals, and highlight the urgent need for a transition to low-carbon energy resources. It includes a sample calculation demonstrating the correlation between decarbonization and energy savings. This study serves as a guide for policymakers, business executives, and industry stakeholders in effectively pursuing sustainable decarbonization strategies.

2.1 Introduction

Decarbonization in the industrial sector is crucial for achieving sustainable energy futures and mitigating climate change impacts. The industrial sector is a major contributor to global greenhouse gas (GHG) emissions. This sector is also pivotal in developing low-carbon solutions, including renewable electricity generation facilities, clean vehicles, and energy-efficient processes [1]. The urgency of reducing industrial emissions is underscored by the Paris Agreement's target to limit global warming to below 2°C, ideally 1.5°C,

DOI: 10.1201/9788743811312-2

which necessitates an 80% to 95% reduction in GHG emissions by 2050 compared to 1990 levels [2].

The industrial sector accounts for approximately 28% of global GHG emissions, of which 90% are carbon dioxide (CO_2) emissions. Between 1990 and 2020, these emissions from the industrial sector increased by 65% (averaging about 2.0% annually), outpacing the overall global GHG emissions growth [3]. Addressing industrial GHG emissions is challenging due to the sector's role in global prosperity, contributing about a quarter of global gross domestic product (GDP) and employment. The sector's GHG emissions stem from the production of essential materials like cement, steel, ammonia, and ethylene, which are important to our daily activities [2].

Decarbonizing the industrial sector requires a multifaceted approach, addressing both direct and indirect emissions. For example, in the United States, the industrial sector's direct process emissions from iron, steel production, and metallurgical coke production reached 40.7 MMT CO_2 equivalent (eq.) in 2022. Similarly, emissions from cement production totaled 41.9 MMT CO_2 eq. the same year [4]. Reducing these emissions involves a variety of decarbonization options, including fuel switching, carbon capture and storage (CCS), and technological changes like electrification or oxygen-fired kilns [5].

Achieving meaningful decarbonization in the industrial sector faces numerous technical and economic barriers. About 45% of CO_2 emissions in focus sectors such as cement, steel, ammonia, and ethylene production result from feedstock processing, which can only be reduced by changing feedstocks or processes. These sectors also have a significant demand for high-temperature heat, necessitating substantial changes in production processes and furnace designs to accommodate alternative energy sources like hydrogen [3]. The need to transition to low-carbon energy resources in the industrial sector is urgent, given the projected growth in industrial output through 2050. Meeting the Paris Agreement targets will require a large-scale reconfiguration of industrial activities, affecting the energy and resources required for production, production processes, and facility control systems (see Figure 2.1) [2].

This study provides a comprehensive overview of the tools and transformative technologies available for industrial decarbonization. It highlights the gaps, limitations, and potential recommendations, focuses on the best practices toward low-carbon emissions, and aligns with global climate goals. Guidance is provided to policymakers, business executives, and industry stakeholders in prioritizing and pursuing decarbonization strategies effectively and sustainably.

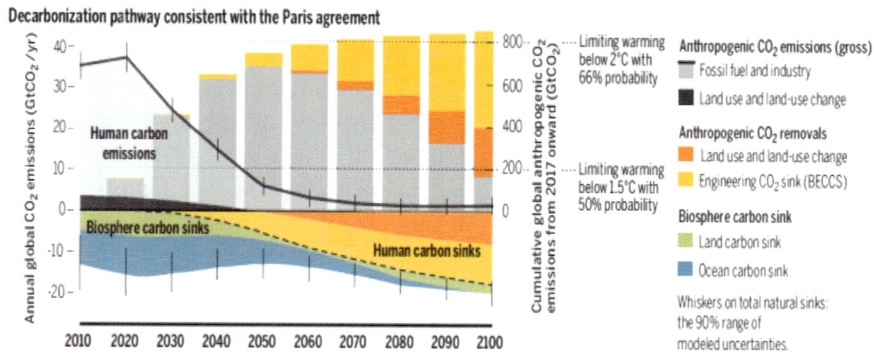

Figure 2.1 Decarbonization pathway to achieve the goals of the Paris Agreement. Note: a deep decarbonization scenario requires immediate reductions in CO_2 emissions, targeting net-zero emissions by 2050. This scenario, slightly more ambitious than previously proposed, relies on enhanced carbon sinks through bioenergy carbon capture and storage, land management, and use of natural sinks. It aligns with a high probability of keeping global warming below 2°C, aiming for a median temperature rise of 1.5°C by 2100, with CO_2 levels at 380 ppm [6].

2.2 The Importance of Decarbonization

Decarbonization involves reducing carbon dioxide emissions, primarily in the energy sector, by transitioning to renewable energy sources and improving energy efficiency. Such reductions are essential to limit the effects of climate change. Technologies and policies have crucial roles in achieving GHG emission reductions. The options include generating secondary energy carriers, targeting end-use energy sectors, understanding sector interdependencies, and perhaps instituting a carbon tax. However, instituting this process is challenging due to high capital requirements, competition among sectors, and inconsistent environmental policies [7].

In recent years, the need for decarbonization has gained prominence due to the increasing intensity of weather phenomena, concerns about climate change, and the anthropogenic nature of these changes, largely attributed to the combustion of fossil fuels [8]. The energy sector's decarbonization efforts, particularly the use of renewable energy (RE), have been an important focus. Countries like Paraguay, Norway, and Costa Rica have achieved high shares of RE in electricity generation, demonstrating the feasibility of major shifts to renewable resources. However, the complexity of decarbonization involves numerous challenges and interdependencies, especially in secondary energy carriers and end-use energy sectors [9].

2.2.1 Industrial decarbonization

The industrial sector is a cornerstone of global wealth and prosperity, contributing to national wealth and employment. However, it is also a major source of GHG emissions and accounts for about 28% of global GHG emissions. From 1990 to 2020, industrial GHG emissions increased significantly, which highlights the urgency of developing and implementing energy transition pathways for industrial processes [3].

Industrial decarbonization is important for meeting international climate goals, such as those outlined in the Paris Agreement (also called the Paris Accords or Paris Climate Accords). Industries must reduce GHG emissions by 80% to 95% of 1990 levels by 2050 to limit global warming to 2°C. This necessitates upgrading or replacing industrial facilities with lower-emission technologies, which requires planning and substantial capital investments. The potential for industrial decarbonization is vast, but achieving it requires overcoming numerous economic and technological barriers [2].

2.2.2 Challenges in industrial decarbonization

Challenges to industrial decarbonization efforts can be categorized as technical and process-related, economic and competitive pressures, investment and infrastructure needs, longevity of industrial sites, and load factors and variability. Each of these challenges is described in greater detail as follows.

Technical and process challenges: Reducing emissions in industries like cement, steel, ammonia, and ethylene production requires changes in feedstocks and processes. Approximately 45% of the GHG emissions in these industries are generated during feedstock processing, necessitating a shift from traditional methods to more sustainable alternatives [2].

Economic and competitive pressures: Decarbonizing the ammonia, ethylene, and steel manufacturing industries faces economic hurdles, as these industries are highly price-competitive. Adopting low-carbon processes may increase production costs, leading to competitive disadvantages in global markets.

Investment and infrastructure needs: Achieving industrial decarbonization will require substantial infrastructure investment, estimated to be roughly $21 trillion globally by 2050. This includes both capital and operational costs for industrial sites, excluding additional costs such as the buildout of zero-carbon electricity generation capacity [2].

Longevity of industrial sites: Industrial plants typically have long lifespans, often exceeding 50 years. Implementing decarbonization measures at these existing sites demands capital-intensive retrofits or rebuilds, adding to the complexity and costs of the transition [10].

Local factors and variability: The effectiveness and feasibility of decarbonization technologies vary based on local factors such as the availability of low-cost, zero-carbon electricity, sustainably produced biomass, and carbon storage capacity. Such factors determine the suitability and cost-effectiveness of various decarbonization options [9].

While decarbonization is important for addressing climate change, its implementation in the industrial sector is hindered by complexities. It requires collaborative efforts from governments, industries, and communities to overcome technical, economic, and infrastructural barriers. Such efforts are vital for achieving a sustainable energy future and mitigating the impacts of climate change.

2.3 Decarbonization Strategies

Transitioning towards a sustainable energy future necessitates innovative strategies to reduce carbon emissions. Examples include hydrogen-enriched natural gas processes, carbon capture and storage, plus technological innovations that are industry-specific. Policies and programs that provide pathways to decarbonization are also needed.

2.3.1 Hydrogen-enriched natural gas

Hydrogen-enriched natural gas (HENG) presents a promising pathway by leveraging existing natural gas infrastructure to distribute hydrogen as an energy carrier. This process utilizes an extensive network of natural gas pipelines and moderates the challenges associated with hydrogen storage and transportation [1,11].

The catalytic thermal decomposition of methane gas (CH_4), a principal component of natural gas, into hydrogen (H_2) and solid carbon is a carbon-free process that aligns with decarbonization objectives.

Unlike conventional steam methane reforming (SMR), the catalytic decomposition of methane (CDM) does not produce CO_2 as a byproduct. Instead, it generates carbon black, which is a valuable commodity in various industries. This process offers dual benefits—the production of both clean hydrogen fuel and a commercially valuable form of carbon [12]. The efficiency of the CDM process is enhanced by the choice of catalysts. Research has shown that nickel (Ni) and iron (Fe)-based catalysts, especially when supported on alumina (Al_2O_3) and silica (SiO_2), exhibit high efficiencies and environmental compatibility. These catalysts facilitate the decomposition of methane at high yields, with conversion rates ranging between 70% and 80% at operating temperatures of 600°C to 700°C [11].

Recent advancements include the development of wet-impregnated, sub-micron catalysts that offer cost-effective and thermally stable solutions up to approximately 800°C. The scalability of these reactions to pilot-scale operations has been explored using ASPEN simulations, revealing the potential for competitive hydrogen production costs compared to existing methods [11]. Incorporating HENG into the energy mix entails considerations regarding the existing natural gas infrastructure's adaptation, safety protocols, and the calorific value adjustments. The gradual introduction of HENG can facilitate a smooth transition, leverage the strengths of the current system, and progressively integrate hydrogen as a sustainable energy source.

Techno-economic assessments underscore the potential of HENG in achieving a lower levelized cost of hydrogen (LCOH) and minimal GHG emissions compared to traditional hydrogen production technologies. The successful implementation of HENG can contribute to decarbonizing the energy sector, aligning with global efforts to mitigate the impacts of climate change (see Figure 2.2) [11].

2.3.2 Carbon capture and storage

In the quest for a sustainable and decarbonized energy future, CCS technologies emerge as critical components in the portfolio of climate change mitigation strategies. These technologies address the inevitable residual emissions from industrial processes that are not easily abated through electrification or renewable energy integration [1]. CCS has a well-established technological basis, historically utilized in oil refining and natural gas processing; it is now pivoting towards broader industrial applications. The processes are diverse, with pre-combustion, post-combustion, and oxyfuel combustion techniques as the primary methods for carbon dioxide capture. Each method demonstrates a unique capability to integrate with existing industrial frameworks and the potential to adapt to future energy landscapes [1,2].

Implementing industrial-scale CCS is not without challenges. The energy-intensive nature of capturing and compressing CO_2 introduces an efficiency penalty, potentially elevating operational costs [13]. It increases

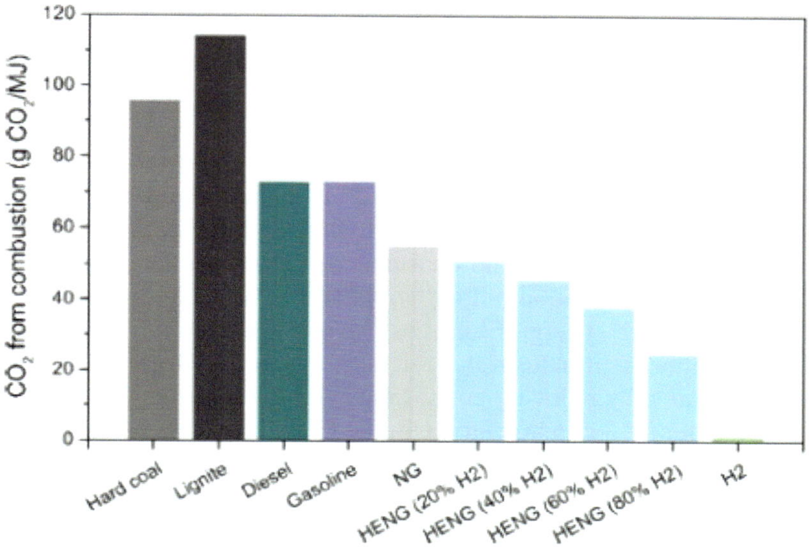

Figure 2.2 Emissions of CO_2 per megajoule of energy produced from burning fuels [11].

air pollution if not managed with additional pollutant capture measures [13]. Despite these challenges, CCS is a commercially ready technology, as evidenced by numerous operating projects with substantial CO_2 capture capacities. The key to the widespread adoption of CCS lies in creating a supportive policy environment that provides the necessary infrastructure and economic incentives. This support could range from carbon pricing mechanisms to subsidies and standards that promote CCS technology deployment to help ensure a transition to a low-carbon industrial sector [1,14].

2.3.3 Policies and the roles of industry

Attaining decarbonization targets in the industrial sector can be advanced by developing comprehensive roadmaps for local and regional governments. These roadmaps are instrumental in defining a long-term direction for decarbonization, facilitating coordination among industrial companies, utilities, and infrastructure owners. By setting a defined trajectory, they can unlock investments in technologies with longer payback periods, ensure the judicious deployment of resources, including zero-carbon electricity, and promote the adoption of supply- and demand-side energy measures. To this end, government policies have a crucial role in crafting regulatory frameworks and financial incentives that are aligned with decarbonization goals. These policies may encompass direct incentives for companies to engage in decarbonization efforts or adapt the financial obligations of energy providers to favor green initiatives [2].

Industrial corporations need to conduct thorough reviews of their operations to assess the availability and viability of decarbonization resources such as low-cost zero-carbon electricity, hydrogen, biomass, and CCS capacities. These assessments should consider the unique circumstances of each site, fostering collaboration among governments, utilities, and other stakeholders to uncover synergies and cost-saving opportunities. For example, companies situated in industrial clusters might share infrastructure for carbon storage, thereby reducing costs. Innovation has an important role in the decarbonization of the industrial sector, necessitating concerted efforts from governments, companies, and research institutions to scale up promising technologies that can lower transition costs and improve efficiency. The collective efforts required for industrial decarbonization mirror the collaborative endeavors that have spurred growth and cost reduction in renewable energy technologies [1,2].

2.3.4 Strategies for iron, steel, and glass production

Iron and steel production, pivotal to the global economy, accounts for large quantities of carbon dioxide emissions due to its reliance on carbon-intensive processes. Transitioning away from coal to charcoal in blast furnaces is a mature and economically feasible way of reducing GHG emissions. Similarly, electric arc furnaces can be powered by renewable electricity to lower their carbon footprint. Direct reduced iron (DRI) production, traditionally reliant on natural gas, can instead utilize biogas or potentially hydrogen, although the latter remains in the pilot phase [2]. Carbon capture and storage technologies offer a means to decarbonize exhaust gases from conventional steelmaking. For deeper emission cuts, innovations such as electrical reduction of iron ore and top gas recycling could be implemented, albeit these require CCS to realize their full decarbonization potential [1,2].

The production of glass, another energy-intensive process, necessitates high-temperature heat for the melting and processing of raw materials. Current decarbonization strategies include the utilization of bio-based feedstocks such as biodiesel or sugar, which can reduce both the production and end-of-life emissions. Furthermore, glass recycling has a crucial role in decreasing the demand for virgin materials and the energy use and emissions associated with glass production. In conjunction with innovative processes and materials, such as new polymer types created through bio-based or electrochemical production, these strategies can diminish the carbon footprint of glass manufacturing processes [2].

2.4 Tools and Transformative Technologies for Decarbonization

Decarbonization in the industrial sector creates multifaceted challenges that demand a suite of tools and transformative technologies tailored to reduce carbon intensity. The spectrum of decarbonization strategies reveals a technological landscape in which innovations range from nascent research-phase options to fully developed applications. Industrial companies are investigating energy efficiency, the use of biomass as a mature technology, and the integration of low-carbon electricity, hydrogen, and carbon capture and storage into their operations. Electrification of heat, hydrogen as fuel or feedstock, and biogas in ammonia production represent adoption-phase options that are ready to apply with minor process modifications. Meanwhile, research-phase technologies like electrochemical-based monomer production from biomass and development-phase options such as electrical steam-cracking furnaces are undergoing industrial-scale tests to validate their commercial viability and decarbonization potential [2].

The transition towards these low-carbon technologies is dependent on their technical readiness and economics; factors like the price of zero-carbon electricity are important. Decarbonization costs are closely linked to the cost of capital and energy, steering the industrial sector's choice among various decarbonization options. For instance, current electricity prices can make CCS the most economical choice for industries; however, a decrease in electricity prices may shift the preference towards electrification and hydrogen-based technologies. Companies are advised to review their asset portfolios, considering access to resources like zero-carbon electricity or hydrogen, and anticipate strategic investments in innovation. Concurrently, governments and regulators can facilitate this transition by establishing roadmaps for industry decarbonization, aligning regulations, and supporting the development of carbon storage infrastructure and renewable energy resources [6].

The potential for biomass to be used as a fuel or feedstock introduces another dimension to decarbonization strategies. Using biomass closes the loop on the carbon cycle. Biomass can be more cost-effective than electrification or hydrogen use, particularly in cement production, and at higher electricity prices in steel production. Biomass offers the advantage of addressing emissions both in the production process and at the product's end-of-life; for example, in the incineration of plastics derived from ethylene. However, the availability of sustainably produced biomass is limited. While it is a viable option, it cannot be the sole solution to industry-wide decarbonization [6].

Demand-side energy measures present opportunities for emissions reduction. Substituting conventional industrial products with lower-emission alternatives, such as replacing cement with timber in construction, offers a pathway to substantial GHG emission reductions. Additionally, the increased circularity of products, through recycling or reuse, can lead to energy and feedstock savings. For example, recycled steel can be less energy-intensive compared to producing virgin steel. This illustrates the potential for demand-side innovations that contribute to decarbonization [2].

2.5 Cost Saving Potential Through Decarbonization

Industrial Assessment Centers (IAC), a program managed by the U.S. Department of Energy under the Office of Manufacturing and Energy Supply Chains (MESC), conduct audits to identify energy-saving opportunities across various industries [15]. The IAC at Oklahoma State University specifically offers energy and productivity assessment services to small and medium-sized manufacturers in Oklahoma, Arkansas, Kansas, and the North Texas region [16].

For our research, we analyzed the IAC database to extract records containing decarbonization-related assessment recommendations. The primary assessment recommendation codes (ARCs) for decarbonization were identified as follows: 1) 2.911, use solar to make electricity/generate heat; 2) 2.7491, insulate glazing, walls, ceiling, and roofs to reduce heating, ventilating and air conditioning (HVAC) loads; and 3) 2.914, use biofuels for energy generation. We have compiled data on recommendations made by all IACs from 2014 to 2023, along with the potential energy savings for each instance. These data are presented in Table 2.1 [15].

2.5.1 Correlation of decarbonization with energy efficiency

Decarbonization encompasses various strategies aimed at reducing CO_2 emissions. The three assessment recommendations (ARs) mentioned are energy efficiency strategies that contribute to reducing energy consumption. When implemented, they indirectly reduce CO_2 emissions, given the direct relationship between energy use and CO_2 emissions. These three strategies are of particular interest because they support decarbonization goals across both domestic and industrial sectors, aligning with the 2030 decarbonization scenario [17].

Table 2.1 Industrial Assessment Center database for assessment recommendations 2.911, 2.7491, and 2.914 (from 2014 to 2023) [15].

Year	AR 2.611 Electricity savings (kW/year)	AR 2.611 Dollar savings ($/year)	AR 2.7491 Electricity savings (kW/year)	AR 2.7491 Energy savings (MMBtu/year)	AR 2.7491 Dollar savings ($/year)	AR 2.914 Electricity savings (kW/year)	AR 2.914 Dollar savings ($/year)
2014	3,263,404	276,535	76,797	10,249	10,249	0	0
2015	2,872,276	309,539	0	2,104	2,104	0	0
2016	5,577,566	625,850	205,306	26	990,025	3,960,000	990,000
2017	15,837,850	1,512,446	126,262	315	315	0	0
2018	15,045,196	1,455,763	891,020	5,147	5,147	0	0
2019	39,115,084	3,256,001	870,166	6,685	6,685	0	0
2020	24,106,278	2,237,019	1,421,220	885	1,670,704	176,705	1,669,819
2021	49,823,370	3,674,632	34,976	539	474,669	1,246,452	474,130
2022	138,638,116	9,878,699	1,147,965	14,119	246,478	30,263	232,359
2023	231,797,260	18,494,595	640,448	5,945	653,545	26,266	647,600

Table 2.2 Input data for an assessment recommendation.

Input data for savings calculations	Units
Electricity cost	per kWh
Electric demand cost	per kWh
Natural gas cost	per MMBtu
Average temperature maintained in summer	°F
Average temperature maintained in winter	°F
Total area to be insulated	ft^2
Thermal resistance of the existing wall/window/ceiling	hr.-ft^2°F/Btu
Thermal resistance of the proposed wall/window/ceiling	hr.-ft^2°F/Btu
Average heating degree days (HDD) of that area	°F
Average cooling degree days (CDD) of that area	°F
Plant heating efficiency	% AFUE
Heating load factor (C_H)	%
Cooling load factor (C_C)	%
Coefficient of performance (COP) of cooling unit	
Installed cost of ceiling insulation	$

2.5.2 Sample assessment recommendation

To understand the correlation among these assessment recommendations and decarbonization efforts, we present the necessary calculations for one of the three ARs. We selected AR 2.7491, insulate glazing, walls, ceiling, and roofs to reduce HVAC loads. To establish a strong evidentiary foundation for energy savings, we analyzed the inputs and calculations within the overall framework of the energy audit evaluation recommendations (see Table 2.2). This involved consulting previous IAC reports and focusing on insulation for assessment recommendation purposes. An earlier IAC report from Oklahoma State University proposed ceiling insulation, which was used as a reference to develop a comprehensive assessment of an AR that can be standardized for future insulation-related recommendations.

The basic energy-saving calculation assumes that heating is provided by natural gas and cooling by electricity. This calculation procedure follows:

Change in *U*-value from insulation (BTU/ hour-ft^2-°F)
= [(1 / Existing *R* value) - (1 / Proposed *U*-value)
= change in *U*-value of ceiling]

Avoided heat loss: Entire heating season (Energy savings, MMBtu/year)
= [(Change in *U*-value) (Area, ft^2) (Heating degree days, *HDD)*
(24) (Load factor, *CH*)] / [(Plant heating efficiency)
(BTUs per unit of heating fuel)]

Reduction in heat infiltration: Entire cooling season (Energy savings, kWh/year)
= [(Change in *U*-value) (Area ft^2) (Cooling degree days, CDD) (24) (Load factor, *CC*)] / [(3,412 BTU/kWh) (Coefficient of performance, COP)]

Energy cost savings
= (Energy savings due to avoided heat loss) (Natural gas cost) + (Reduction in heat infiltration in summer) (Electricity cost)

These calculations do not incorporate savings from electric demand. Electric demand savings vary depending on an industrial facility's operating hours and electric requirements. To accurately account for demand savings, it is essential to log data on the HVAC loads to determine the exact peak electric demand and its contribution to the monthly demand.

2.6 Conclusions

This chapter emphasizes the urgent need for industrial decarbonization to combat climate change. It highlights the industrial sector's contribution to global GHG emissions, especially CO_2, and the challenges faced in achieving decarbonization. Various technologies like hydrogen-enriched natural gas, carbon capture and storage, and innovative process solutions for iron, steel, and glass production are discussed. The authors stress the importance of collaborative efforts among governments, industries, and communities to overcome technical, economic, and infrastructural barriers. It also offers a sample calculation procedure that demonstrates the correlation between decarbonization and energy savings. This analysis serves as a valuable guide for policymakers, business executives, and industry stakeholders. It encourages them to prioritize and effectively pursue sustainable decarbonization strategies to align with global climate goals.

References

[1] Rissman, J., Bataille, C., Masanet, E., Aden, N., Morrow, W., Zhou, N., Elliott, N., Dell, R., Heeren, N., Huckestein, B., Cresko, J., Miller, S., Roy, J., Fennell, P., Cremmins, B., Koch Blank, T., Hone, D., Williams, E., de la Rue du Can, S., … and Helseth, J. (2020). Technologies and policies to decarbonize global industry: Review and assessment of mitigation drivers through 2070. *Applied Energy*, 266, 114848. https://doi.org/10.1016/j.apenergy.2020.114848.

[2] Pee, A., Pinner, D., Roelofsen, O. and Somers, K. (2018). Decarbonization of industrial sectors: the next frontier. http://dln.jaipuria.ac.in:8080/jspui/bitstream/123456789/10762/1/Decarbonization-of-industrial-sectors-The-next-frontier.pdf.

[3] International Energy Agency (2023, March 2). CO_2 emissions in 2022—analysis. https://www.iea.org/reports/co2- emissions-in-2022.

[4] U.S. Environmental Protection Agency (2024, February 8). Inventory of U.S. greenhouse gas emissions and sinks: 1990-2022 [reports and assessments]. https://www.epa.gov/ghgemissions/inventory-us-greenhousegas-emissions-and-sinks-1990-2022.

[5] Pisciotta, M., Pilorgé, H., Feldmann, J., Jacobson, R., Davids, J., Swett, S., Sasso, Z. and Wilcox, J. (2022). Current state of industrial heating and opportunities for decarbonization. *Progress in Energy and Combustion Science*, 91, 100,982. https://doi.org/10.1016/j.pecs.2021.100982.

[6] Rockström, J., Gaffney, O., Rogelj, J., Meinshausen, M., Nakicenovic, N. and Schellnhuber, H. (2017). A roadmap for rapid decarbonization. *Science*, 355(6331), 1,269–1,271. https://doi.org/10.1126/science.aah3443.

[7] Avşar, C., Tümük, D., Yüzbaşıoğlu, A. and Gezerman, A. (2022). Focusing on the Merseburg process: Benefits on industrial decarbonization and waste minimization. *Environmental Technology Reviews*, 11(1), 148–155. https://doi.org/10.1080/21622515.2022.2119171.

[8] Wesseling, J., Lechtenböhmer, S., Åhman, M., Nilsson, L., Worrell, E. and Coenen, L. (2017). The transition of energy intensive processing industries towards deep decarbonization: characteristics and implications for future research. *Renewable and Sustainable Energy Reviews*, 79, 1,303–1,313. https://doi.org/10.1016/j.rser.2017.05.156.

[9] Papadis, E. and Tsatsaronis, G. (2020). Challenges in the decarbonization of the energy sector. *Energy*, 205, 118,025. https://doi.org/10.1016/j.energy.2020.118025.

[10] Ji, S., Lee, B. and Yi, M. (2021). Building life-span prediction for life cycle assessment and life cycle cost using machine learning: a big data approach. *Building and Environment*, 205, 108,267. https://doi.org/10.1016/j.buildenv.2021.108267.

[11] Deng, Y., Dewil, R., Appels, L., Van Tulden, F., Li, S., Yang, M. and Baeyens, J. (2022). Hydrogen-enriched natural gas in a decarbonization perspective. *Fuel*, 318, 123,680. https://doi.org/10.1016/j.fuel.2022.123680.

[12] Qian, J., Chen, T., Enakonda, L., Liu, D., Mignani, G., Basset, J. and Zhou, L. (2020). Methane decomposition to produce COx-free hydrogen

and nano-carbon over metal catalysts: a review. *International Journal of Hydrogen Energy*, 45(15), 7,981–8,001. https://doi.org/10.1016/j.ijhydene.2020.01.052.

[13] Supekar, S. and Skerlos, S. (2015). Reassessing the efficiency penalty from carbon capture in coal-fired power plants. *Environmental Science and Technology*, 49(20), 12,576–12,584. https://doi.org/10.1021/acs.est.5b03052.

[14] Global Status of CCS Report: 2019. Global CCS Institute. https://www.globalccsinstitute.com/resources/publications-reports-research/global-statusof-ccs-report-2019, accessed 7 December 2023.

[15] Industrial Assessment Center [dataset]. https://iac.university.

[16] Industrial Assessment Center—Oklahoma State University (2021, August 27). https://ceat.okstate.edu/det/iac/index.html.

[17] Sofia, D., Gioiella, F., Lotrecchiano, N. and Giuliano, A. (2020). Cost-benefit analysis to support decarbonization scenario for 2030: a case study in Italy. *Energy* Policy, 137, 111,137. https://doi.org/10.1016/j.enpol.2019.111137.

3

Software Analysis of Solar PV System Efficacy

Stephen A. Roosa

There are many times when an energy engineer is tasked with developing a solar project using virtual tools. This chapter provides an overview of how an energy engineer can perform preliminary site investigations of solar projects by virtually assessing their generation and economic potential without the need to physically visit the site. It discusses ways to perform a solar photovoltaic analysis of a sample site using virtual surveying and the National Renewable Energy Laboratory's PVWatts, a web-based tool used to estimate the electrical generation from a solar photovoltaic system. Two case studies for hypothetical sites are provided using PVWatts to assess the solar potentials of orientation options and selected module racking systems. One example considers a new solar array mounted on the roof of an existing commercial building; the other assesses a canopy-mounted system for an electric vehicle charging station. The case studies are designed to compare the electrical production of the solar PV systems based on the orientation and type of racking system. The use of solar thermal systems for water heating is also considered.

3.1 Introduction

Estimating the potential electrical generation from the installation of a solar photovoltaic (PV) system for a commercial or industrial site can be challenging. The production varies due to factors such as the site location, weather conditions, and orientation. There are times when an energy engineer is tasked with developing an estimate of a solar PV system's output without

DOI: 10.1201/9788743811312-3

physically visiting the site—certainly a challenging assignment. There are innumerable potential configurations. Solar PV arrays vary in size and are location sensitive. They can be green field-located, floated on ponds or lakes, mounted on rooftops, used as awnings to shade fenestration, or installed on canopies. Canopy structures are often used in existing automobile parking areas and customized to supply power for electric vehicle charging stations.

Information that the designer requires would include the site location, physical site layout, what buildings and structures are nearby, and electrical utility locations. Ultimately, an estimate of the solar potential of a preliminary solar PV configuration will be needed for proof-of-concept. If the engineer is unable to physically visit the development site, basic data may need to be accumulated virtually. Virtual analysis will often use a platform such as Google Earth to gather information about the site. Preliminary analysis of the generation potential might use the PVWatts calculator, developed by the National Renewable Energy Laboratory (NREL), to estimate the electrical generation potential of a solar array [1]. It may also be necessary to consider the site's solar thermal potential and accommodate the installation of solar hot water heating.

3.2 The Problem of Estimating Solar Generation

Solar power is an intermittent energy resource. Maximum electrical generation potential is only available when the sun is shining. The resource is not available at night. While solar power can be used to generate electricity using thermal energy, solar PV cells can directly convert sunlight to electricity. Silicon solar cells use visible daylight in the red to violet spectrum (near the infrared range) with wavelengths of 400–700 nm for electrical generation.

The solar constant is the average density of solar radiation measured outside Earth's atmosphere and at Earth's mean distance from the sun; it is 1.360 kW/m^2 or 0.13 kW/ft^2. The amount of solar insolation that penetrates the atmosphere and reaches the surface of a solar collector can be much less; it varies widely based on non-controllable conditions such as season, time-of-day, cloud cover, air pollution, ambient conditions, etc. The length of time each day that sunlight is available varies throughout the year. Substantially more daylight is available during the summer months than in the winter months in non-tropical regions. In winter months, snow and ice might accumulate on the surface of the collectors, which reduces module efficiency until it is removed either by natural conditions or physical intervention. Overheating can reduce panel efficiencies. Because the electricity comes from the flow of electrons inside an electrical circuit, heat waves cause the electrons to vibrate too actively, which reduces voltage. Nearby trees may need to be trimmed to prevent an array from being shaded. Development in urban areas may lead to high-rise construction on a site adjacent to a solar array, causing it to be shaded during parts of the day, which reduces power generation. These are reasons why solar power is an intermittent renewable resource.

Maintenance is required to maximize the array output. Controllable conditions that affect the output of solar collectors include repairing or replacing damaged panels and removing dust from the collector surface. Effective strategies include performing physical assessments, remote monitoring, and performing routine maintenance. Infrared thermography can be used to help identify damaged solar PV modules.

3.2.1 Solar module efficiencies

A key parameter used for estimating solar array output is system efficiency. Solar cell efficiency is typically defined as the percentage of the solar energy falling on the cell's surface that is converted into electrical energy. Module (panel) efficiency is the overall efficiency of the combination of photovoltaic cells assembled into a module. However, the true efficiency measure of a solar PV module is its energy yield as determined by laboratory testing. Energy yield is the actual output in kilowatt hours (kWh) as measured by recording equipment over time and verified by independent test facilities that certify energy yield performance.

Common categories of solar PV modules include silicon and thin-film photovoltaics. Silicon PV modules are one of the most prevalent commercial technologies and are classified as either monocrystalline silicon (c-Si) or polycrystalline silicon (mc-Si). Monocrystalline silicon photovoltaic cells are typically made using the Czochralski process, in which single-wafer silicon cells are cut from cylindrical ingots of highest-grade refined silicon; they have higher efficiencies, longer life, provide better low-light and low-temperature performance, and are more costly. Polycrystalline silicon cells are made from raw silicon that is melted and poured into a square mold; cooled they can be cut into perfectly square wafers consisting of many small crystals. Modules that use polycrystalline silicone cells are less efficient than monocrystalline panels, but are less affected by high temperatures, have better high-temperature performance, and are less costly. Cell efficiencies for monocrystalline cells are a maximum of 26.8%; for multi-crystalline cells, the maximum is 23.3%. Thin-film PV modules are the least efficient but can be applied to different types of surfaces (e.g., vinyl, glass, or stainless steel).

Common technologies are cadmium telluride (CdTe), copper indium gallium selenide (CIGS), gallium arsenide multijunction (GaAs), and amorphous silicon (a-Si). All PV cell technologies natively produce electricity in the form of direct current (DC). Inverters are needed to create AC power.

Solar PV cells operate at their peak efficiencies at temperatures of 25°C (77°F). During high temperatures, the panel's temperature increases, leading to increased resistance within the PV cells [2]. The resistance increases the amount of heat generated, leading to a further reduction in efficiency [2]. The decline in performance becomes more evident in areas with hot and humid climates, where temperatures repeatedly exceed 40°C (104°F). Research has shown that most solar panels that use silicone technologies exhibit a roughly linear rate of decline in output over time [3]. They have a median degradation rate of about 0.5% per year, with the mean in the range of 0.8–0.9% per year; the rate could be higher in hotter climates with mounting configurations that lead to sustained higher temperatures (i.e., roof-mounted systems) [3]. A degradation rate of 0.5% implies that production from a solar panel will decrease at a non-compounded rate of 0.5% per year. This means that in year 20, the module is producing approximately 90% of the electricity it produced in the year it was initially installed. Degradation rates tend to be higher for multi-crystalline silicon (mc-Si) solar PV modules than for monocrystalline silicon solar PV modules. When performing a life-cycle cost assessment, an estimated degradation rate needs to be factored into the economic analysis.

Solar cells are assembled into groups of panels or modules. When configured into panels, the module efficiency is less than the cell efficiencies. Figure 3.1 provides module efficiencies for various types of solar PV panels that have been standardized and confirmed by independent, recognized test labs (i.e., NREL, AIST, JRC-ESTI, and Fraunhofer-ISE). The measurements are based on standard test or reporting conditions as defined by the global reference spectrum for flat-plate devices and the direct reference spectrum for concentrator devices as listed in standards IEC 60904-3, edition 2, or American Society for Testing and Materials (ASTM) G173 [4].

The amount of electricity generated by a solar array varies based on the number and type of solar modules used, their efficiencies, panel age, physical orientation, and mounting systems. For solar photovoltaic arrays, module efficiency equates to space and labor efficiency; higher module efficiency means that less collector surface area and installation labor are required to generate the amount of electricity needed. Output also varies based on the type and orientation of the selected mounting system. Typical types of mounting systems for solar arrays include field, roof, or canopy fixed tilt

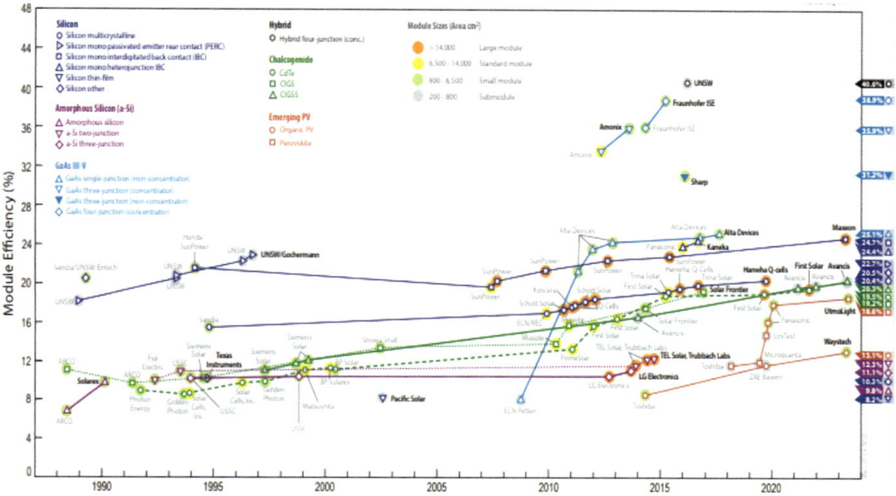

Figure 3.1 NREL Champion solar PV module efficiencies [4].

systems, and single-axis tracking. Dual-axis tracking systems are available but are rarely used.

3.2.2 Estimating solar array electrical output

One of the problems we find in practice is that the actual electrical generation of installed solar arrays often falls short of design expectations. Fluctuations in the amount of light incident on the panel, such as at dawn and dusk, cloud cover, passing storms, and debris, are often cited. Lack of maintenance, such as allowing dust, snow, or ice to accumulate on the panels, can cause solar arrays to underperform. Corroded connections between the panel and the inverter can reduce electrical output. While weather conditions are often cited as the uncontrollable and principal cause, other factors, often occurring in combination, are usually the culprit causing efficiency losses. The output of solar arrays is often overestimated due to:

- Using cell efficiencies rather than the tested module or panel efficiencies.
- Failure to account for module degradation.
- Excluding system losses (e.g., from inverters, converters, etc.)
- Unaccounted losses in turn-around cycles from battery energy storage systems.

- Accepting the rated DC output in kWh as usable AC power.
- Panel overheating (often caused by not elevating the panels above the mounting surface).
- Maintenance issues such as cleaning or snow removal.
- Unidentified damage to the solar PV modules.

Energy storage systems offer a heuristic example of the magnitude of the problem. Lithium-ion (L-I) batteries are often sold as having a 95% turn-around efficiency. A more holistic approach would calculate conversion losses and auxiliary power consumption [5]. Surprisingly, when operating in an installed configuration for low-power operating systems, the losses in the power electronics are greater than the battery losses [6]. Conversion round-trip efficiencies upon installation have been measured in the range of 70% to 80% [5]. If a system is storing just 20% of the electricity for later use with a 75% functional turn-around efficiency, then the added system losses may total 5%. Systems can be designed to use DC power only (i.e., supply DC-generated electricity to DC devices and loads) and thus avoid higher levels of loss.

When attempting to assess the electrical generation of a given solar array, the software platforms and online assessment tools often handle key variables differently. Some allow solar model efficiencies to be inserted directly, while others ask for a categorical type of module. Many allow a system efficiency factor to be entered, which includes derating, but may not allow entry of specific inverter efficiencies. Software tools provided by equipment manufacturers tend to estimate system output based on optimal conditions, which rarely occur in practice. The software user needs a fundamental understanding of how key parameters impact the estimation of solar array output and how they are applied.

It is important to consider the interrelationships among the system components and how system efficiencies may be impacted. Using a conservative system derating factor is one way to help ensure that the electrical generation of the system is not overestimated, given the widely variable design uncertainties. If estimating output over a number of years, a degradation rate needs to be included in the analysis.

3.3 Case Examples

For this exercise, two case examples are provided that are often applied in energy engineering practice. The first one considers a commercial solar array using monocrystalline modules, which is mounted on an existing roof; the second considers an array used to supply electricity for electric vehicle charging stations. For both discrete examples, panel efficiencies, module characteristics, the array size, orientation, and system losses are held constant. This allows a comparative assessment and analysis of selected racking systems and their differences based on electrical production. The hypothetical sites selected for each case are in the U.S. northern hemisphere in the metropolitan area of Louisville, Kentucky (Latitude 38.25, Longitude -85.74). The solar array azimuth for these configurations is oriented at 180° (facing due south). This first case assumes that there is no connected energy storage system, all electricity is used by the facility or exported to the host grid, and the unit value (kWh) of the electricity produced is constant. The second case has a supplemental energy storage system. As solar irradiance levels are site-specific, case study data and output would vary, perhaps widely, if another site elsewhere were selected.

3.3.1 Roof-mounted solar PV array

The project scope for this example case study considers various orientations and mounting options for a roof-mounted solar array. There are cost-saving benefits to placing solar arrays on an existing roof area: 1) the location precludes public access which improves security; 2) fencing is not required; 3) no foundations need to be constructed; 4) drainage systems are already in place; 5) water for cleaning is available from the existing building; 6) the panels shade the roof which reduces building air conditioning loads; and 7) the site likely has existing electrical supply and metering. Most importantly, this option is space-efficient as no land is needed for development. For most locations, the permitting process is less burdensome.

A new modular solar monocrystalline solar PV system with panels arrayed on the existing roof of a hypothetical commercial campus is used in this example. The system array, as configured, uses 1,064 (45-inch × 90-inch) 550 W solar panels configured in 19 identical groups; each group contains 56 solar PV panels. Seven groups of solar PV panels are located on single-story buildings, and 12 groups are located on an adjacent multi-story building (see Figure 3.2). The rated electrical generation capacity of the system is 585.2 kW.

The modeled installation might use either a ballasted racking system or be mechanically attached to the roof structure, but is not installed directly flush with the roof. This allows airflow below the solar array. Modules could be fixed in position or use tracking systems.

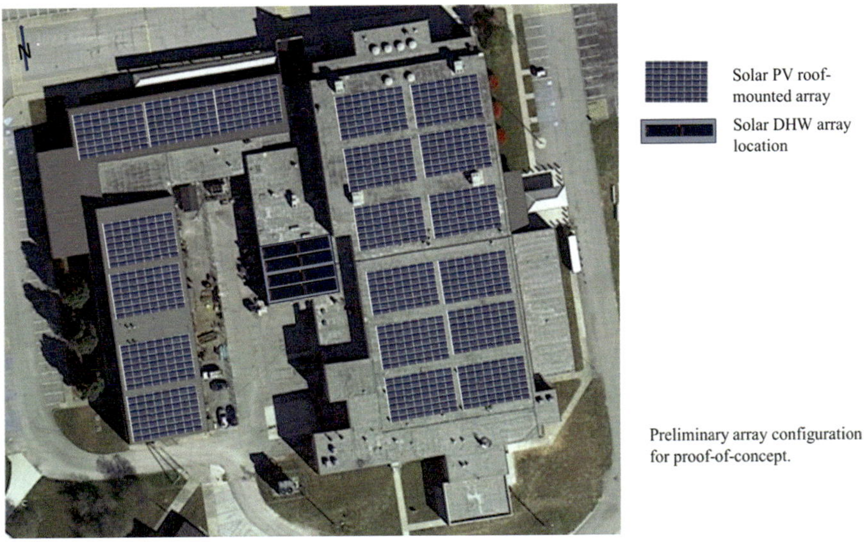

Figure 3.2 Sample campus solar PV and thermal (DHW) arrays.

Before installation of the new solar array, it is important that the type of roof and its remaining life be assessed. It may be necessary to apply a white reflective coating or replace the roof with a new reflective surface if it is nearing the end of its economic life. This would help reduce heat concentrations near the surface of the roof, improve reflectivity, and enhance module efficiency. Engineers would not recommend installing a new solar PV system with a potential 25-year life on a roof that was going to need to be replaced in 5 or 10 years. Substantial labor costs would be incurred dismantling the solar array to enable roof replacement, packaging it for storage, and then reinstalling it after a new roof was installed. In roof areas where the solar PV modules and mounting systems are proposed, an assessment of the added bearing/loading capacity of the roof structure may be required. In most cases, the added loads will not be so excessive as to preclude installation.

While there are many array orientation options, the optimum array orientation in the northern hemisphere would normally face and slope toward due south (or due north in the southern hemisphere) unless there were other reasons to adjust the orientation. For this case assessment, the array uses a fixed, roof-mounted configuration with panels tilted either 10° or 38° or otherwise uses a single-axis tracking system. A tilt of 10° would improve electrical output compared to a flat module, reduce the potential for wind damage, be more conveniently accessible for cleaning, allow for drainage from rain, and could be installed at a low cost. A fixed tilt at an angle approaching the latitude (in this case, 38°) is known in the industry to generally increase output. However, damage from high wind events would likely be greater than if mounted with less tilt. Racking systems will be more costly, and rows of collectors would need to be properly spaced to prevent shading. Single-axis tracking systems that move the arrays from east to west during the day to optimize collector orientation tend to further increase system electrical generation. However, the increased costs for tracking systems can be prohibitive; maintenance costs also increase because of the operable components and controls required. Dual-axis tracking applications for roof-mounted arrays usually maximize output but are rare due to high comparative costs; this option was not assessed.

The question facing the energy engineer is which standard racking system design would generate the most electricity? What is the value of the electricity produced with each type of mounting configuration? With this type of data, the energy engineer can apply the estimated installation costs of the system and determine which is the most cost-effective.

For the case analysis, the configuration would include all components necessary to make the system functionally operational, including a DC/AC inverter, charge controller, and other controls to interconnect with the building's electrical supply system. A new bi-directional electrical meter would be installed as part of the host utility interconnection; this would allow all electricity (AC) to be used either by site loads (behind the meter) or transferred to the local grid (beyond the meter) via a utility interconnection.

The analysis assumes the use of premium single-sided PV modules, system losses of 23%, an inverter efficiency of 96%, and a coverage ratio of 0.8. The analysis is for the first year of operation and assumes that all the electricity tabulated, given the assumed parameters, is usable.

A comparison of the results of the analysis using PVWatts is found in Table 3.1. For the modeled location, solar insolation for the 10° fixed tilt array annually averages 4.58 kWh/m^2/day, and the 38° fixed tilt array averages 4.88 kWh/m^2/day. For the array with single-axis tracking systems, solar insolation for the 10° tilt array annually averages 4.94 kWh/m^2/day; the 38° fixed tilt array averages 5.10 kWh/m^2/day. The single-axis tracking systems capture more sunlight by optimizing module orientation. The fixed tilt array sloped 10° generates an estimated 669,743 AC kWh annually. The fixed tilt array sloped 38° generates an estimated 718,409 AC kWh annually. The output improves with the single-axis tracking systems: 736,846 AC kWh annually for an array sloped 10° and 762,744 AC kWh annually for an array sloped 38°.

Table 3.1 shows the dollar value of the electricity produced (kWh) by the various racking system scenarios for electricity prices of $0.10/kWh, $0.12/kWh, and $0.14/kWh. The economic value of the electricity generated (assuming all electricity is consumed by connected loads) using $0.12/kWh is $80,369/year for the 10° fixed tilt array, which offers the least electrical generation. This compares to $91,529/year for the single-axis tracking system sloped at 38° with an 11.4% increase in estimated annual income. The greatest output and cash inflows are available when the single-axis system is used. If the facility is on an electrical demand tariff and summer peaking, additional income may result from the added output during the summer months. A separate analysis of the demand profile and an assessment of daily electrical generation would be required to decide on the best option.

For a typical 12-month period, the roof-mounted solar array with single-axis tracking, sloped at 10° when compared to the roof-mounted, fixed-tilt array also sloped at 10° increases electrical output (kWh) by 10.0%. Using the roof-mounted array with single-axis tracking sloped at 38° rather than the fixed-tilt array of the same slope increases electrical output by only 6.2%. An economic analysis would compare the incremental costs of the more expensive option to its incremental savings to obtain the return on investment for the added costs. For this case, it would be difficult to justify the added initial cost of a single-axis tracking system after the increased operations and maintenance costs are included, regardless of the slope of the array. However, such an analysis would be necessary for each individual site and each project's specific conditions and configuration.

3.3.2 Solar array for an electric vehicle charging station

The project scope for this example considers building a solar array for new electric vehicle charging stations. The hypothetical green field site selected for study is undeveloped and adjacent to an existing parking lot (see Figure 3.2). Site preparation includes removing vegetation, leveling, grading, spreading a gravel base, and installing drainage piping as needed.

A newly designed canopy structure will be constructed on the site. A racking system for a monocrystalline silicon solar photovoltaic panel array will be installed. The design will allow for parking spaces under the canopy structure for electric vehicles while charging. This turn-key solar PV system includes DC-to-AC microinverters, a charge controller, and electricity connections. It will supply electricity to charging stations with a metered utility interconnection. Overhead power lines are assumed. The installation for the solar PV charging station will include four Level 2 charging units (240 V),

Table 3.1 Roof-mounted solar array orientations, mounting systems, generation, and production values.

	Roof-mounted Fixed Tilt Array				**Roof-mounted Array With Single Axis Tracking**			
Month	**10° Fixed Tilt Ave. Solar Radiation (kWh/m²/day)**	**10° Fixed Tilt Output/ Month (AC kWh)**	**38° Fixed Tilt Ave. Solar Radiation (kWh/m²/day)**	**38° Fixed Tilt Output/ Month (AC kWh)**	**10° Tilt Ave. Solar Radiation (kWh/m²/day)**	**10° Tilt Output/ Month (AC kWh)**	**38° Tilt Ave. Solar Radiation (kWh/m²/day)**	**38° Tilt Output/ Month (AC kWh)**
January	2.56	34,301	3.57	48,192	2.75	37,832	3.73	50,799
February	3.23	39,059	4.09	49,646	3.45	42,589	4.20	51,640
March	3.94	50,941	4.35	56,295	4.16	54,624	4.46	58,533
April	5.22	62,974	5.20	62,622	5.55	68,104	5.36	65,730
May	6.06	74,427	5.54	67,927	6.46	80,687	5.72	71,471
June	6.53	75,948	5.75	66,741	7.02	83,164	5.98	70,909
July	6.33	74,703	5.70	67,075	6.93	83,419	6.03	72,725
August	6.25	74,016	6.11	72,216	6.76	81,651	6.41	77,454
September	5.21	60,325	5.60	64,718	5.71	67,496	5.96	70,240
October	4.29	53,538	5.26	65,832	4.74	60,322	5.63	71,168
November	2.94	37,375	3.97	50,662	3.20	41,504	4.16	53,776
December	2.39	32,136	3.43	46,483	2.58	35,454	3.52	48,299
Total	4.58	669,743	4.88	718,409	4.94	736,846	5.10	762,744
Production Value	$0.10/kwh	$66,974		$71,841		$73,685		$76,274
	$0.12/kwh	$80,369		$86,209		$88,422		$91,529
	$0.14/kwh	$93,764		$100,577		$103,158		$106,784

each capable of simultaneously charging two EVs, parked on opposite sides of the charging unit. Charging units will be mounted on a concrete base elevated above finished grade. Charging unit locations will be designed to minimize the distance to the charging ports of the EVs.

A 20 kW lithium-ion battery energy storage system will be provided to store excess electricity generated by the canopy solar PV installation during periods of high solar insolation. A challenge with the use of L-I batteries for certain applications is that multiple charge-discharge cycles over the course of a day may cause them to degrade somewhat, which shortens their operating life-cycle [7]. They are also unable to discharge at full power for more than four to six hours [7]. They will be mounted at an accessible height so that they can be monitored and maintained as needed. The control configuration will allow the batteries to be charged directly from the solar array when excess power is available. The primary components of the installation include:

- Foundation and structure for the stand-alone solar charging canopy.
- The racking system and solar PV arrays are installed on the canopy structure.
- Multiple Level 2 (240V) plug-in electric vehicle charging units, each capable of charging two vehicles simultaneously.
- A stationary lithium-ion battery energy storage system.
- Electrical interconnections, components, and electronic hardware (e.g., inverters, charge controllers, devices, etc.).
- Intelligent software to enable remote data and operational monitoring.
- A bi-directional utility meter and underground electrical cabling, conduit, and connections.
- Other components as specified by the selected system configuration.
- Photosensor-controlled light-emitting diode (LED) floodlights to illuminate access ramps and areas under the canopy at night.

The solar PV canopy structures (i.e., structural steel) will be elevated to allow the electric vehicles (EVs) to drive through and park underneath the canopy when charging batteries. The supporting pitched or saw-tooth array configuration of the panel structure will be oriented to the south, with the slope of the fixed racking system optimized for maximum electric output (see Figure 3.3, which shows a concept design). The panel or module configuration of the

Figure 3.3 Canopy solar for electric vehicle charging stations.

passive emitter and rear contact monocrystalline panels in the solar PV array will be determined based on an assumed array using 144, 45" × 90" panels rated at 550 W each with a surface area totaling approximately 4,050 ft^2 (376.3 m^2). The rated 79.2 kW array will also power eight new weatherproof LED flood lights with internal batteries and daylight sensors to illuminate the paved access ramps to the charging stations at night.

The racking system for this case study is the elevated structural canopy (i.e., fixed tilt, open rack). The solar PV array is not subject to as much heat compared to a roof-mounted system, as there is a free flow of air underneath and over the solar panels. The analysis compares three potential design orientations with panels sloped either 10°, 25°, and 38° to azimuth. The analysis assumes the use of premium single-sided solar PV modules, system losses of 18%, an inverter efficiency of 96%, and a coverage ratio of 0.4. For the modeled location, solar insolation for the 10° fixed tilt array averages 4.58 kWh/m^2/day, the 25° fixed tilt array averages 4.79 kWh/m^2/day, and the 38° fixed tilt array averages 4.75 kWh/m^2/day.

The results of the analysis are provided in Table 3.2 and enable a comparison of the output of the three different orientations for the canopy-mounted, fixed tilt array. The analysis assumes that all tabulated AC electricity produced is usable. The fixed tilt array sloped 10° generates the least electricity, an estimated 97,388 AC kWh annually. The fixed tilt array sloped 25° generates the most electricity, an estimated 102,599 AC kWh annually. The fixed tilt array sloped 38° generates an estimated 101,8233 AC kWh annually.

Table 3.2 shows the dollar value of the electricity produced (kWh) by the various racking system scenarios for electricity prices of $0.10/kWh,

Table 3.2 Canopy solar array orientations, generation, and production values.

	Canopy-mounted Fixed Tilt Array					
Month	**10° Fixed Tilt Ave. Solar Radiation (kWh/m²/day)**	**10° Fixed Tilt Output/Month (AC kWh)**	**25° Fixed Tilt Ave. Solar Radiation (kWh/m²/day)**	**25° Fixed Tilt Output/Month (AC kWh)**	**38° Fixed Tilt Ave. Solar Radiation (kWh/m²/day)**	**38° Tilt Output/Month (AC kWh)**
January	2.55	5,005	3.13	6,181	3.46	6,820
February	3.22	5,693	3.72	6,595	3.96	7,015
March	3.93	7,410	4.21	7,937	4.24	7,990
April	5.21	9,148	5.26	9,238	5.05	8,856
May	6.05	10,806	5.84	10,421	5.38	9,600
June	6.52	11,022	6.17	10,424	5.58	9,428
July	6.52	10,852	6.05	10,366	5.54	9,504
August	6.32	10,751	6.25	10,764	5.95	10,252
September	5.20	8,770	5.50	9,264	5.48	9,223
October	4.29	7,793	4.90	8,910	5.18	9,399
November	2.94	5,448	3.54	6,608	3.87	7,212
December	2.39	4,690	2.96	5,871	3.29	6,524
Total	4.57	97,388	4.79	102,599	4.75	101,823
Production Value	\$0.10/kwh	\$9,739		\$10,260		\$10,182
	\$0.12/kwh	\$11,687		\$12,312		\$12,219
	\$0.14/kwh	\$13,634		\$14,364		\$14,255

$0.12/kWh, and $0.14/kWh. The economic value of the electricity generated (assuming all electricity is consumed by connected loads) using $0.12/kWh is $11,687/year for the 10° fixed tilt array, which offers the least electrical generation. This compares to $12,312/year for the fixed tilt array sloped at 25° and $12,209/year for the fixed tilt array sloped at 38°.

Of the three racking system orientations assessed, the fixed tilt array sloped at 25° offers the best performance and generates the most income, 5.3% more than the 10° fixed tilt array. Increasing the slope of the array to 38° would cost more, be more likely to sustain damage if high wind speeds were experienced, and provide no added economic benefits. If the facility has an electrical demand tariff and is summer peaking, additional income from electrical demand may accrue from the added output during the summer months. For the assessed configurations, additional electricity, sourced from the host utility, will be needed for consistent and full charging of the EV batteries.

3.4 Solar Water Heating

When installing a solar PV system, hot water for domestic purposes can be heated by using the electricity generated to supply power directly to electric water heaters. However, this option is comparatively inefficient and may consume large amounts of electricity. It often creates a variable load; the electricity used to maintain the temperature of water stored in the tanks may augment intermittency. Solar thermal water heating is often a lower-cost and more efficient alternative because the solar energy from the sun is converted directly into heat. Principal factors determining the cost-effectiveness of solar thermal systems are the ability to use the solar energy when it is available, the specific nature of the application, local climate, system efficiency, and the cost of the traditional fuel being offset by the energy collection system [9].

Solar thermal systems can be active or passive; they are used to generate heat, electricity, or for combined heat and power applications. Passive systems are unmechanized. Active solar thermal heating systems move heated fluid (air or a liquid) into the interior of the building to a heat storage system, and the heat is released when needed or dissipates [10]. Fans or pumps move the fluid through collectors to be heated, then to the interior of the building or heat storage system, and finally back to the collector to be reheated [10]. To heat and supply water for a domestic hot water system, active solar water heating designs typically provide a tank for storing solar-heated water. The collectors can be designed to be concentrating or non-concentrating. Unlike solar PV panels, the efficiencies of solar thermal panels are unaffected by heat, and efficiencies do not degrade over time, assuming they are properly maintained.

Perhaps the most ubiquitous applications for solar thermal systems are for domestic water heating or to heat swimming pools. For domestic hot water (DHW) heating, two types of these systems are predominant. Most of these systems use collectors, a storage tank, piping, circulating pumps, and a controller. A solar PV panel may be used to provide electricity to the pumps and controller. Solar-heated water from the collector is passively or actively (pumped) transported and stored in a hot water tank. Indirect solar hot water heating systems use similar components but use a secondary heat transfer solution and a heat exchanger usually immersed in the tank. The thermal transfer fluid, typically propylene glycol, is pumped to a heat exchanger to heat water in the storage tank.

Often, a commercial building's primary DHW heaters are in a main mechanical room. The water heaters are typically fired using electricity or natural gas. Due to lower energy costs, it is often difficult to justify replacing existing water heaters that use natural gas unless they are at or near the end of their economic lifecycle. However, electric water heaters used for commercial applications are more expensive to operate and will often increase electrical demand (kW) costs.

When electric water heater units are operating correctly, the immersed electric elements are staged to maintain the hot water temperature setting in the tank. Controls include a water heater controller mounted on the existing storage tank. As more employees now work from home, many commercial buildings have lower rates of occupancy, which causes the demand for hot water to also be lower. When a DWH system is over-sized for the requirements, electricity is wasted, creating inordinate costs for electrical energy. For many single-shift commercial operations, hot water is primarily required during the daylight hours when the building is fully occupied; hot water use after hours is minimal, and the demand for hot water matches times when solar energy is available for heating. In such cases, it is reasonable to consider a new solar thermal system installation to satisfy the DHW needs of the building occupants using primarily solar energy.

This type of solar energy and cost-saving application would involve replacing the existing DHW system with a new solar thermal system. The scope of work would require the demolition, removal, and disconnection of the existing electric hot water heating system, ancillary equipment, piping, storage tank, controls, etc. Then, a new active, indirect circulation, solar thermal DHW system would be installed. This would require the installation of a solar thermal collector array. In locations that are not subject to freezing, a direct system could be installed as an alternative.

3.4.1 Configuration for solar hot water heating

For this case, south-oriented, non-concentrating solar thermal collectors will be installed on the roof, above, or proximate to an existing central mechanical room to minimize piping and costs. The site selected is the same as in the analysis for solar PV (see Figure 3.2). This active system will be configured to heat a secondary compound, which serves as the heat transfer fluid. This configuration provides heat to the transfer loop connected to the heat exchangers located in the hot water storage tanks. Using a heat transfer fluid prevents the system from freezing.

Modular water heaters can be staged and interconnected to support the variable demand for DHW. The new solar thermal DHW system will include the necessary roof-mounted collectors, equipment, piping, valves, storage tanks, and controls for a complete and operational system. Dedicated solar photovoltaic modules will support the system's circulating pumps and dedicated controls. When no sunlight is available, the system's circulating pumps and controls will be interconnected with electricity supplied by a nearby electrical panel. Each hot water storage tank would have backup electric heating elements based on the specific design of the system, as suggested by the manufacturer. The system would be modular with valves installed to enable bypass of a single unit in the event of component malfunction.

3.4.2 Analysis and design considerations

Solar thermal applications such as domestic hot water heating require a separate economic analysis based on the specific design configuration. There are locations where a solar PV system may be uneconomical to install, yet a solar thermal system may prove cost-effective. One way to obtain an initial estimate of the approximate heat input from the solar collector system is to obtain the monthly kWh from an online software tool, convert the estimated kWh generated to its heat content (i.e., 1kWh = 3,412 BTU, or 1kWh = 3.6 megajoules), and apply a reasonable system efficiency factor.

A better way is to use a custom online software tool. The Federal Energy Management Program (FEMP) has a free online solar hot water system calculator designed to help estimate which size of solar hot water system will work best for an application based on the solar resource and the cost-effectiveness of that size [8]. The optimal system size is often one that provides 60% to

90% of the load, depending on the solar resource [8]. If hybrid modules (i.e., those that generate electricity and heat for DHW) are used, the analysis is more complex.

There are design considerations when the option of installing a solar DHW is not pursued; enabling the site to be solar thermal-ready for a future roof-mounted system should be considered. This can be accomplished by leaving roof space unused above or near an existing mechanical room that houses the building's hot water system. This allows a collector array for a solar thermal domestic hot water system to be installed in the future without necessitating the removal of part of a roof-mounted solar PV array (see Figure 3.2).

3.5 Conclusion

There are many ways to incorporate solar energy systems in existing commercial buildings. The amount of electricity generated by a solar array varies based on the number and type of solar modules used, their efficiencies, panel age, physical orientation, maintenance, and mounting systems. For the practitioner, it is important to consider the interrelationships among the components of a solar PV system and how system efficiencies may be impacted. This chapter considered roof-mounted solar PV arrays, canopy solar PV arrays used for electric vehicle charging, and solar thermal systems, particularly those used to heat domestic hot water. The NREL PVWatts online software tool was used to compare and analyze the racking systems and their orientations for the case studies presented.

Installing solar arrays on roofs, in open fields, and on specially designed canopy structures are mainstream applications. There are benefits to installing solar PV and thermal systems on the roof areas of existing buildings. Care must be taken in the analysis and best practices applied to avoid overestimating the quantities and value of the energy produced. For the case of similarly sized, roof-mounted systems, this study held certain variables constant; it compared selected racking system configurations for fixed-pitch and single-axis tracking solar PV arrays. For the location and orientations studied, the fixed pitch array sloped 10° generated the least amount of electricity; the racking system with a 38° slope using a single-axis tracking system generated the most electricity.

To assess variations of open rack systems, a canopy-mounted solar PV system for electric vehicle charging stations was used as a case example. Of the orientations studied, the fixed tilt array sloped at 25° provided the best system performance and generated the greatest amount of income.

Solar water heating is often economically justifiable. Solar thermal systems can be active or passive; collectors can be concentrating or non-concentrating. There are design considerations and numerous configurations that can be used to heat water for domestic purposes. An online software tool from the FEMP has a calculator designed to help estimate which size of solar hot water system will work best for specific applications. Solar thermal applications for domestic hot water heating systems require a separate economic analysis based on the design configuration. If not being installed as part of a project, it is advisable to allocate roof space for future use.

Eventually, the solar arrays will require decommissioning at the end of their useful life. When installing new solar PV and thermal systems, component recycling needs to be considered [10]. It is becoming increasingly important to identify and use project components that can be more easily dismantled, sorted, and recycled.

3.5.1 Opportunities for future research

By holding certain variables constant, this study provides a basic assessment of a few examples of how orientation and types of racking systems impact the electrical generation of solar PV systems. It is limited by the applications considered and is by no means exhaustive. The are innumerable opportunities for future study and reporting. Additional site locations, collector orientations, and various types of racking systems are worthy of analysis. Different types of solar panels, including bi-facial PV and hybrid (solar thermal and PV) modules, would likely provide improved efficiencies and are worthy of analysis.

References

[1] National Renewable Energy Laboratory (2023). PVWatts calculator, API Version 8.0. https://pvwatts.nrel.gov, accessed 19 November 2023.

[2] Newpowa (2023, November 16). How temperature impacts solar panel efficiency: exploring the relationship between temperature and output. https://www.newpowa.com/a/blog/how-temperature-impacts-solar-panel-efficiency-exploring-the-relationship-between-temperature-and-output, accessed 31 July 2025.

[3] Jordan, D., Kurtz, S., VanSant, K. and Newmiller, N. (2016, February 7). Compendium of photovoltaic degradation rates. https://onlinelibrary.wiley.com/doi/full/10.1002/pip.2744, accessed 2 December 2023.

[4] National Renewable Energy Laboratory (2023). Photovoltaic device performance calibration services. https://www.nrel.gov/pv/pvdpc, accessed 29 July 2025.
[5] Schimpe, M., Naumann, M., Troung, N., Hesse, H., Santhanagopalan, S., Saxon, A. and Jossen, A. (2018, January 15). Energy efficiency evaluation of a stationary lithium-ion battery container storage system via electro-thermal modeling and detailed component analysis. *Applied Energy*, 210, 211–229.
[6] GenixGreen (2022, May 30). What is the efficiency of lithium-ion batteries? https://www.genixenergy.com/battery-knowledge/what-is-the-efficiency-of-lithium-ion-batteries.html#:~:text=Typically%2C%20the%20losses%20in%20the%20%22power%20electronics%22%20are,measured%20in%20the%20range%20of%2070%25%20to%2080%25, accessed 2 December 2023.
[7] Roosa, S. (2023, July). Energy storage is becoming more powerful. *International Journal of Strategic Energy and Environmental Planning*, 5(4), 30–40.
[8] Federal Energy Management Program (2023, June 20). Solar hot water system calculator. https://www.energy.gov/femp/articles/solar-hot-water-system-calculator, accessed 22 January 2026.
[9] Caunt, G., Meep, S. and Roosa, S. (2018). Renewable energy, chapter 16, in *Energy Management Handbook*, 9th ed., Roosa, S. (editor), 464.
[10] Roosa, S. (2023, September). An asset manager's conundrum: sunsetting solar farms. *International Journal of Strategic Energy and Environmental Planning*, 5(5), 31–49.

4

Wind Turbine Generators—They Just Keep Getting Bigger

Stephen A. Roosa

The wind turbine industry is evolving. One notable trend is to design, develop, and install ever-larger wind turbine generators. Is bigger really better? When using machines of larger capacities, their efficiencies increase. Fewer are required to generate the electricity needed. This chapter discusses the shift in the wind power industry to producing larger capacity machines for use in the development of both land-based and offshore wind farms.

In the U.S., renewable energy resources—biomass, geothermal, hydropower, solar, and wind—provide nearly 30% of total electrical generating capacity. Together, they comprise the largest portion of electrical generation using clean energy. One notable trend is to use wind turbine generators with larger capacities and higher efficiencies in new and existing wind farms. This means that fewer machines are required to generate the electricity needed. This chapter discusses the status of wind power in the U.S. and the primary ways wind energy is harnessed to generate electricity. The growth of the U.S. wind farm developments is limited by technical, economic, maintenance, and developmental challenges.

This chapter further considers the reasons why manufacturers are racing to build larger wind turbine generators, discusses development limitations, and identifies the electrical power capacities of the largest ones that have been installed. While there are no theoretical limits, the size of wind turbine generators is limited by technical, economic, and developmental challenges. The basic mechanics of wind turbines are detailed. The importance of making on-site operational performance assessments after prototypes have been placed into operation is emphasized. It is clear that there are opportunities

DOI: 10.1201/9788743811312-4

for efficiency improvements and growth in the wind power industry that will drive future innovation and create the ability to produce machines with larger capacities.

4.1 Introduction

Though economic constraints hindering the development of wind farms must be overcome, wind power is a renewable energy resource that offers many important benefits. Wind power provides mechanical capabilities and generates clean electricity. Unlike other forms of electrical generation, it coexists with agricultural and other land uses. Wind power is becoming more economically competitive when compared to fossil fuel-fired means of electrical generation. It provides needed revenues, creates local employment, and increases the property tax base in rural areas. Perhaps more importantly, in some regions of the world, locally available wind resources offer the lowest cost of producing electricity. The expansion of electrical generation using wind power is viewed by some countries as a strategic solution to help them meet their environmental and sustainability objectives. This is because harvesting wind power does not produce air pollution nor emit greenhouse gas (GHG) emissions.

Wind power requires optimum site selection and is an intermittent renewable energy resource. Overcoming the challenge of variable electrical generation when base load power is often required remains a problem for the industry. This issue can be resolved by interconnecting to a host electric grid, using energy storage to smooth the power produced, or by developing microgrids.

Despite the inherent issues with using wind power for electrical generation, the industry is rapidly increasing its generation capabilities and capacity. It was only about 20 years ago when the standard wind turbine generators (WTGs) used for commercial wind farms were rated at 1.5 to 2.0 MW. Today's super-sized horizontal axis wind turbines (HAWTs) are taller than some of the skyscrapers found in nearby cities. Bigger is not necessarily better; building larger turbines entails resolving numerous logistical, engineering, and manufacturing challenges. While land-based wind farms have different requirements depending on the site, offshore wind power developments often deploy the newest, tallest, and most efficient turbines [1].

4.2 The World's Wind Power Production is Increasing

The world's growing use of wind and solar power for electrical generation has been particularly impressive since 2020 (see Figure 4.1). This is especially true in European and Asian countries due to recent wind power capacity

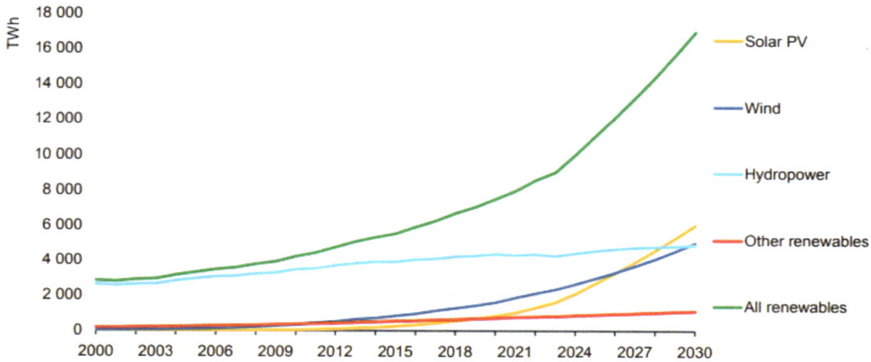

Figure 4.1 Global electricity generation by renewable energy technologies (2000–2030) [3].

additions. Several European countries have achieved relatively high levels of wind energy penetration in their electricity grids [2]. Denmark supplies 48% of its grid electricity using wind power, Ireland (33%), Portugal (27%), Spain (23%), the UK (21%), and Germany (20%) [2]. Denmark has substantial investments in offshore wind farms. Trains in the Netherlands operate on electricity, which is totally supplied by harnessing wind, some purchased from neighboring countries. The South American country of Uruguay obtains 94% of its electricity from wind power.

Regardless of regional declines in global capacity additions, the world's new land-based wind installations rebounded 65% to 107 GW in 2023. This increase came mostly from expanded wind power development in China and, to a lesser extent, India. While China has seen success in offshore wind farm development, elsewhere in the world, the industry is experiencing increasing costs, abandoned leases, project delays, and limited supply chain investment that hinders development. Some countries are belatedly discovering that offshore windfarms require specific policies, supply chains, and infrastructure that must be in place to support their development. Despite these struggles, the world's wind-based electrical generation is expected to surpass hydropower by 2030 (see Figure 4.1) [3].

Harnessing wind energy is one of the world's fastest-growing segments in the energy sector [4]. Wind generated 6.6% of the world's electricity in 2021 [4]. There are reasons for making wind turbines bigger. One is to quickly add renewable capacity to national electric grids or support remote microgrids [4]. Larger capacity wind machines reduce the geographic footprint needed for wind farm development. When using WTGs of larger capacities, fewer machines are required to generate the electricity needed. Their greater height and expansive swept areas provide the ability to harvest more consistent winds that are unobstructed by variable landscape conditions or nearby structures. For example, a single 15 MW HAWT when properly sited replaces a least ten older 1.5 MW machines. As a result, a developing industry trend is to reduce the total number of installed machines and use larger, more efficient, and higher capacity WTGs that increase total electrical production.

4.3 The U.S. Experience

In the U.S., renewable energy—biomass, geothermal, hydropower, solar, and wind—provides nearly 30% of total electrical generating capacity. More electricity is now produced from wind power than from any other form of renewable generation. This is due to increases in U.S. fleet capacities and recent declines in electrical production from hydropower caused by drought conditions in the western regions of the country. Another reason is economic—U.S. capacity-weighted, utility-scale installation costs for wind power declined by 2022 to $1,451/kW while solar PV costs $1,588/kW, or 9.4% more.

The U.S. relies primarily on land-based wind farms, which are less costly to develop than those located offshore. Some turbines from the early 1990s are still in operation, and many older units are being replaced with larger ones with higher generation capacities. While development locations at sea are inexpensive to lease, they are more difficult to access, and development can be hindered by variable ocean conditions.

The growth in U.S. electrical generation using wind power has been particularly impressive, beginning in 2006 (see Figure 4.2); total generation has since more than doubled [5]. For 2022, wind turbines were the source of 10.25% of total U.S. utility-scale[1] electricity generation [6]. The U.S. generated 42.5 terawatt-hours (TWh) of electricity from wind in 2023—less than 2022 due to less wind availability—yet enough to power the equivalent of more than 39 million homes [5]. However, in 2024, wind turbine output rebounded and produced a record total of 47.7 TWh—a substantial increase compared to 2023. During the spring of 2024, net electricity generation from wind power exceeded that of coal for two consecutive months. The country now has over 90,000 WTGs across 45 states.

Generating electricity is a leading form of providing power using renewable energy in the U.S. By 2023, 25 states generated at least 10% of their total in-state electricity from wind and solar combined [5]. The 2023

---

[1] Utility-scale is defined as those with at least one megawatt (1,000 kilowatts) of electrical generation capacity.

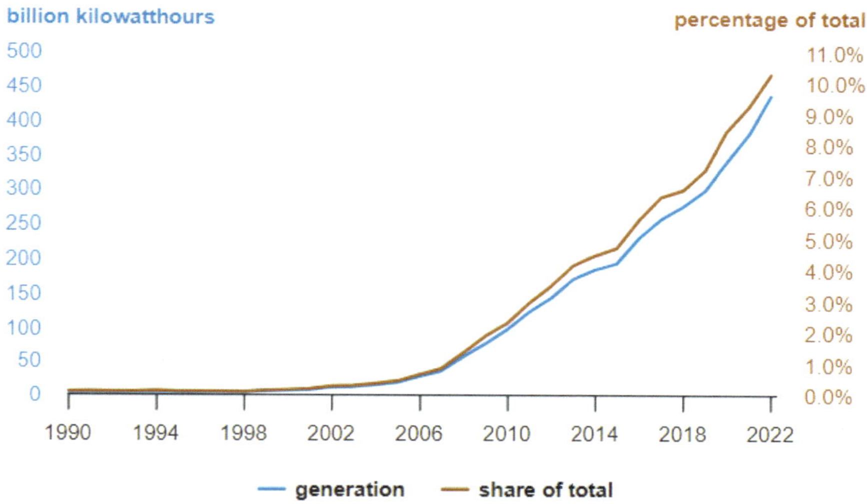

Figure 4.2 U.S. Electricity generation using wind power (1990–2022) [4].

total generating capacity from wind power in the U.S. stood at 83,408 MW, a 130% increase from 2014 through 2023. It accounted for 12% of the nation's total electrical generating capacity from all utility-scale sources (including renewables, nuclear, and fossil fuels) [5]. The leading U.S. states in 2023 with appreciable capacity growth in the last decade were Texas with 26,658 MW (up 190%), Oklahoma 8,466 MW (up 224%), Iowa 7,241 MW (up 130%), Kansas 6,074 MW (up 205%), Illinois 4,347 MW (up 123%), and New Mexico 3,597 MW (up 443%) [5].

New land-based wind installations are the mainstay in the U.S. By the end of 2024, the U.S. had an estimated total capacity of 150.1 GW (see Figure 4.5, which shows installed capacities by state for the first quarter of 2024) from utility-scale onshore and offshore wind installations—an increase of 8 GW from 2022. The states of Texas, Oklahoma, and Iowa generate over 30 billion kWh annually using wind power (see Figure 4.5). This growth mirrors wind development elsewhere; the world's wind-based electrical generation is expected to surpass that of hydropower by 2030 [6].

4.3.1 Utility-scale wind farms

Siting considerations are key to a successful wind farm. Sustained and consistent wind profiles are needed to ensure a thriving development. The Tehachapi

Figure 4.3 Tehachapi Pass, California wind farm (photo credit: Stephen Roosa).

Pass wind farm in Mohave, California, was among the first large, land-based wind farms in the U.S. The Tehachapi wind resource area exports electrical energy to other parts of California. It has about 710 MW of installed capacity produced by roughly 3,400 wind turbines; total site generation is about 800 million kWh annually. Pictures of the wind farm taken by the author during a site visit are shown in Figure 4.3. The mountain pass where the wind farm is located provides geographic conditions that create a Venturi effect for air flowing between the ocean and desert, thus increasing wind speeds. An initiative to upgrade the transmission system in southern California began in 2008 and was completed in 2016 [7]. California's Tehachapi Renewable Transmission project involved the construction of approximately 173 miles (278 km) of new and upgraded high-voltage (69 kV to 500 kV) transmission lines with a capacity of 4,500 MW. The primary intent of this $2.1 billion project was to carry electricity from wind farms and other generating units in southeastern Kern County to Los Angeles and San Bernardino counties [7]. This project enabled regional wind and solar power development.

Yet another large U.S. wind farm is the Horse Hollow Wind Energy Center in Texas. It consists of 291 GE 1.5 MW wind turbines and 139 Siemens 2.3 MW wind turbines arrayed over about 47,000 acres (19,020 hectares). At the time of its completion in 2006, it was the world's largest wind farm. The project has a combined electricity-generating capacity of about 735 MW.

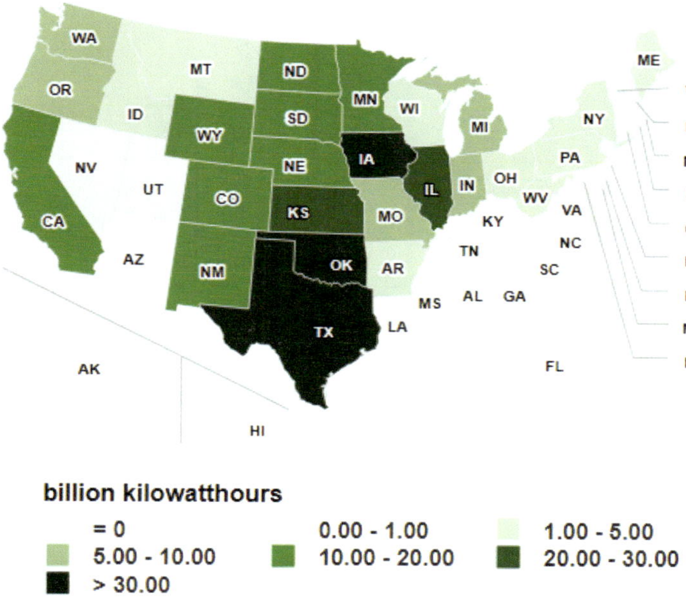

Figure 4.4 U.S. utility-scale wind electricity generation by state, 2022 [8].

A last land-based example, the Meadow Lake Wind Farm, is an 801 MW wind development near Brookston and Chalmers, Indiana, owned and operated by EDP Renewables North America. It produced 2,209,034 MWh in 2022. The facility has six completed operational phases with a total of 414 turbines. It is a prominent feature on both sides of Interstate 65 when driving through northwestern Indiana. The turbines are exceptionally well-maintained. The author once drove through the wind farm on a windy day and found only two machines that were not spinning. When complete, its nameplate capacity is planned to be over 1,000 MW—making it one of the largest wind farms in the world.

The U.S. has only a few offshore installations in operation, but more are planned. The Block Island Wind Farm, located 3.8 mi (6.1 km) offshore from the state of Rhode Island in the Atlantic Ocean, has five turbines with a total capacity of 30 MW. It was the first offshore wind farm in coastal U.S. waters. Coastal Virginia Offshore Wind is an offshore wind power project located about 43 km (27 mi) off the coast of Virginia Beach, Virginia, on the outer

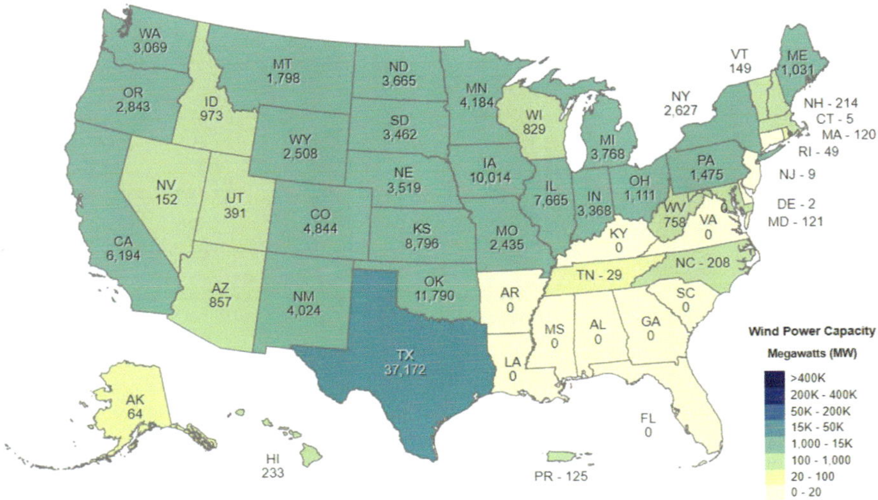

Figure 4.5 U.S. total installed wind capacity by state, Q1 2024 [9].

continental shelf. The initial phase of this 12 MW pilot project, constructed in 2020, used six 2 MW wind turbines. As of December 2022, wind project developers reported plans for about 7,500 MW of nameplate wind electricity generation capacity in waters off the coasts of Maryland, Massachusetts, New Jersey, Rhode Island, and Virginia, and about 20 MW in Ohio state waters on Lake Erie [8].

4.4 Mechanics That Limit the Capacities of WTGs

Today's commercial wind turbine generators are classified as either vertical or horizontal axis machines. Vertical axis wind turbines (VAWT) are omnidirectional machines, so the problem of orienting the machines into the wind is resolved by their configuration. Despite their robust designs using blades fixed in position, they can extract a maximum of only 15% of the wind's energy—a key limitation to their broader use. The two main types are the Savonius (drag-type) and the Darrieus (lift-type) vertical axis wind turbines. The largest installed prototypes are rated at about 1.5 MW. The modern, lift-type, three-blade, horizontal axis wind turbines (HAWTs) with longer airfoils and larger swept areas are simply more efficient than vertical axis machines. HAWTs are designed to automatically orient themselves upwind or toward the oncoming winds to optimize their capability of generating electricity. The pitch of their blades is adjusted to help regulate the speed of rotor rotation.

4.4.1 Gauging the efficiency of HAWTs

For horizontal axis wind turbines, increasing the diameter of the swept area of the rotating blades increases the power extracted from the wind. For HAWTs, Betz's Law (a.k.a., Betz's limit) indicates the extractable power from the wind, independent of the design of the turbine in open air flow. It states that the theoretical maximum extractable power is a function of the swept area of the blades, the wind velocity, air density, and the power coefficient (a measure of efficiency) of the specific machine being studied. Betz's Law establishes an upper limit—no HAWT can capture more than 16/27 or 59.3% of the wind's kinetic energy [10]. This is called the power coefficient (*Cp*) and is usually defined as a maximum factor of *Cpmax* = 0.593. Unfortunately, wind turbines cannot operate at this upper theoretical limit. The power (*P*) available from a wind stream can be estimated using Equation 4.1.

$$P = 1/2\ \rho A v^3 \tag{4.1}$$

Wherein, ρ = air density, A = swept area; and v = velocity (m/s).

While lower-rated WTGs are somewhat less efficient, the larger utility-scale horizontal axis wind turbines achieve a maximum theoretical efficiency of 75% to 80% of the Betz limit. Actual operating efficiencies are less as they are constrained by the height of the tower, the mechanical configuration of the equipment (see Figure 4.6), efficiency losses, and blade designs. Rarely do WTGs operate at rated capacity. A HAWT when well-matched for a given site will operate 60% to 80% of the time and provide full rated power only about 10% of the time. On an average day, a typical machine will generate about 30% to 35% of what it would generate if it were operating continuously at full power.

Once we incorporate various engineering requirements of a wind turbine—particularly strength and durability—the actual power coefficient is less than the Betz Limit. *Cp* values of 0.35–0.45 are common for the best-designed wind turbines. After accounting for the other factors in a complete wind turbine system (i.e., the gearbox, bearings, generator efficiencies, etc.), less power from the wind is converted into usable electricity. Hence, the machine's power coefficient (*Cp*) needs to be factored into the equation to better estimate the capacity. An estimate of the extractable power available (P_{avail}) from the wind in watts (W) using a specific horizontal-axis wind turbine can be estimated using Equation 4.2.

$$P_{avail} = 1/2\ \rho A v^3\ Cp \qquad (4.2)$$

For example, assume that the following data is provided for a specific horizontal-axis wind turbine generator: blade length (l) = 52 m, wind speed (v) = 12 m/second, air density (ρ) = 1.23 kg/m^3, and a power coefficient (Cp) = 0.4. Note that the blade length used in this equation is the distance from the center of the rotor hub to the tip of a blade, which indicates the radius (r) of the rotational swept area (A). Therefore, $A = \pi r^2 = 3.1416 \times 52$ m = 8,495 m^2. Using Equation 4 2:

$$P_{avail} = 0.5 \times 1.23\ \text{kg/m}^3 \times 8{,}495\ \text{m}^2 \times 12^3\ \text{m/s} \times 0.4 = 3{,}611{,}123\ \text{W or } 3.61\ \text{MW.}$$

If the Cp value was 0.2, the P_{avail} would half as much (0.2/0.4 × 3.61 MW) or 1.805 MW.

4.4.2 Basic configurations

Most HAWTs are direct drive, variable speed, and blade pitch-regulated. The long, narrow design of the airfoils (typically with three and sometimes two rotor blades) functions like an airplane wing. The rotating speeds can be lowered or increased by changing the pitch of the blade and its orientation to the wind. Because of the aerodynamic design and length of the blade, the rotation speed at the tip of the airfoil during normal operation is a multiple of the wind speed.

WTGs designs are configured either with or without gearboxes. Direct-drive generators do not use a gearbox to generate electricity (see Figure 4.6, left). They generate power using a giant ring of permanent magnets that spin with the rotor to produce an electric current as they pass through stationary copper coils [11]. The large diameter of the ring allows the generator to create power when turning at the same speed as the blades (8–20 rotations per minute) [11]. For WTGs with gearboxes, the drivetrain is comprised of the rotor, main bearing, main shaft, gearbox, and generator (see Figure 4.6, right) [11]. The drivetrain converts the low-speed, high-torque rotation of the turbine's rotor (blades and hub assembly) into electrical energy [11].

Once installed, the larger WTGs have an on-board digital computer control system that accumulates data and optimizes the output based on recorded

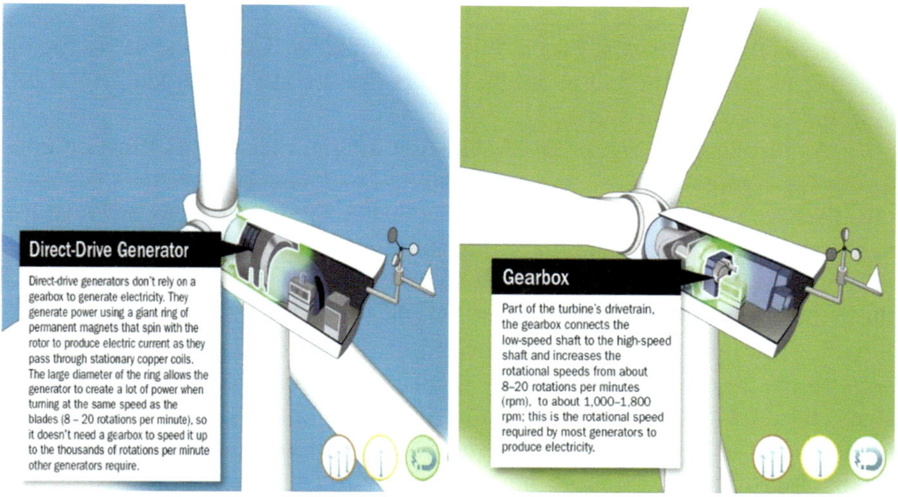

Figure 4.6 Wind turbine generator configurations, direct drive vs. gearbox [11].

wind speeds and ambient weather conditions. Digital sensing, control, and wireless communication systems enable remote data accumulation, system observation, management, and diagnostics. For power transmission, transformers receive electricity at one voltage and increase or decrease the voltage as needed. A wind power plant will use a step-up transformer to increase the voltage, which decreases the power losses that happen when transmitting large amounts of current over long distances with transmission lines [11]. A nearby substation links the transmission system to the distribution system that delivers electricity to the electric grid.

4.4.3 Engineering and cost limitations

While there is no theoretical physical limit to the height of wind turbine generators, there are engineering and cost constraints that must be overcome as the machines get larger. While inadequate maintenance and premature gearbox failures are recurrent issues, blade damage is the most common problem that causes failures. As blade length increases, so do the demands for greater stiffness and strength to prevent issues like stall, vibration, and blade breakage [12]. Blade fatigue is manageable, but of greater concern is the erosion caused when blades collide with raindrops, sea spray, sleet, or hail [13]. When operating, the rotational speed of the blades multiplies the damage from repeated impacts. Blade damage is often visible because cracks and pits will appear on the leading edges of the blades. If unrepaired, structural damage to the blades will occur. With modern HAWT designs, the blade tip speed is usually limited to about 90 meters per second (just under 200 mph) to help avoid blade erosion [13].

As turbines get bigger and blades get longer, their rotors must be configured to rotate at slower speeds. Designed to rotate 10–20 times per minute, there are consequences when slowing the rotor speed. To produce the same amount of power, the blades must deflect the wind to a greater extent; this greatly increases the forces on the other turbine components [13]. As the wind turbine blades increase in length, they also tend to become more flexible. Because the blades are upwind of the tower, during extreme wind conditions, it becomes increasingly difficult to ensure that the blades do not strike the turbine tower while rotating [13]. Braking systems are designed to slow or stop hub rotation when such conditions occur. Finally, the longer the blades, the more difficult they are to manufacture, transport, install, and maintain.

Over land, component transport of WTG tower sections and blades can be limited by the design of available roadways and overpasses. For offshore wind farm developments, this requires bigger port facilities, larger and specially designed ships, and taller cranes that can operate reliably in variable ocean conditions [13]. Accommodating the various engineering design and component transportation challenges for offshore installation increases the production and delivery costs.

Once installed and operational, the taller the wind turbines, the more difficult they are to maintain. The larger the WTGs, the more complex the control systems and mechanical components. Safe access and emergency egress are primary concerns. For land-based machines, turbine cowboys can access the towers at the base and ascend the ladders to the nacelle. This can be challenging when carrying tools or other equipment. Offshore wind farms can be more difficult to access under variable weather and sea conditions. Maintenance personnel must be ferried to the tower base by boat. Some nacelles are designed to be accessed by helicopter. This may be difficult under certain conditions—large WTGs are strategically located in places with consistent and higher wind velocities.

4.4.4 Capacities of utility-scale WTGs

Once rare but no longer unusual in its capacity, the world's largest horizontal axis WTG was once the SeaTitan rated at 10 MW, designed by AMSC, an American energy technologies company; it uses a direct-drive turbine with a 190 m rotor diameter and hub height of 125 m [14]. The Sway Turbine ST10 offshore wind turbine, designed and developed by the Norwegian technology company Sway, is also rated at 10 MW. It is equipped with a rotor diameter of 164 m, has a 2-rpm nominal speed, and blades 67 m in length [14]. The turbine features a direct drive, permanent magnet, ring-style generator with an ironless stator core [14]. The turbine was developed between 2005 and 2012 with an investment of €20 million ($27.4 million) and is suitable for both fixed and floating foundations [14].

Afterwards, WTGs of larger capacities began moving to assembly line production. The world's expanding manufacturing infrastructure has enabled the construction of ever-larger HAWTs. These broader manufacturing capabilities are being driven by the lower costs of installing WTGs on land and in-country requirements for domestic content in their production. Components for WTGs are being manufactured worldwide, but mainly in China, the European Union (EU), India, and North America (see Figure 4.7). Most are for onshore developments. While manufacturing in China and the EU includes producing offshore WTGs, the U.S. and India have focused mostly on the production of machines to be placed in onshore wind farms.

On a world-scale production basis, component manufacturing is often mismatched—especially for those used by onshore wind farms (see Figure 4.7). This creates a supply chain conundrum. In 2023, global onshore wind turbine manufacturing capacity increased by 6 GW for towers, 22 GW for blades and 23 GW for nacelles [3]. Capacity at the least developed stage of the supply chain (towers) reached 134 GW. As of September 2024, manufacturing projects under development comprised about 10 GW of towers, 20 GW of blades, and 55 GW of nacelles [3]. Offshore wind global manufacturing capacity in 2023 expanded by 6 GW for blades and 9 GW for nacelles. Almost all of the new manufacturing plants coming online are in China.

Many of the largest HAWTs are atypical or only have a few units in production. Individual components are custom-designed and manufactured for the specific prototype. The Swedish company Vestas completed its V236-15.0 MW production prototype, which started producing electricity in January of 2023; mounted at a height of 280 m, it has a blade length of 116 m and a rotor diameter of 236 m [4]. The company at present has no plans to build anything with greater capacity. In October 2024, a 15 MW onshore WTG (the SI-2701150) developed by SANY Renewable Energy was installed in Tongyu, Jilin Province, China. It has a design life of 25 to 30 years [15]. With a rotor diameter of 270 m and a blade length of 131 m, it has a maximum swept area of 57,256 m^2 (616,159 ft^2) [12]. With its swept area boosting wind capture ability, optimized aerodynamic blade design (using a

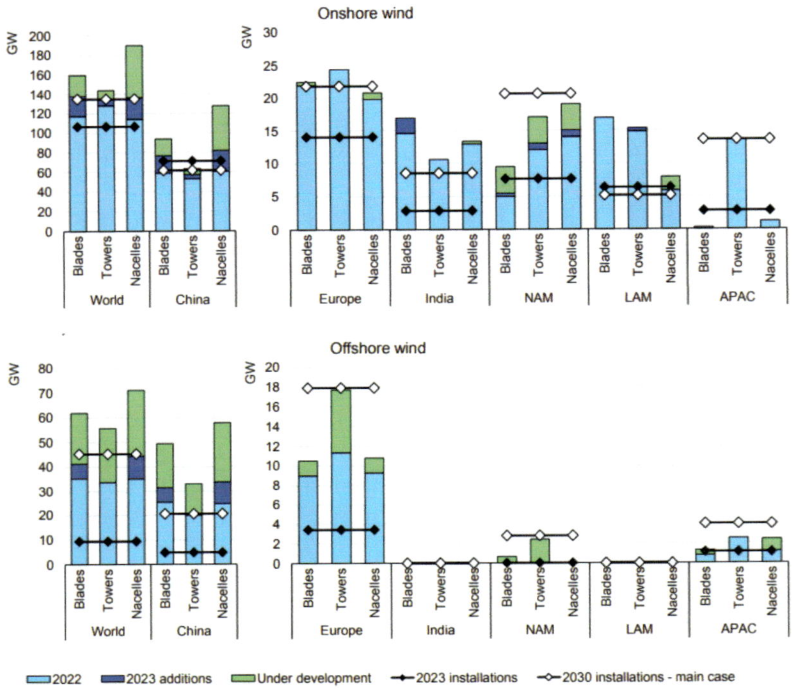

Figure 4.7 Wind equipment nameplate manufacturing capacity by region, component, and WTG installation, 2022–2030 (NAM, North America; LAM, Latin America, APAC, Asia-Pacific excluding China and India) [3].

larger thickness and blunt trailing edge), and control technologies to enhance efficiency, a single unit can generate electricity to power 160,000 households annually[2] [12]. The drivetrain includes a dual tapered roller bearing integrated main shaft support system that ensures high load-bearing capacity and stability, making it more resilient in variable operating conditions [12]. The Siemens SG 14-222 DD turbine model has a 14 MW capacity, reaching up to 15 MW using what the company calls its power boost function. The model installed in Østerild, Denmark, features a 222 m diameter rotor and 108 m blades [16].

WTGs just keep getting bigger and have higher capacities. The Chinese company MingYang Smart Energy (MySE) recently installed

[2] The total number of households a WTG can supply varies based on location. Such estimates require context. In China the average electricity use of a household is 756 kWh/year. A U.S. household consumes an average of about 10,800 kWh/year.

its first MySE 16-260 with a rotor diameter of 260 meters (853 feet) and its swept area of 53,902 m^2 (580,196 ft^2); it is among the most powerful wind turbines so far, offering 16 MW of power, which is enough to supply approximately 36,000 homes [17]. It was installed on a wind farm off the coast of Fujian Province in China. The company has a 16 MW production machine (MySE 16.0-242) with a blade length of 118 m and a rotor diameter of 242 m [4]. Each unit can power 20,000 homes over a 25-year service life [4].

MySE lays claim to the largest WTG installed to date—the MySE18.X-20 offshore prototype in Hainan, China—which has a power output up to 20 MW [18]. It is designed to be more lightweight, modular, and reliable. With its massive wind rotor diameter of 260–292 m (853–958 ft), it has a maximum wind sweeping area of 66,966 m^2 (720,816 ft^2)—more than 12 NFL football fields [18]. When sited in locations with average wind speeds of 8.5m/s, it generates 80 million kWh annually. Its reliability and survivability in high winds remain to be proved. Installed in August 2024, it survived typhoon Yagi in September without damage. However, pieces of the wind turbine's blades broke off the following December and fell while the rotor continued to spin. The company has announced plans for a larger offshore turbine, capable of delivering 22 MW of power.

4.5 The Importance of Performance Observations

The output of wind turbine generators is enhanced when they are properly sited. While both land-based and offshore locations are available, offshore locations benefit from more consistent wind flows that are not disrupted by geographic features or nearby buildings.

Actual performance observations are important to estimate the efficiency and power generated by the WTGs at different wind speeds. This is particularly true for prototypes installed at a specific site. Assessments of wind turbine operation are data-driven and require on-site operation of the machines after installation and commissioning. The 14 MW Siemens SG 14-222 DD in Denmark first generated electricity in December 2021 and produced 359 MWh within a 24-hour period; the turbine maker said this was the most power a single wind turbine has ever produced over this duration [16]. The previous record was held by the General Electric (GE) Haliade-X offshore wind turbine prototype installed at the Port of Rotterdam, which generated 312 MWh of electricity in November 2020 [16]. Better assessments can be made when multiple prototype WTGs are in closer proximity and operate in similar environmental conditions.

It will be interesting to see what happens when these larger, utility-scale, wind turbine prototypes are manufactured at production scale and installed in commercial offshore wind farms. The ability to make direct performance observations of multiple similar machines is likely to improve during the next few years. This will enable engineers and scientific researchers to study recorded data and compare the efficiency and capabilities of these large WTGs when operating under real-world and highly variable conditions.

Numerous opportunities for operational assessments will soon be available. GE Wind Energy's Haliade-X (14 MW) prototype is planned for commercial deployment by the Vineyard Wind and Ocean Wind projects in the U.S. and on Doffer Bank in the UK. Siemens Gamesa will supply 72 units of its flagship 14 MW SG 14-236 DD wind turbines to the 1 GW Danish Thor offshore wind farm based at Thorsminde, Denmark, near the Nissum Fjord [19]. The SG 14-222 DDs feature a rotor diameter of 222 m and swept area of 39,000 m^2 (419,792 ft^2). This design has been configured for sustainable deconstruction. After reaching end-of-life, the decommissioning process enables the separation and recycling of the blade materials. These units are also planned to spin in wind farms being developed at Moray West off the coast of Scotland (882 MW), the Hai Long projects offshore Taiwan, and the Dominion Energy Coastal Virginia Offshore Wind Project (2,640 MW) in U.S. waters [16]. Because of the larger number of similar WTGs being installed and the numerous site locations, operating data will be valuable to optimize engineering decisions regarding operational adjustments, siting, and design improvements.

4.6 Conclusion

There is a strong industry commitment to delivering larger WTGs in greater numbers. As a result, the costs for wind power are expected to further decline in the future. Like automobiles, the price for wind machines declines precipitously after assembly line production of a particular model occurs. The production of components for WTGs is happening worldwide. After 2,000 to 3,000 units of a specific model are manufactured, costs can decline by as much as 20% to 30%. At some point in the manufacturing process, most engineering flaws have been resolved, and the output of each specific model is optimized.

As production and operation costs for WTGs decline and the costs for grid-supplied electricity increase, the economic feasibility of wind power improves. Many naysayers who once shunned wind power now believe a closer look is justified. Onshore wind power development is already competitive with conventional fossil fuel-fired generation in some regions of the world. Offshore windfarms make use of no-cost real estate unavailable to fossil plants and their supportive infrastructure—they simply cannot be constructed in offshore locations. The global weighted-average lifecycle cost of energy for offshore wind power dropped from being more than twice as expensive as the least costly fossil fuel alternative in 2010 to just 17% more expensive by 2022 [20]. During that period, the average lifecycle cost of electricity produced by wind power fell from $0.197/kWh to $0.081/kWh [20].

We will see wind farms in the future being developed with multiple WTGs rated at over 10 MW. While much of this expansion is occurring in China, the annual global supply chain potential has reached almost 35 GW for towers, over 400 GW for blades, and almost 45 GW for nacelles—all are much greater than the 9 GW of total installations in 2023 [3]. China's share in the global offshore supply chain reached 60% in 2023; with the commissioning of projects under development, it is expected to expand further by 2030 [3]. China's manufacturers are pushing the limits on scale and striving not to be outdone by their domestic competitors. Dondfang Electric Corporation, a Chinese state-owned manufacturer of power generators, has unveiled a typhoon-resistant 26 MW offshore WTG that has rolled off the production line; its height equals that of a 63-story residential building [21]. After it is installed, it will be the world's largest in both size and capacity, given its swept area of 53,000 m^2 [21].

Whether bigger wind generators are better is a question whose answer requires observation, study, and assessments. As these larger wind energy conversion systems are completed and become operational, we will gain the ability to study the economic viability and environmental impacts of these installations more fully [22]. While technical and engineering challenges remain, getting the large capacity wind machines into assembly line production and installed in offshore wind farms is the probable next step [22]. No doubt some of the larger installations will be hindered and delayed by permitting, financing, interconnection requirements, and site development constraints. Ultimately, many will prevail and become operational.

There is enormous worldwide potential to generate electricity using wind turbines, whether located on land or offshore. The larger the capacity of the machines, the fewer that are needed. Tomorrow's wind farms will gradually use fewer WTGs. Needing fewer turbines on land-based wind farms allows power plant developers more flexibility in siting—creating opportunities to avoid culturally and environmentally sensitive areas [23]. Increasing wind power capacities and developing offshore wind farms are trends that will shape the future of wind power developments [24]. Because of the consistently strong winds at sea, turbines can produce energy more reliably and efficiently than onshore wind turbines [18]. Typhoon-proof turbines will become the standard; already, some HAWTs are capable of withstanding winds up to 220 kph (137 mph). We will also see hybrid renewable energy systems that use wind power, greater grid integration, and creative applications of energy storage.

As larger WTGs come into production and their costs decline, wind power will have an increasingly important role in meeting the world's hunger for renewable energy [24]. When wind resources are available and WTGs are properly sited, they can generate large amounts of electricity near densely populated coastal regions [25]. Due primarily to accessibility and logistics, onshore wind farm developments predominate. Economic challenges remain, especially for offshore wind power development. While costs for land-based wind farms are declining, offshore wind farms have a global average development cost of \$230 per MWh produced—their costs have increased somewhat in the past couple of years to more than triple the average of \$75/MWh for onshore facilities [25].

Wind power is a mature technology that provides pollution-free electrical generation [26]. As a result, it has a key role in meeting regional sustainable development goals. Wind turbines are crucial for the generation of clean energy at a time when the world is transitioning away from fossil fuels [27]. From a policy perspective, wind power will form a considerable part of government strategies to incentivize local electrification. Wind power offers a technically viable renewable energy alternative in an era dedicated to decarbonization. No greenhouse gas emissions are emitted in the process of producing electricity. Emissions are limited to those in the manufacture of components and their delivery, site preparation, and assembly—which in sum are much less than the carbon footprints of developing new fossil fuel-fired plants. To alleviate environmental concerns about decommissioning, future WTG components will be designed to be reusable (foundations and towers) or recycled (blades, metals, and electronic parts).

Harnessing wind energy provides opportunities for continued innovation. In an era when the commercial growth in electrical demand is concentrated in U.S. states such as Arizona, Texas, and Virginia, which are experiencing rapid development of large-scale facilities thirsty for electricity (i.e., manufacturing and data centers), electricity generated using wind power is increasingly valuable. Surprising technological innovations in wind power generation are just over the horizon—where our wind farms can be found. In some remote areas of the world that lack grid interconnections, wind power will offer the first source of electrification by producing electricity for local microgrids. With such large capacities for power production, we may decide that our newest and biggest blade runners are big enough—at least for now.

References

[1] Morison, R, (2023, January 16). Wind turbines are big enough for now, Vestas CEO says. https://www.bloomberg.com/news/articles/2023-01-16/wind-turbines-are-big-enough-for-now-vestas-ceo-says, accessed 6 December 2024.

[2] Venditti, B. (2022, June 3). Animation: the world's biggest wind turbines. https://www.visualcapitalist.com/visualizing-the-worlds-biggest-wind-turbines, accessed 1 May 2023.

[3] International Energy Agency (2024, October). Renewables 2024, analysis and forecast to 2023. https://origin.iea.org/reports/renewables-2024, accessed 1 November 2024.

[4] Roosa, S. (2021). *Fundamentals of Microgrids*, 100-101. CRC Press: London.

[5] Climate Matters (2024, April 10). A decade of U.S. wind growth. https://www.climatecentral.org/climate-matters/a-decade-of-us-wind-growth-2024, accessed 4 January 2025.

[6] U.S. Energy Information Administration. Renewable wind. https://www.eia.gov/kids/energy-sources/wind, accessed 4 January 2025.

[7] Wikipedia. Tehachapi Renewable Transmission Project. https://en.wikipedia.org/wiki/Tehachapi_Renewable_Transmission_Project, accessed 6 January 2025.

[8] U.S. Energy Information Administration (2023, February). Electric Power Monthly, Table 1.14.B., preliminary data, accessed 4 January 2025.

[9] U.S. Office of Energy Efficiency and Renewable Energy. U.S. installed and potential wind power capacity and generation. https://windexchange.energy.gov/maps-data/321, accessed 4 January 2025.

[10] Burton, T. et al. (2001). *Wind Energy Handbook*. John Wiley and Sons. ISBN 0471489972.

[11] U.S. Department of Energy. How a wind turbine works. https://www.energy.gov/eere/wind/how-wind-turbine-works-text-version, accessed 5 August 2025.

[12] Malayil, J. (2024, October 10). 15 MW: world's largest onshore wind turbine set to power 160,000 homes yearly. *Interesting Engineering*. https://interestingengineering.com/energy/china-installs-worlds-largest-onshore-wind-turbine, accessed 6 December 2024.

[13] Hogg, S. (2023, January 3). Wind turbines are already skyscraper-sized. Is there any limit to how big they will get? https://techxplore.com/news/2023-01-turbines-skyscraper-sized-limit-big.html, accessed 6 December 2024.

[14] Power Technology (2014, January 1). The world's 10 biggest wind turbines. https://www.power-technology.com/features/featurethe-worlds-biggest-wind-turbines-%80%99s%2010%20biggest%20wind%20turbines%201%20SeaTitan,6.2M126%20...%208%20Siemens%20SWT-6.0-154%20...%20More%20items, accessed 5 August 2025.

[15] SANY (2024, October 16). SANY renewable energy successfully installs the 15-MW onshore wind turbine, press release. https://www.sanyglobal.com/press_releases/3922, accessed 2 November 2024.

[16] Durakovic, A. (2022, October 10). Siemens Gamesa offshore wind turbine prototype breaks 24-hour output record. https://www.offshorewind.biz/2022/10/10/siemens-gamesa-offshore-wind-turbine-prototype-breaks-24-hour-output-record, accessed 4 January 2025.

[17] Nield, D. (2023, July 24). The largest and most powerful wind turbine ever built is now operational. https://www.sciencealert.com/the-largest-and-most-powerful-wind-turbine-ever-built-is-now-operational, accessed 11 December 2024.

[18] Ghoshal, A. (2024, September 4). Colossal 20 MW wind turbine is the largest on the planet (for now). https://newatlas.com/energy/world-largest-offshore-wind-turbine-20-mw-mingyang, accessed 6 November 2024.

[19] Memija, A. (2023, January 12). Siemens Gamesa 14 MW turbines for Danish offshore wind farm. https://www.offshorewind.biz/2023/01/12/siemens-gamesa-14-mw-turbines-for-danish-offshore-wind-farm, accessed 6 December 2024.

[20] International Renewable Energy Agency (2023, August). Renewable power generation costs in 2022. ISBN: 978-92-9260-544-5.

[21] Memija, A. (2024, October 14). World's first 26 MW offshore wind turbine rolls off production line. https://www.offshorewind.biz/2024/10/14/worlds-first-26-mw-offshore-wind-turbine-rolls-off-production-line, accessed 6 December 2024.

[22] Roosa, S. (2023, September). Big wind—how big is big enough? *International Journal of Strategic Energy and Environmental Planning*, 5(5), 5–11.

[23] U.S. Office of Efficiency and Renewable Energy (2024, February 8). Wind farms of the future will be more powerful and quieter. https://www.energy.gov/eere/wind/articles/wind-farms-future-will-be-more-powerful-and-quieter, accessed 4 January 2025.

[24] Windcycle Energy. The future of wind energy: predictions and trends.

[25] Groom, N. and Chestney, N. (2024, November 18). Global offshore wind industry poised to miss big targets as obstacles mount. https://www.reuters.com/sustainability/global-offshore-wind-industry-poised-miss-big-targets-obstacles-mount-2024-11-18/, accessed 11 December 2024.

[26] Proles, F. (2018). Hybrid electrical generation for grid independent oil and gas well fields, in chapter 6, Roosa. S. ed., *International Solutions to Sustainable Energy, Policies and Applications*. Fairmont Press: Lilburn, Georgia.

[27] Mishr, R. (2024, December 13). Typhoon-proof blades of 'world's most powerful' wind turbine break off in China. https://www.msn.com/en-us/money/companies/typhoon-proof-blades-of-world-s-most-powerful-wind-turbine-break-off-in-china/ar- AA1vO1VG?ocid=msedgdhp&pc=HCTS&cvid= a25a22d4fe7b49d1ae99015d1d53a383&ei=17, accessed 14 December 2024.

5

Clean Energy Technologies

Stephen A. Roosa

This chapter discusses the status of selected advanced clean energy technologies. It considers hydropower, tidal energy, floating solar installation, and the situation with nuclear energy. Hydropower is a proven, mature, predictable, price-competitive, and scalable technology. Hydropower has a key role in providing renewable and emissions-free generation of electricity. There is growing interest in harnessing tidal power to generate electricity. One of the key benefits of tidal power is that tides are highly predictable. Floating solar farms offer the potential to take advantage of unusable or underused lake surface areas. By using solar PV systems over stable areas of water, they provide opportunities to generate revenue. Nuclear energy technologies promise reliable and resilient energy supplies. They are not dependent on the transportation of large volumes of fossil fuels to the plant site or on variable weather conditions. Nuclear power plants can produce large amounts of energy in a limited land area due to their high energy efficiency and power density.

5.1 Status of Hydropower in the U.S

Hydropower was one of the first sources of energy used for electricity generation [1]. Since the energy source is provided freely by nature, there is no combustion process, and there are no costs to deliver the resource to the site. Until 2019, hydropower was the leading U.S. source of renewable electricity generation [1]. In 2022, hydroelectricity accounted for about 6.2% of total U.S. utility-scale electricity generation [1]. In 2023, hydropower generated about 27% (245 TWh) of total U.S. renewable electricity.

DOI: 10.1201/9788743811312-5

Table 5.1 The 10 largest U.S. hydropower plants in terms of average net electricity generated [2].

Hydropower plant	State plant is located	2021 Average net generation (MWh)	Total U.S. hydropower generation (%)
Grand Coulee	Washington	19,550,777	7%
Robert Moses – Niagara	New York	14,186,130	5%
Chief Joseph	Washington	11,092,216	4%
John Day	Oregon	9,041,083	3%
Robert Moses – St. Lawrence	New York	6,906,420	3%
The Dalles	Oregon	6,613,185	2%
Rocky Reach	Washington	5,935,038	2%
McNary	Oregon	5,369,726	2%
Wanapum	Washington	4,820,651	2%
Bonneville	Oregon	4,659,483	2%
Total	N/A	88,174,709	32%

Due to the density of water and the use of gravity to increase pressure, hydropower is one of the most efficient ways to convert kinetic energy into electricity. Utility-scale hydropower plants operate at efficiencies greater than 90% and have a minimum of 1,000 MW of net generating capacity. These plants are typically configured to use Francis or Kaplan turbines, but those with less than 75 m of head often use Pelton turbines. In the U.S. a turbine can weigh as much as 172 tons and turn at a rate of 90 revolutions per minute (rpm).

There are about 1,450 conventional and 43 pumped-storage utility-scale hydropower plants operating in the U.S. The total U.S. conventional hydropower fleet includes 2,252 hydropower plants with a total generating capacity of 80.6 GW. Pumped-storage hydroelectric plants store water by pumping it to a secondary reservoir located at a higher elevation. The water is released from the upper reservoir to power hydro turbines located at a lower elevation below the upper reservoir [1]. Due to its low cost and large storage capacities compared to other types of energy storage, pumped-storage is the main technology used for hydropower applications. It is used primarily to ensure baseload electric generation, help meet demand peaks, and smooth power from other renewable resources.

More hydropower is generated in the Northwest and Rocky Mountain regions than in other regions of the country. The 10 largest hydropower plants are in the states of New York, Oregon, and Washington; together they provide almost one-third of the U.S. average net generation (see Table 5.1). The largest U.S. hydropower facility, and the largest U.S. electric power plant

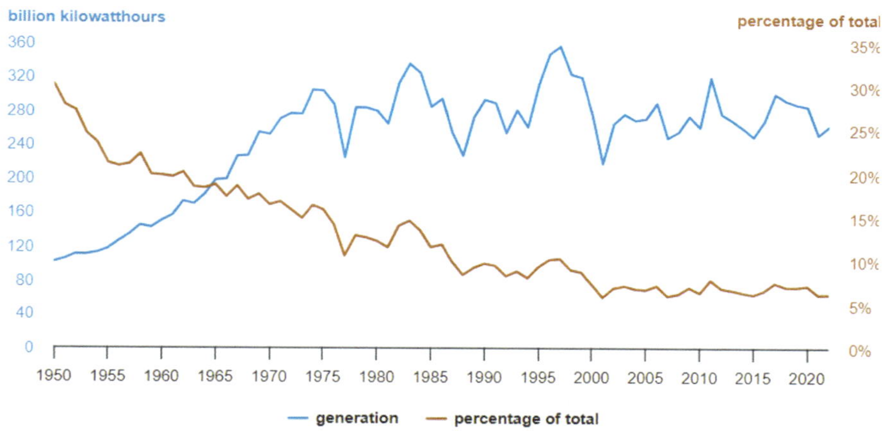

Figure 5.1 Hydroelectric generation and share of total U.S. electric generation from 1950 to 2022 [4].

in electric generation capacity, is the Grand Coulee hydropower dam on the Columbia River in Washington State, with 6,765 MW total generation capacity [1]. It generates over 19.5 million MWh of electricity annually and supplies power to eight states, including parts of Canada [2]. This is substantially larger than the generating capacity of Plant Vogtle in Waynesboro, Georgia, the largest nuclear plant in the U.S., which is rated at 4,464 MW [3]. While both hydro and nuclear energy provide baseload electricity, nuclear power is not subject to unpredictable weather conditions (i.e., droughts) and provides higher operational capacities [3].

The generation of electricity from U.S. hydropower resources has remained stable since the 1970s, fluctuating from 230 to 360 billion kWh annually (see Figure 5.1). As the demand for electricity in the U.S. has increased, its share of electric generation as a percentage of total generation has declined (see Figure 5.1).

The total precipitation that drains into the rivers and streams of a geographic watershed determines the amount of water available for producing hydropower [1]. For about the last 25 years, the southwestern U.S. has experienced its driest period in the last 1,200 years. Since 2020, the drought conditions have become severe [2].

When reservoir levels drop below certain thresholds, hydropower plants must reduce electricity generation or risk completely shutting down operations [2]. 2022 was a particularly tough year for the U.S. hydropower industry. In December 2022, Lake Mead (Hoover Dam in Nevada and Arizona), the

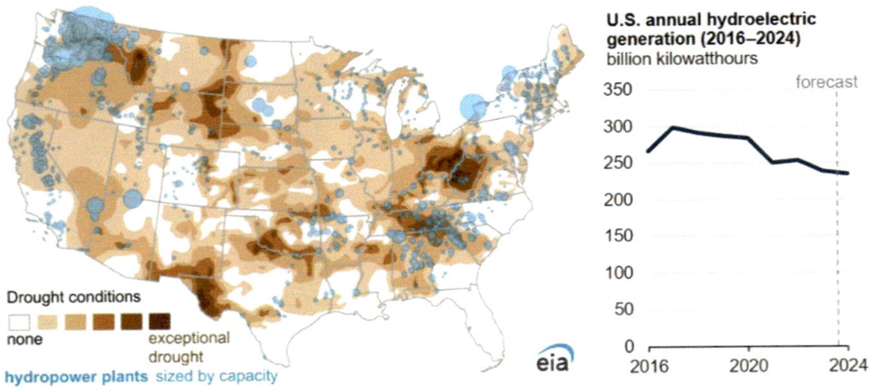

Figure 5.2 U.S. drought conditions and hydropower capacity in September 2024 [5].

largest reservoir in terms of storage capacity, and Lake Powell (Glen Canyon Dam in northern Arizona) recorded their lowest levels of water storage in the last 30 years of operation [2]. On December 4, 2022, the reservoir at Lake Mead was filled to only 28% capacity, and Lake Powell was at 23% [2].

Perhaps the most concerning issue with the use of hydropower to generate electricity is that it is hydrology-dependent. Drought conditions, in some areas classified as extreme, continued to be widespread and persistent in regions of the U.S. as of September 2024 (see Figure 5.2). This lack of rainfall causes lower water levels in reservoirs; ultimately, less water is available to be released through turbines to generate electricity. Widespread drought conditions reduce the generation capacities of hydropower plants, lower total hydropower production, and increase the potential for electricity shortages.

In 2023, the U.S. generated approximately 239.9 terawatt hours (TWh) of hydroelectric power, a decrease from 254.8 TWh in 2022 [6]. Washington was the leading U.S. state in terms of conventional hydroelectricity production in 2023, with more than 60 TWh produced, followed by California with 32 TWh [6]. By November 2024, due to rainfall from atmospheric rivers that impacted areas of the far western U.S. states, conditions had somewhat improved. For example, the water level in Lake Mead was at 1,065 feet (324.6 m) of mean sea level, which was above its historic 2022 lows but far short of its full pool level of 1,229 feet (344.6 m) as shown in Figure 5.3 [7].

While no major high-head hydropower projects are planned in the U.S., there are some encouraging trends. At the end of 2022, the U.S. hydropower development pipeline included projects to construct 117 new facilities with a combined capacity of 1.2 GW; however, only eight are presently under

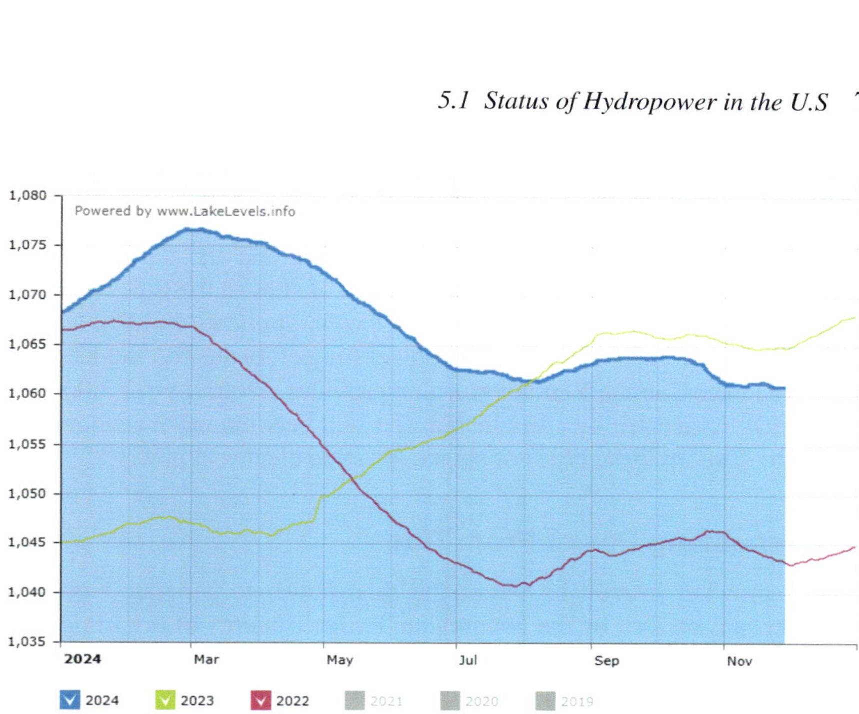

Figure 5.3 Lake Mead water levels 2022–2024 [7].

construction [8]. Non-powered dam retrofits account for 95% of all proposed new hydropower capacity in the U.S. [8]. In addition to the new facilities, 23 active hydro upgrade projects, upon completion, will increase the capacity of the existing fleet by 254 MW [8]. Pumped storage hydropower capacity increased 1.4 GW over the past decade, with 97% of this increase due to upgrades to the existing fleet—most occurring in the western regions of the U.S. Total energy storage capacity now exceeds 553 gigawatt-hours [9]. These plants accounted for 96% of utility-scale energy storage capacity. At least 11 hydropower plants in the U.S. have added or are planning to add battery capacity to their facilities, a promising trend for hydropower in the future [9].

Hydropower has many benefits. It is a proven, mature, predictable, price-competitive, and scalable technology. Hydropower has a key role in providing renewable and emissions-free generation of electricity. Unlike fossil fuel-fired plants, hydropower plants do not require a combustion process to generate electricity and do not consume large volumes of water. They operate at high efficiencies and do not produce unusable heat. For such reasons, we are seeing increased interest in the use of hydropower to generate electricity. By the end of 2022, there were 3,909 hydropower projects in the global development pipeline with a combined total capacity of 557 GW [8].

In addition to large utility-scale projects, future capacity additions will likely include the use of enhanced turbine technologies, run-of-the-river and tidal applications, greater use of energy storage, and the development of micro-hydropower plants. There are thousands of sites in the U.S. with existing dams that are available for micro-hydropower applications [10]. A recent example is the Mother Ann Lee hydropower project located on the Kentucky River in Mercer County, Kentucky, which provides electricity for 2,000 residences [10]. As electricity costs increase, it will become more economically feasible to develop these types of sites to provide electrical power.

5.2 Harnessing Power From Tides

There is growing interest in harnessing tidal power to generate electricity. Tidal energy is created by the potential and kinetic energy of tides that occur naturally due to the orchestrated movements of the Earth, Sun, and Moon. Tidal power is classified as a form of renewable hydraulic energy and can be harnessed to generate electricity. Water is 830 times denser than air, and its movement due to gravitational forces contains enormous amounts of potential kinetic energy.

Finding the best applications for tidal power poses challenges. Unlike wind and solar power, one of the key benefits is that tides are predictable. Given almost any coastal location on Earth, we can estimate when the tides will flow into and out of an estuary. With observation and direct measurement, we can also determine the flow rates at various times of the day.

Potential tidal power sites, when located near coastal populations, lower the cost of developing new and extensive electrical transmission systems. Sites in the U.S. that have the potential to harness tidal power have been identified in the waters of Florida, Hawaii, Maine, New Jersey, New York, Oregon, South Carolina, and Washington. Some of these sites are located near populated areas where there is a need for additional electricity.

What types of methods do we have available to utilize power from the tides? Several of the configurations are like those we use to generate electricity with wind power technologies. Others mirror the configurations of low- and medium-head hydroelectric dams. The technologies used to generate electricity include:

- Horizontal axis turbines and vertical axis turbines;
- Oscillating hydrofoils that use hydraulic fluid;
- Venturi effect designs that concentrate the flow into a funnel to drive a turbine;

- Tidal fences that look like turnstiles in caissons; and
- Tidal barrage harvesting systems.

Tidal barrages are a type of tidal range technology. They are like hydroelectric dams but take advantage of locations with high-amplitude tides to harvest their potential energy. Barrages are constructed across an estuary to hold large bodies of water and use bidirectional turbines to generate electricity. Constructed to restrict the movement of tidal flows in and out of an estuary or tidal bay, barrages direct water through strategically located turbines and generate electricity regardless of the direction that water is flowing. This technology holds a dominant share of the globally installed capacity of total ocean renewable energy technologies [11].

To determine the best ways to harness tidal power, various technologies need to be considered. Potential sites must be selected and discreetly assessed. A specific study of the tidal flows at the selected site is often required. Assessments of the impacts on navigation and the environment are necessary. Engineering studies are needed to determine which technologies to apply and the scale of deployment. The technology selected for a given site must be designed to adapt to the tidal resources available. After a proposed engineering design is completed, the costs need to be estimated, and an economic analysis of the project must be performed. In this way, economic and environmental costs may be lowered and optimized.

Tidal current energy converters (TCECs) have made substantial progress in commercial development in the last decade [11]. There are now many examples of successful tidal power projects, most of which are found in the northern hemisphere (see Figure 5.4) [12]. Getting it right usually depends on choosing the best sites and the appropriate technologies. Once constructed, tidal power projects have longevity; the Rance Tidal Power Station (240 MW) in France has been in operation since 1966 [13]. However, most operational tidal projects are small in scale. Examples include the Jiangxia Tidal Power Station in China (1980/3.2 MW), the Eastern Scheldt Barrier Tidal Power Plant in the Netherlands (2015/1.3 MW), the Uldolmok Tidal Power Station in South Korea (2009/1.5 MW), and the Meygen plant in the UK (2017/6 MW) [13]. The largest one currently in operation is the Sihwa Lake Tidal Power Station in South Korea (2011/254 MW) [13]. At least 17 major projects are underway in the European Union that will provide over 160 MW of electric capacity with an estimated investment cost of $1.2 billion [14]. This works out to an average capacity cost of about $7.5 million/MW.

Sustainable energy solutions require assessments of the benefits and issues associated with the production of electricity. When properly developed,

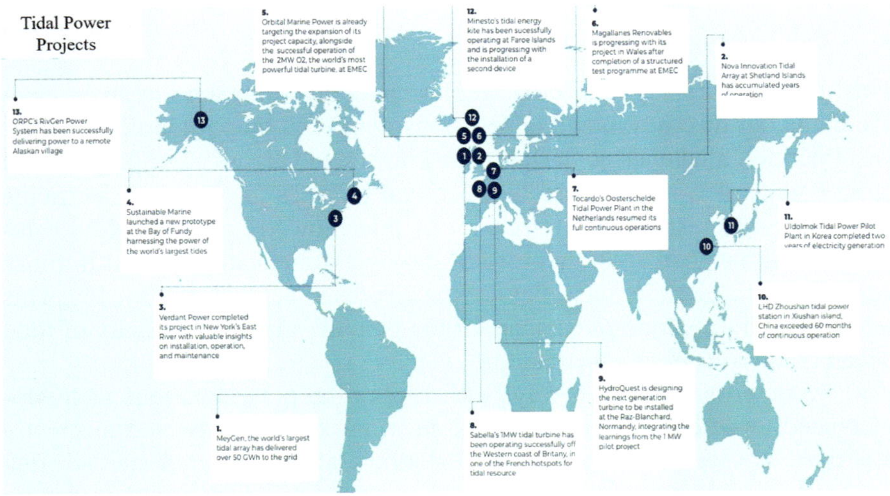

Figure 5.4 Tidal power projects that are active or under development [12].

tidal power projects provide many benefits. The energy resources used to operate the TCECs are available at no cost and delivered to the site by natural processes. Their electrical output is not influenced by weather. Tidal power provides a cost-effective alternative to using fossil fuels to generate electricity at relatively high efficiencies [11,15]. Production is quiet, predictable, and capacities tend to be higher than most other renewables. Because the energy resource is continuously available and the technologies used have relatively few moving parts, electrical production is reliable and resilient—particularly when compared to the web of technologies and mechanical components required to refine fossil fuels, deliver them to the site, and combust them in a controlled fashion for Rankine cycle power plants [15]. The longer lifespan of tidal power facilities makes them more competitive in the long term [15]. Once a tidal power system is operating, they are relatively inexpensive to maintain. From an environmental perspective, the electricity produced by tidal power does not create air or water pollution nor produce greenhouse gases [15].

In the U.S., tidal energy represents an estimated 220 TWh/year of resource potential, which is equivalent to 5.4% of the U.S. total electrical energy demand of 4,096 TWh in 2024 [16]. If only a portion of these resources is captured, they could provide affordable and locally available clean power to rural, remote island and shoreline communities [17]. However, the cost of tidal power is substantial due to the engineering required, site preparation costs, upfront costs of building plants, and expenses associated with maintaining machinery. The average commercial tidal energy project costs as much \$280/MWh [18].

Tidal projects may cause ecological issues such as changes in sediment, shrinking, or damage to tidal flats [19]. For example, barrages that block off estuaries may redirect the movement of water and have unanticipated effects on their ecosystems. Many animals, especially migratory birds, rely on these tidal flats and may not be able to utilize other similar habitats [19]. Regional regulatory issues associated with any potential environmental impacts must be considered and addressed.

"Our oceans and rivers represent a huge potential source of renewable energy," said Alejandro Moreno, Acting Assistant Secretary for the U.S. Office of Energy Efficiency and Renewable Energy [17]. The tidal energy industry requires long-term and substantial funding to move from testing, siting, project prove-out, and commercialization. The complexity of installing devices, navigating permitting processes, and difficulties connecting to local power grids have proven to be barriers to advancing the broader use of tidal energy [17].

The global potential to generate electricity using tidal energy is estimated to be on the order of 500 GW [18]. More research on tidal technologies is needed to overcome their geographical limitations, costs, and ecological impacts [18].

5.3 Floating Solar Photovoltaics

Utility-scale solar PV often requires large parcels of land; however, land-constrained developing countries, especially in densely populated regions, may prioritize land use for agricultural, forestry, or other needs [20]. There are alternatives when land areas are unavailable or costly to access.

Floating photovoltaics (FPV), also referred to as floating solar power or floatovolatics, is an important option. FPV is a new application that was first patented in 2008 [21]. Arrays of solar PV modules float on bodies of water by being mounted on buoys or floats. The arrays are placed in ponds, lakes, reservoirs, or irrigation ponds. The better sites are found in inland, near-shore locations near existing distribution and transmission infrastructure. Many of these installations are found in unused areas on artificial bodies of water, such as hydroelectric dam reservoirs, wastewater treatment ponds, or drinking water reservoirs [21]. Optimum sites are ponds and lakes in industrial areas that do not attract many visitors, have steady water levels, and are not subject to heavy winter snowfalls or wave action [22]. This placement of solar PV arrays on bodies of water avoids the issue of competition for land. Figure 5.5 shows the basic configuration and key components of FPV systems.

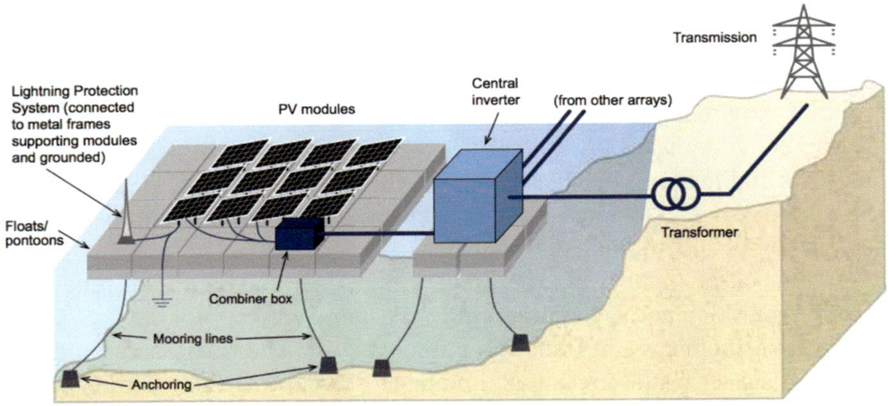

Figure 5.5 Schematic of a typical floating solar PV system [23].

There are many benefits that support the development of FPV. Chief among these are the avoided costs of placing the solar arrays over lake areas, which are less expensive to rent than land areas. The potentially unusable or underused surface areas are converted into ones that allow for revenue-generating uses [20]. Costs to prepare land areas for installation, such as creating access roads, clearing trees, and vegetation, are diminished or avoided. A key benefit is that the efficiency of the solar PV modules is enhanced and their longevity is maximized because of the regulating effect of water bodies, which stay cooler than land areas. This is notably true in summer months when solar energy production is greater and ambient temperatures are higher. The voltage at the maximum power will decrease as the solar module surface temperature increases [24]. The convective heat transfer rates are higher than the radiative heat transfer rates [24]. Module overheating is avoided. Shade, which can prevent sunlight from reaching the collector surfaces and reduce solar PV output, is rarely a problem. However, in regions where water is scarce, the solar PV arrays cover lake surfaces and thus reduce loss of water due to evaporation. The arrays can also be installed near existing electrical power plants, such as near hydropower dams, reducing the substantial costs of extending power lines. The cost of security fencing can be avoided.

There are costs and impacts associated with FPV installations. Like any solar PV project, each potential site requires a custom engineering assessment and project design. In some cases, total development costs are higher than land-based systems. The solar panels require special coatings to prevent deterioration and the potential for rust. Anti-rust materials must be used.

Floatation is designed to be buoyant and often uses polyethylene that can support two and a half times its weight [21]. Extreme and cyclic loads due to wind and waves can cause failures. To limit movement, arrays must be properly anchored. Effective mooring systems are needed and must be tailored to local water movements and wave action to ensure stability. FPV systems may be disruptive to aquatic life and must be placed in locations where they do not pose hazards to navigation. An environmental impact assessment may be required for each site. Finally, due to the location of the solar arrays over water bodies, access for maintenance must be by boat. Costs for periodic visual and infrared inspections can be reduced by using aerial drones.

There are numerous examples of FPV installations. Floating solar has predominantly been installed in countries such as China, Japan, and the U.K. and is quickly gaining popularity in the U.S., especially in California and New Jersey [21]. Among U.S. examples are the Healdsburg Floating Solar Farm, California, with a capacity of 4.8 MW [21]. This installation sits on two ponds and provides 8% of Healdsburg's annual electricity needs [21]. It uses bifacial solar panels, capturing the overhead sunlight and the rays that the water reflects. The world's largest installation is the Dezhou Dingzhuang Floating Solar Farm, China, with a capacity of 320 MW, which is located near the Dezhou thermal power stations in Shandong Province. The system was completed in 2020 and is connected to a wind farm and an 8 MWh battery storage system; it generates 550 million kilowatt-hours (kWh) of electricity annually [21,25].

The World Bank estimated there are 6,600 large bodies of water worldwide suitable for floating solar, with a technical capacity of over 4,000 GW if only 10% of their surfaces were covered with solar panels [22]. Of this total, 1,260 GW of this potential capacity is in North America [22]. The private sector leads the development and deployment of FPV technologies, with the research community showing increased interest [20]. The FPV market is expected to grow globally from $13.8 million in 2015 to $2.7 billion by 2025 [20].

Floating photovoltaic systems are now commercially available, and their deployment is accelerating worldwide [20]. As the industry grows, new and innovative technologies for FPV structures are being developed [26]. China's first floating solar power project located entirely in a seawater environment is operating in Qingdao City [27]. This project has a capacity of 7.5 MW and is expected to generate 16.7 million kWh of electricity annually [28]. The innovative design allows the photovoltaic panels to float and move with the tides, maintaining a minimal gap between the panels and the water surface [28]. It uses anti-corrosion floats, barnacle-resistant supports, and a robust underwater anchoring system [28].

The levelized cost of solar PV installations is declining. The main driver for the continued growth of FPV is that unit investment costs are approximately \$1.2/Wp and are anticipated to fall to \$0.65/Wp by 2050 [26]. Continued research concerning FPV systems must advance emerging global applications, plus offer multiple benefits to stakeholders, such as diversity of generation, increased investment opportunities, and economic development [20]. Efforts must be made to determine ways to lower costs. Real-world experience is needed in the development of mooring systems, which have been identified as a main cause of system failures due to the inherent complexity of station-keeping [26]. One promising application is the use of FPV at pumped-storage hydropower sites. The upper reservoirs are often engineered for stability and designed to maintain minimum water levels. "A hybrid floating solar and hydropower system can have a more stable power output throughout the year," said Prateek Joshi, an energy researcher at the National Renewable Energy Laboratory [27]. "Overall, we can reduce the variability of generation" [27]. Electrical transmission infrastructure is usually nearby. With greater development experience and reduced costs, floating photovoltaics systems are poised for broader deployment.

5.4 Nuclear Power Plants

Nuclear power plants (NPPs) are viewed as a way for electricity suppliers to meet the increasing demand for clean energy. Nuclear energy technologies promise reliable and resilient energy supplies. They are not dependent on the transportation of large volumes of fossil fuels to the plant site or on variable weather conditions, as is the case with some types of renewable energy resources.

Compared to other alternatives, nuclear power has advantages over contenders to replace traditional coal and oil sources. NPPs can produce large amounts of energy in a limited land area due to their high energy efficiencies and power density. Relying on a relatively ubiquitous mineral for a fuel source, the technology provides electrical generation at higher capacities than fossil fuel-fired processes. The use of nuclear power is viewed as a solution to reducing the atmospheric greenhouse gas emissions associated with utility-scale fossil-fuel-fired generation. A benefit of NPPs is that they provide a means of producing electricity that does not emit carbon dioxide except for that attributed to plant construction and development.

Nuclear energy technology offers seemingly infinite supplies of energy and provides base-load electrical power generation at high operating capacities [29]. Many of today's nuclear power plants function as utility-scale

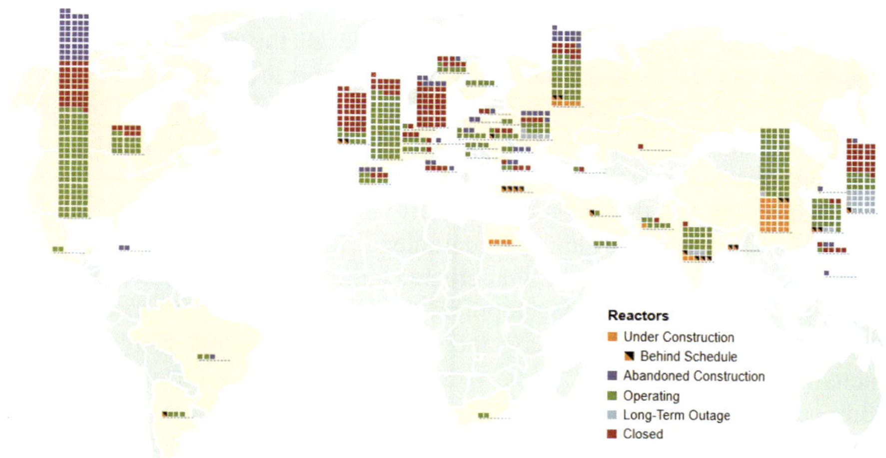

Figure 5.6 Nuclear power reactors around the world [33].

generation facilities and account for almost 10% of the world's total electrical energy production. Nuclear plants use enriched uranium-235 as the fuel source and fission reactors to generate electricity. Most utility-scale reactors in the U.S. are either boiling water reactors (BWRs) or pressurized water reactors (PWRs) [30]. Both systems use Rankine cycle processes to boil water to make steam [30]. The steam must be cooled after it is directed through a turbine to produce electricity, which necessitates its signature cooling towers [31].

All electrical power plants, regardless of type or energy resource used, have a finite lifecycle. Nuclear power plants became an option for the generation of electricity beginning in the mid-1950s. Their lifecycle is typically 40 to 60 years. Seen as a high-capacity and reliable means of generating power, the nuclear power industry in some regions of the world is growing, most notably in China [32]. In May 2025, there were 814 nuclear reactors worldwide, with 94 in the U.S. and 59 in China (see Figure 5.6) [33]. Of these, 471 are in commercial operation; in total, they have a net installed capacity of 377,221 MW [34].

As nuclear technologies advance, gauging the economics of nuclear power is perplexing. Ways to assess the economic performance of nuclear plants are important as they are used to justify nuclear plant development and operations. Cost calculation procedures allow comparative assessments with other energy sources, which indicate the economic efficiency of power plants [31]. While there are various methods for calculating the operating and construction costs for NPPs, few include all the relevant factors, and often the lifecycle costs for nuclear baseload plants are underestimated [31]. As a result, the formation of rate structures for electricity sold to consumers is challenging; in the different regions of the world, electricity costs sourced from nuclear power are widely variable.

Costs for the development of utility-scale nuclear reactors are substantial, and fuel sources are not guaranteed. Reactors that supply power to electric grids have limitations that create challenges for assessments of their costs. Design, permitting, development, and prove-out may require a decade or more, and construction delays increase costs [35]. Other viable concerns include refined uranium's price volatility, how the converted uranium is sourced (about 40% of the refined global supply comes from the Russian Federation), the ultimate disposition and storage of spent fuel rods, and public concerns about safety and the potential of meltdowns [35]. The costs for plant decommissioning are often not included in lifecycle cost evaluations.

Another more common concern is that traditional nuclear plants withdraw and consume vast volumes of water, more than coal-fired plants of similar capacities [35]. If plentiful water supplies are unavailable, a nuclear power station may be forced to temporarily shut down. This is the primary reason why nuclear plants are located near large natural bodies of fresh water.

When new electrical generation is needed quickly, the time frames required to develop large nuclear plants preclude them from being considered as a solution. The reasons for this include environmental concerns, long lead times for design and development, and high construction costs. Unresolved environmental problems include their impact on water quality and concerns regarding the storage of nuclear waste. Public opposition to the development of new nuclear plants is common.

For NPPs, there are a few recent U.S. examples of newly constructed plants. Costs are often misrepresented or unavailable. Despite efforts to make nuclear generation more economical, building new reactors in the U.S. is the most expensive way known the generate electricity at scale. Two new reactors commissioned in 2025 in Georgia cost over $34 billion and now supply the world's most expensive electricity [30]. Ratepayers in the service territory ultimately bear the burden of the high electricity costs required to amortize the financial investment.

The demand for electricity in the U.S. is growing due to the need for new data centers, electric vehicle battery charging, and greater use of air conditioning systems for buildings. Will nuclear energy help meet this increase in demand? Unfortunately, the number of nuclear reactors that are being retired in the U.S. is greater than new ones that are being constructed. A likely interim solution is that existing plants might be refurbished, upgraded, and reinspected to enable extended operating licenses. There is hope that advanced nuclear technologies, such as thorium or fusion reactors, might be commercially demonstrated and become available. In the future, small nuclear reactors might be mass-produced at a lower cost. However, until the key issues of environmental impacts and high development costs are resolved, it is unlikely that many large-scale, commercial nuclear plants will be constructed in the U.S.

5.5 Conclusion

The chapter considered hydropower, tidal energy, floating solar installation, and the situation with nuclear energy. These technologies have key roles in providing renewable and emissions-free generation of electricity. All provide routes to regional decarbonization. Hydropower has many benefits. It is a proven, mature, predictable, price-competitive, and scalable technology. There are numerous technologies available to harness tidal power. Floating photovoltaics are a potential alternative when land areas are unavailable or costly to access. Compared to other alternatives, nuclear power has advantages over contenders to replace traditional coal and oil sources. Nuclear energy offers seemingly infinite supplies of energy and provides base-load electrical power generation at high operating capacities. The development and deployment of these advanced electrical generation technologies require supportive policies, programs, and infrastructure.

References

[1] U.S. Energy Information Administration (2023, April 20). Hydropower explained. https://www.eia.gov/energyexplained/hydropower, accessed 29 November 2024.

[2] Bhutada, G. (2022, December 1). The 30 largest U.S. hydropower plants, National Public Utilities Council. https://decarbonization.visualcapitalist.com/the-30-largest-u-s-hydropower-plants, accessed 29 November 2024.

[3] Georgia Power. Alvin W. Vogtle electric generating plant. https://www.georgiapower.com/about/energy/plants/plant-vogtle.html, accessed 29 November 2024.

[4] U.S. Energy Information Administration, Monthly Energy Review, Table 7.2.a. (2023, March) and *Electric Power Monthly*, Table 1.1, preliminary data for 2022 (2023, February).

[5] U.S. Energy Information Administration (2024, November 7). Drought conditions reduce hydropower generation, particularly in the Pacific Northwest. https://www.eia.gov/todayinenergy/detail.php?id=63664, accessed 7 November 2024.

[6] Statista (2024, September 1). Net generation of conventional hydroelectric power in the United States from 2000 to 2023. https://www.statista.com/statistics/183461/generation-of-energy-from-hydropower-sources-in-the-us-since-2000, accessed 30 November 2023.

[7] LakesOnline.com (2024 November 29). https://mead.uslakes.info/level.asp, accessed 29 November 2024.

[8] U.S. Department of Energy (2023). U.S. hydropower market report. Executive summary. https://www.energy.gov/sites/default/files/2023-10/U.S.%20Hydropower%20Market%20Report%202023%20Edition%2C%20Executive%20Summary.pdf, accessed 30 November 2024.

[9] U.S. Department of Energy. Hydropower market reports. https://www.energy.gov/eere/water/hydropower-market-reports, accessed 30 November 2024.

[10] Roosa, S. and Jhaveri, A. (2018). Carbon reduction technologies, chapter 21, in Roosa, S. (ed.) *Energy Managerment Handbook,* 9th edition, page 563. The Fairmont Press, Inc.: Lilburn, Georgia.

[11] Li, G. and Zhu, W. (2023, June). Tidal current energy harvesting technologies: a review of current status and life cycle assessment. *Renewable and Sustainable Energy Reviews*, 179, 113269.

[12] International Energy Agency (2023). Tidal current energy developments highlights. https://tethys-engineering.pnnl.gov/sites/default/files/publications/oes-2023-tidal-current-energy-spreads.pdf, accessed 7 December 2024.

[13] Wikioedia, The Free Encyclopedia (2021, April 14). List of tidal power stations. https://en.wikipedia.org/wiki/List_of_tidal_power_stations#:~:text=The%20following%20table%20lists%20tidal%20power%20stations%20that,%20%201968%20%204%20more%20rows%20, accessed 4 May 2021.

[14] Crosse, G. (2023, January 10). Projects, pipelines and power: around the world's tidal projects. https://www.power-technology.com/features/tidal-power-development-projects/?cf-view, accessed 5 December 2024.

[15] Max (2023, April 2024). Advantages of tidal energy: a useful guide. https://www.linquip.com/blog/advantages-of-tidal-energy, accessed 4 December 2024.

[16] DiSavino, S. (2024, April 9). U.S. power use forecast to reach record highs in 2024 and 2025. https://money.usnews.com/investing/news/articles/2024-04-09/us-power-use-forecast-to-reach-record-highs-in-2024-and-2025, accessed 29 January 2025.

[17] U.S. Department of Energy, Office of Energy Efficiency and Renewable Energy (2022, October 18). U.S. Department of Energy announces $35 million in funding to advance tidal and river current energy systems. https://www.energy.gov/eere/articles/us-department-energy-announces-35-million-funding-advance-tidal-and-river-current, accessed 5 December 2024.

[18] Lai, C. (2022, August 5). Tidal energy: advantages, disadvantages, and future trends. https://earth.org/what-is-tidal-energy, accessed 2025, January 27.

[19] Clark, N. (2006, March 27). Tidal barrages and birds. *International Journal of Avian Science*, 148, 152–157.

[20] National Renewable Energy Laboratory. International applications for floating solar photovoltaics. https://docs.nrel.gov/docs/fy19osti/73907.pdf.

[21] Mehra, A. (2022, August 26). Floating solar panels (floatovolatics): what to know. https://www.energysage.com/community-solar/floating-solar-what-you-need-to-know, accessed 30 June 2025.

[22] Tirone, J. (2023, August 3). Floating solar panels turn old industrial sites into green energy goldmines. Bloomburg. https://www.bloomberg.com/news/articles/2023-08-03/floating-solar-panels-turn-old-industrial-sites-into-green-energy-goldmines, accessed 30 June 2025.

[23] Solar Energy Research Institute of Singapore, University of Singapore.

[24] Maima, C., Diniz, A., Bonfim, S. and Kazmerski, L. (2024). Evaluation of the electrical parameters and performance of floating PV generators. *Renewable Energy and Environmental Sustainability*, 9(5).

[25] Bellini, E. (2022, January 3). World's largest floating PV plant goes online in China. https://www.pv-magazine.com/2022/01/03/worlds-largest-floating-pv-plant-goes-online-in-china, accessed 1 July 2025.

[26] DNV (2022, April). The future of floating solar: drivers and barriers to growth. https://assets.super.so/9bc769b9-2607-4333-8ea9-2c761b0a0aa6/files/ff29e89b-e9c7-4a42-8d5d-e3f88e44b496.pdf, accessed 1 July 2025.

[27] Simon, M. (2025). What will the rise of floating solar panels mean for wildlife? https://www.msn.com/en-us/news/technology/what-will-the-rise-of-floating-solar-panels-mean-

for-wildlife/ar-AA1Hj0Xa?ocid=msedgntp&pc=U531&cvid 28cb4255c&ei=54=d863f9f37f414c1cf8d0522.

[28] CGTN (2025, July 4). China launches first offshore floating PV project in Qingdao. https://news.cgtn.com/news/2025-07-04/China-launches-first-offshore-floating-PV-project-in-Qingdao-1EJA3Wuff3y/p.html, accessed 5 July 2025.

[29] Rosen, L. (2013, April 7). The future of microgrids. http://hplusmagazine.com/2014/01/07/the-future-of-micrgrids, accessed 2 August 2019.

[30] Roosa, S. (2023, September). Nuclear energy—lessons from Georgia's new power plant. *Alternative Energy and Distributed Generation Journal*, 6(3), pages 5–12.

[31] Mamalyga, V., Roosa, S., Yiang, Q., Yan, J. and Varava, O. (2025, September). Nuclear power plants in the U.S. and China: procedure for determining and comparing operating and construction costs. *International Journal of Strategic Energy and Environmental Planning*, 7(5), 10-26.

[32] Roosa, S. (2025, September). Decommissioning nuclear power plants. *Alternate Energy and Distributed Generation Journal*, 7(3), 5–9.

[33] World Nuclear Industry Status Report (2025, February 18). 70+ years of nuclear power, world nuclear power reactors 1951-2025. https://www.worldnuclearreport.org/reactors.html#tab=iso, accessed 22 February 2025.

[34] International Atomic Energy Agency (2025). The database on nuclear power reactors. https://pris.iaea.org/pris, accessed 17 May 2025.

[35] Roosa, S. (2025, June). Reinventing the electric grid. *Alternative Energy and Distributed Generation Journal*, 7(2), 24–38.

6

Nuclear Energy—Lessons from Georgia's New Power Plant

Stephen A. Roosa

Electricity is a key enabler of contemporary life. The opportunities for nuclear power as a clean energy resource seem limitless. Nuclear energy has the potential to solve many of the world's most pressing energy problems. It can provide nearly unlimited supplies of energy and, more importantly, offer base-load electrical power generation. Some see nuclear energy as key to meeting decarbonization goals.

Nuclear energy accounts for almost 10% of the world's electrical energy production with about 440 operating reactors. The need to achieve ambitious decarbonization objectives has driven increases in nuclear power investments. During the 28th Conference of the Parties held in Dubai in December 2023, twenty of the world's countries, including the U.S., committed to increasing the output of nuclear energy worldwide. The financial risks associated with utility-scale energy projects are mitigated when construction and financing costs are predictable. This chapter considers the costs and benefits of the first new nuclear power plant in the U.S. in over 30 years, which is now operating in Waynesboro, Georgia.

6.1 Power from Uranium

Nuclear energy is derived from uranium. Being sourced from a mineral, it is not classified as either a renewable energy resource or a fossil fuel. It is often called a clean energy resource (grouped with renewables) because it produces little to no greenhouse gas emissions or air pollution during electricity generation [1]. It offers a reliable and consistent source of energy that

DOI: 10.1201/9788743811312-6

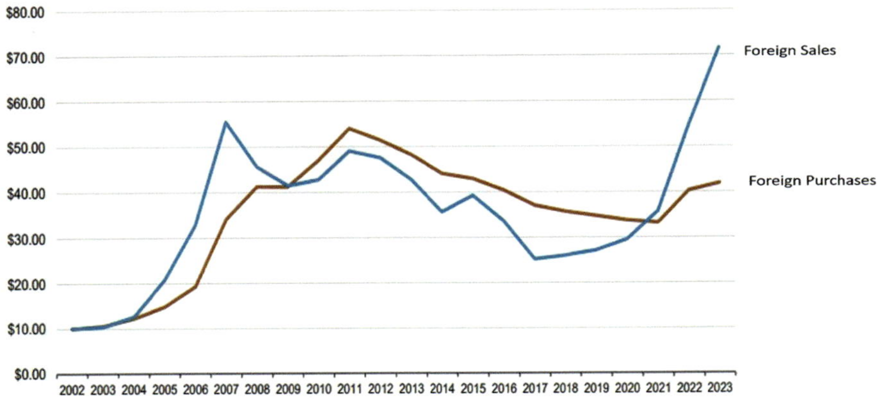

Figure 6.1 Weighted-average price of foreign purchases and sales of uranium from 2002 to 2023, $/lb. (Data Sources: U.S. Energy Information Administration: *Uranium Industry Annual reports, 2002*. Form EIA-858, *Uranium Marketing Annual Survey, 2003–2023* [4].

can help meet the growing demand for electricity [1]. Nuclear power plants operate at high capacities, meaning they can produce electricity continuously over long periods of time, regardless of weather conditions or other environmental factors [1].

In the last 50 years, uranium has become one of the world's most important energy minerals. Uranium is a ubiquitous and dense metal found in rocks all over the world. Used almost exclusively in the energy industry, uranium is mined and concentrated using methods like many other metals. The cumbersome process of fuel fabrication involves bundling the enriched uranium into rods, which are used in nuclear reactors as the fuel source [2]. Fuel rods are then shipped to the power plant to be readied for use. After being used only once, spent fuel rods at U.S. plants must be stored on site—at least for the next 10,000 years or so. While there are technologies to reprocess uranium, the only functioning plant in the world with this capability is the Seversk plant; it is owned by the Russian group Rosatom and located in the Tomsk region of Siberia [3].

Uranium is a commodity. Its price varies with market conditions and is dependent on whether it is purchased from or sold to a foreign source (see Figure 6.1). The volatility in uranium prices is influenced by supply constraints and demand dynamics, production costs, regulations, and geopolitical events. The standard contract unit is 250 pounds of U-3O8, which is traded on the New York Mercantile Exchange. In January 2020, the price

Table 6.1 Leading nuclear electric generation countries in 2021.

Country	Nuclear electricity generation capacity (million kilowatts)	Nuclear electricity generation (billion kilowatthours)	Nuclear share of country's total electricity generation
United States	95.49	778.15	19%
France	61.37	360.70	68%
China	53.26	407.52	5%
Russia	27.73	222.44	20%
South Korea	24.43	150.52	26%

Source: U.S. Energy Information Administration, International Energy Statistics, as of August 7, 2023

was $24.50/lb; the price increased to $88.96/lb by 26 February 2026, almost tripling in the previous five years but less than its peak of over $100/lb in early 2024. The price volatility of uranium does not significantly alter the economics of nuclear plants because the fuel costs are small compared to the cost of the electricity they produce.

6.2 The Realities of Nuclear Power Plants

Nuclear power plants use uranium as the principal fuel. Nuclear reactors generating electricity in the U.S. fall mainly into two categories: boiling water reactors (BWRs) and pressurized water reactors (PWRs) [2]. Both systems boil water to make steam (BWRs within the reactor and PWRs outside the reactor); this steam must be cooled after it is directed through a turbine to produce electricity [2].

In 2021, 33 countries had nuclear electricity generation plants. The U.S., France, and China are countries that rely heavily on electricity generated by nuclear energy (see Table 6.1). In the U.S., it provides 18.9% of the electricity produced with a capacity factor of 92.4%. As of August 2023, the U.S. had 93 operating commercial reactors at 54 plants; it remains the world's leader in electrical generation using this resource.

Most of the uranium used by U.S. nuclear plant operations is imported. In 2022, U.S. uranium supplies were purchased from Russia (12%), Canada (27%), Kazakhstan (25%), Uzbekistan (11%), and Australia (9%). Importing uranium for U.S. nuclear reactors fails to support the goal of national energy independence. In 2024, a new law barred the U.S. from continuing to import uranium from Russia, historically one of its largest suppliers, to jumpstart domestic uranium mining. Table 6.2 summarizes the sources and costs of uranium used by civilian nuclear power reactors in the U.S.

Table 6.2 Uranium purchased by owners and operators of U.S. civilian nuclear power reactors by origin country and delivery year, 2019–2023 (1,000 lbs. of uranium, $s/lb.) [4].

	Deliveries in 2019		Deliveries in 2020		Deliveries in 2021		Deliveries in 2022		Deliveries in 2023	
Origin country	**Purchases**	**Weighted-average price**	**Purchases**	**Weighted-average price**	**Purchases**	**Weighted-average price**	**Purchases**	**Weighted-average price**	**Purchases**	**Weighted-average price**
Australia	8,504	35.39	5,597	39.86	6,712	36.88	3,620	42.08	10,605	51.15
Brazil	0	--	0	0	0	0	0	0	0	0
Bulgaria	0	--	0	0	0	0	0	0	0	0
Canada	10,172	33.06	10,976	35.05	6,908	35.09	11,100	37.22	13,162	43.73
China	0	--	141	24.45	0	0	0	0	W	W
Czech Republic	0	--	0	0	0	0	0	0	0	0
Germany	W	W	0	0	0	0	W	W	0	0
Hungary	0	--	0	0	0	0	0	0	0	0
Kazakhstan	8,760	35.69	10,828	33.37	16,557	34.16	10,019	39.05	10,622	43.64
Malawi	0	--	239	29.01	60	52.25	W	W	W	W
Namibia	2,450	40.40	2,517	35.28	3,214	36.01	W	W	1,546	47.67
Niger	998	41.21	2,050	34.36	1,773	39.08	W	W	1,418	38.57
Portugal	0	--	0	0	0	0	0	0	0	0
Russia	7,365	27.31	8,064	25.73	6,314	22.76	4,781	35.20	6,042	30.86
South Africa	0	--	0	0	1	31.04	W	W	W	W
Ukraine	0	--	0	0	0	0	0	0	0	0
United Kingdom	0	--	666	35.40	0	0	0	0	0	0
Uzbekistan	4,365	38.99	3,940	35.93	2,499	33.74	4,438	39.21	4,887	44.20
unknown	W	W	100	24.36	225	29.70	0	0	0	0
Foreign total	**W**	**W**	**45,367**	**33.53**	**44,263**	**33.40**	**38,467**	**38.45**	**49,239**	**43.95**
United States	W	W	3,567	30.09	2,474	43.04	2,052	50.96	2,386	40.63
Total purchases	**48,328**	**35.59**	**48,934**	**33.27**	**46,736**	**33.91**	**40,519**	**39.08**	**51,625**	**43.80**

W = Data withheld to avoid disclosure of individual company data. -- = Not applicable.

Notes: Totals may not equal sum of components because of independent rounding. Weighted-average prices are not adjusted for inflation.

Data Source: U.S. Energy Information Administration, Form EIA-858, *Uranium Marketing Annual Survey* (2019–23)

The realities of deploying utility-scale nuclear electrical generation are perplexing. Primary concerns include public opposition. People are leery of having a nuclear power plant nearby. Others include issues with the mining process for nuclear fuels, operational safety, water consumption, development costs, and waste storage. Processing uranium requires mining, milling, enrichment, and fuel fabrication; all expose workers to health risks and use substantial quantities of fresh water. Water consumption is often the most important consideration, especially in western parts of the U.S. and in other countries with limited access to water. Mining and processing uranium consumes as much as 14 gallons (53 liters) of water per million Btu (MMBtu) of thermal output. In the U.S., the milling process uses a mix of liquid chemicals; enrichment of the ore is typically performed using gas diffusion. This approach uses more water than the gas centrifuge processes normally used in Europe and elsewhere. Water is also needed in the cooling process. Nuclear plants consume about 400 gallons (1,514 liters) of water per MWh of electricity produced.

Like other thermoelectric power plants, nuclear reactors use once-through or recirculating cooling systems. BWRs and PWRs use comparable amounts of water to produce a unit of electricity. Nuclear plants withdraw and consume more water per unit of electricity produced than coal plants using similar cooling technologies because nuclear plants operate at lower temperatures and turbine efficiencies; they do not lose heat via smokestacks [5]. For example, once-through nuclear plants withdraw 25,000 to 60,000 gallons (94,635 to 227,125 liters) of water per MW of electricity produced and consume 100 to 400 gallons (379 to 1,514 liters) per MW [6]. Recirculating systems withdraw 800 to 2,000 gallons (3,028 to 9,842 liters) of water per MW of electricity produced and consume 160 to 800 gallons (607 to 3,028 liters) per MW [6]. Regardless of the system used, fish, sea turtles, sea lions, and other forms of wildlife can be drawn into the cooling system's intakes. Such impacts on local marine life are often underreported [7]. In 2003, a major environmental catastrophe occurred when one of California's coastal nuclear plants killed about 32 million fish [7].

6.2.1 Financial risks of U.S. nuclear power

The financial risks associated with utility-scale energy projects are mitigated when construction and financing costs are predictable. Unfortunately, it has become increasingly difficult to estimate the costs associated with new nuclear plants in the U.S. A project in South Carolina, one of the country's poorest states, failed in 2017 after $9 billion had been expended on construction; this boondoggle produced zero electricity and left the utility's rate payers with the tab [8].

A new nuclear power plant, the first in the U.S. in over 30 years, is now operating in Waynesboro, Georgia. Bechtel assumed responsibility for the Plant Vogtle project (adding reactors 3 and 4 to the site) in 2017. A substantial portion of this project's anticipated cost of $14 billion was subsidized by the U.S. Department of Energy, which provided more than $12 billion in direct loan guarantees. Delays, labor issues, cost overruns, and inflation during the latter years of construction eventually increased the project cost to over $31 billion [9]. Add in $3.7 billion that the original contractor Westinghouse paid to the Vogtle owners to walk away from constructing the reactors, and the total approaches $35 billion [10]. Seven years behind schedule, the total capital costs for the two new reactors are an astounding $15.8 million per MW—making new utility-scale nuclear power in the U.S. the world's most expensive way to generate electricity. Tom Newsome, the Georgia Public Service Commission director of utility finance, noted in testimony at a public hearing that "the cost increases and schedule delays have completely eliminated any benefit on a life-cycle cost basis" [11]. Cited costs for the two new reactors do include any funds set aside for decommissioning. China's Three Gorges Dam hydropower project, the world's largest electrical generating station of any type, cost the equivalent of $32 billion; it has a generating capacity of 22,500 MW, over 10 times that of Plant Vogtle's units 3 and 4 combined, and provides electrical generation using a renewable energy resource.

Plant Vogtle units 3 and 4, both pressurized water reactors, have a capacity of 1,100 MW each and are expected to generate more than 17 million MW hours of electricity, enough to power an estimated 1.6 million average American homes [9,11]. Once all four generating units are online, the site will become one of the largest suppliers of clean energy in the nation [9]. Safety is paramount; Vogtle's new reactors have advanced safety systems that automatically shut down without any operator action or external power source [12]. The new Vogtle reactors are important because government officials and some utilities are considering nuclear power as a means of alleviating climate change by generating electricity without the use of natural gas, coal, and oil [11].

6.2.2 Small modular reactors

Small modular reactors (SMRs) have the potential to power microgrids in the future; advanced SMRs are currently being developed in the U.S. that can be used for electrical generation, process heat, desalination, or specific industrial applications [13]. They vary in output from a few to hundreds of megawatts and use various types of liquid coolants [14]. SMRs offer advantages that include their small size, variable output capabilities, reduced capital investment, ability to be sited in locations not possible for larger nuclear plants, and the ability to incrementally add power [15].

NuScale Power has the only small modular nuclear reactor design certified for use in the United States [16]. The company was planning to complete a six-reactor demonstration project in 2029 at the Idaho National Laboratory; it would generate enough electricity to power more than 300,000 homes [16]. After about 10 years of work, the project was cancelled in November 2023. The primary reason for cancelling this project was rising costs—the planned \$59/MWh grew to \$89/MWh—and local utility companies were unwilling to commit to buying the project's output [17].

6.3 Nuclear Power as a Clean Energy Alternative

As the world focuses on ways to reduce carbon emissions and develop environmentally friendly energy sources, nuclear energy becomes increasingly attractive, driving up the demand for uranium. Despite other serious environmental issues (i.e., disposal, storage, and management of spent nuclear fuel rods) and the financial risks associated with new plant construction, nuclear generation has a formidable role in providing clean electrical energy generation. Nuclear energy provides roughly 10% of the world's electricity and 4.3% of primary energy production. It has grown in importance as the world seeks to implement technologies that reduce the reliance on fossil fuels and deploy new ways to reduce greenhouse gas emissions [18].

For utility-scale electrical generation, the nuclear reactors at Georgia's Vogtle Plant may initiate a resurgence in nuclear power production in the U.S.—or, they may thwart nuclear development due to their high costs. The construction of reactors 3 and 4 at the Vogtle Plant provides a case study that highlights the faults of the older nuclear technologies. The regulatory hurdles, recurring construction delays, excessive construction costs, and quality control failures that were experienced will likely discourage many electric utility companies from pursuing nuclear energy. We may learn from our experiences in South Carolina and Georgia and seek less expensive ways to generate power.

Opponents of the new Vogtle reactors have argued that power generated from solar and wind is much less expensive than nuclear power and can be brought online more quickly if transmission and distribution systems are available [11]. Despite the drawbacks of conventional nuclear generation, the technology does offer a route to decarbonization for regions lacking other, less costly and time-consuming alternatives.

Small modular reactors provide hope that combined heat and power configurations may be more attractive because of their enhanced safety features. Some U.S states, such as Tennessee, in partnership with the Tennessee Valley Authority, are promoting and setting aside funds for the development of small nuclear reactors (300 MW or less) as an alternative to coal-fired generation [19]. Regardless, a key problem for the nuclear industry is that SMRs remain an unproven technology at scale—not so with renewables such as solar and wind power. Expanding the use of these intermittent generation resources while providing base-load power will be challenging. Perhaps utility-scale microgrids in the future might find ways to combine SMRs and renewable energy technologies into hybrid generation systems.

References

[1] Ajebe, M. (2023, September). Statistical forecasts on Africa's energy consumption, population with access, and GHG emissions. *International Journal of Strategic Energy and Environmental Planning*, 5(5), 65.

[2] Union of Concerned Scientists (2013, July 15). How it works: water for nuclear. https://www.ucsusa.org/resources/water-nuclear#:~:text=BWRs%20and%20PWRs%20use%20 comparable%20amounts%20of%20water, efficiency%2C%20and%20do%20not%20lose%20heat%20via%20 smokestacks, accessed 28 May 2023.

[3] Mouterde, P. and Cessac, M. (2022, December 3). Russia owns the only plant in the world capable of reprocessing spent uranium. https://www.lemonde.fr/en/energies/article/2022/12/03/russia-owns-the-only-plant-in-the-world-capable-of-reprocessing-spent-uranium_6006479_98.html, accessed 29 October 2023.

[4] U.S. Energy Information Administration (2023, June). 2023 Uranium marketing annual report. https://www.eia.gov/uranium/marketing/pdf/2023%20UMAR.pdf, accessed 6 August 2025.

[5] World Nuclear Association (2013). Cooling power plants. London, UK.

[6] Union of Concerned Scientist (2008, July 13). Got water? Nuclear power plant cooling needs. https://www.ucsusa.org/resources/nuclear-power-plant-cooling-water-needs, accessed 30 October 2023.

[7] Gunter, L., Gunter, P. Cullen, S. and Esquire, N. (2001). Licensed to kill—how the nuclear power industry destroys endangered marine

wildlife and ocean habitat to save money. http://www.nirs.org/wp-content/uploads/reactorwatch/licensedtokill/LiscencedtoKill.pdf, accessed 28 July 2025.

[8] Lacy, A. (2018, February 6). South Carolina spent $9 billion to dig a hole in the ground and then fill it back in. https://theintercept.com/2019/02/06/south-caroline-green-new-deal-south-carolina-nuclear-energy, accessed 28 July 2025.

[9] Thibault, M. (2023, August 3). $30B Bechtel-built nuclear plant starts operation in Georgia. https://www.constructiondive.com/news/bechtel-nuclear-power-plant-starts-operation-georgia/689738, accessed 29 July 2025.

[10] Associated Press (2023, August 17). Utilities begin loading radioactive fuel into a second new reactor at Georgia nuclear plant. https://www.usnews.com/news/business/articles/2023-08-17/utilities-begin-loading-radioactive-fuel-into-second-new-reactor-at-georgia-nuclear-plant, accessed 29 October 2023.

[11] Amy, J. (2023, July 7). First U.S. nuclear reactor built from scratch in decades enters commercial operation. https://abcnews.go.com/US/wireStory/american-nuclear-reactor-built-scratch-decades-enters-commercial-101861665, accessed 29 October 2023.

[12] Energy.gov (2019, April 18). 5 things you should know about Plant Vogtle. https://www.energy.gov/ne/articles/5-things-you-should-know-about-plant-vogtle, accessed 29 July 2025.

[13] Roosa, S. (2009). *Fundamentals of Microgrids, Development and Implementation*, pages 254–256. Taylor and Francis: London. ISBN 978-0-367-53539-1.

[14] Roosa, S. (2020). Reimagining nuclear energy. *International Journal of Strategic Energy and Environmental Planning*, 2(5), 5–6.

[15] Office of Nuclear Energy. Advanced small modular reactors (SMRs). https://www.energy.gov/ne/advanced-small-modular-reactors-smrs, accessed 11 November 2023.

[16] McDermott, J. and Daly, M. (2023, November 9). First-of-a-kind nuclear project is terminated in a blow to Biden's clean energy agenda. https://finance.yahoo.com/news/first-kind-nuclear-project-terminated-224710694.html, accessed 12 November 2023.

[17] Meyer, D. (2023, November 9). NuScale shares plunge as it cancels flagship small nuclear reactor project. https://finance.yahoo.com/news/nuscale-shares-plunge-cancels-flagship-173443587.html, accessed 12 November 2023.

[18] Roosa, S. and Jhaveri, A. (2009). *Carbon Reduction: Policies, Strategies and Technologies*, page 37. The Fairmont Press, Lilburn Georgia. ISBN 978-1-4200-8382-8.

[19] Taylor, S. (2024, December 1). Tennessee Nuclear Advisory Council encourages all-in approach. *Nashville Banner*. https://nashvillebanner.com/2024/12/03/tennessee-nuclear-energy-future, accessed 1 August 2025.

7

Energy Storage Capacities are Increasing

Stephen A. Roosa

One of the more exciting developments in electrical generation is the recent focus on energy storage technologies. The deployment of storage technologies for utility-scale electrical energy applications is fast becoming mainstream. For these, pumped hydropower systems have been the mainstay as a means of leveling intermittent electrical loads. Energy storage systems are being reinvented, and the technologies they use are improving. This chapter explores the applications of energy storage systems, the various uses for energy storage, and discusses its applications for renewable energy systems. Several types of battery energy storage systems are discussed, and those used for utility-scale applications are identified.

The chapter concludes by suggesting that energy storage systems offer a wide range of benefits for distributed generation and will become more specialized to meet the needs of future applications. Energy storage technologies will continue to be incorporated with renewable generation systems and will likely utilize hybrid generation configurations.

7.1 Energy Storage

Storing electrical energy is not a new concept. In 1800, Alessandro Volta, an Italian physicist, invented the first actual battery, storing and releasing a charge through a chemical reaction, which came to be known as the *voltaic pile*. In 1859, the French physicist Gaston Planté created the world's first rechargeable battery that was based on lead-acid components.

When Thomas Edison pioneered the first electric plant to provide energy for both lighting and heat at the Pearl Street Station in New York City

DOI: 10.1201/9788743811312-7

in 1882, it used a battery system to provide backup power. The first known applications for pumped storage hydropower (PSH) were found in Italy and Switzerland in the 1890s; PSH was first used in the United States in 1930 [1].

Today's energy storage systems are much more sophisticated yet based on similar principles. We now have a wide variety of different types of storage systems that we can use, all capable of being customized for various applications. Recent market developments that provide impetus for electric utility companies to consider energy storage include the need to level the output of the renewable generation, provide electricity without expanding existing transmission systems, and supply power for electric vehicles.

7.1.1 Ways to store energy

Energy storage is accomplished by devices or physical media that save excess energy to perform a useful operation. These systems are typically configured to enable the storage of thermal or electrical energy. The devices capable of storing energy are called accumulators or batteries.

The capability of storing electricity, a high-quality form of energy, is important because electricity must somehow be consumed or transformed the instant it is generated. Otherwise, the source energy used to generate the electricity is wasted. Storing excess electricity has an important role in programs such as load leveling and planning for decarbonization.

Storing energy allows us to better balance energy supply and demand. The primary types of energy storage systems include mechanical, electrical/electrochemical, chemical, biological, and thermal. Fossil fuels, such as coal and oil, are forms of stored biological energy. The type of energy storage system used depends on the technology and its application. For example, the rotation of flywheels, a form of mechanical energy storage, is used to store energy needed for short periods of time. They can provide short bursts of electricity, especially when needed to meet peak electric loads. To achieve larger capacities, modular energy storage systems are replicated and configured into groupings that are often called banks (i.e., battery banks).

7.1.2 Mechanical energy storage

There are numerous types of energy storage systems that are classified as mechanical energy storage. Pumped storage hydropower remains the most common form of mechanical energy storage for microgrid applications and utility-scale projects when gauged by total installed capacity. Conventional pumped hydropower is configured using two water reservoirs, separated by elevation; the reservoir located at the lower altitude is called the lower

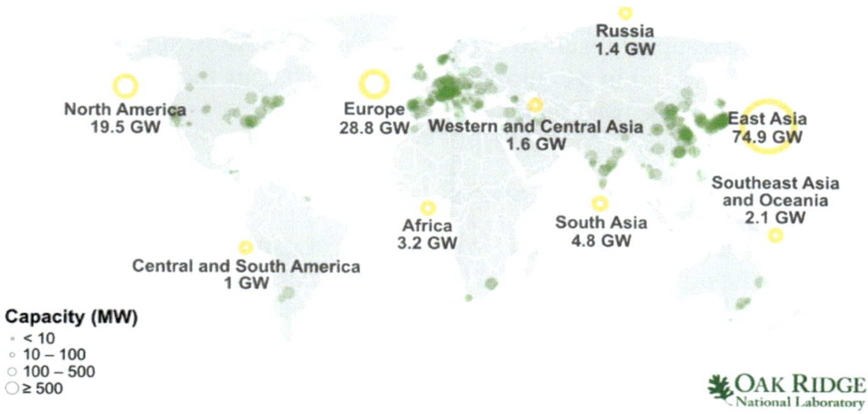

Figure 7.1 Global operational pumped storage hydropower in 2023 [2].

reservoir, and the one at the higher elevation is called the upper reservoir. During off-peak hours or when the electricity generated has low value, water is pumped from a lower reservoir to an upper reservoir. This is accomplished by reversing the turbines if so designed or using another set of pumps. During peak load periods when electricity is costly, the water flow is reversed (from the upper reservoir and directed to turbines) to generate more electricity.

PSH acts like a giant battery, because it can store power (water at a higher elevation) and then release it when needed [1]. PSH currently accounts for 96% of all utility-scale energy storage in the United States [1]. The U.S. has 43 PSH plants comprising 14% of total global capacity; the country has the potential to add enough new PSH plants to more than double its current capacity [2]. Some high dam hydropower plants have a storage capability which can be dispatched as pumped hydropower. The world's PSH fleet is largely concentrated in the U.S., European Union, China, and Japan; these four regions represent 80% of global capacity (see Figure 7.1) [2].

Compressed air energy storage provides another example of mechanical energy storage. Excess energy is used to store and compress air in a storage vessel or underground chamber. This air stored under higher pressures is released in a controlled manner when needed and directed to a turbine to generate power.

Another category of mechanical systems used for energy storage includes gravity energy storage. Rather than using water as in a pumped hydropower process, which also uses gravity, excess energy is directed to a system of heavily weighted objects. When lifted to a higher elevation, the energy is stored as potential energy. Solid block energy storage involves lifting heavy blocks, such as those made from concrete, to a higher elevation using a crane. Advanced rail energy storage systems use weighted rail cars, which are towed up a hillside to accomplish this task. Allowing the weighted objects to be lowered in a controlled manner discharges the stored potential energy, which can be used to generate electricity.

7.1.3 Hydrogen energy storage

Energy can be stored in secondary products. For example, excess electrical energy can be used to manufacture hydrogen, which can then be drawn upon to supply power to a process or consumed as a vehicular fuel. In these applications, hydrogen is an energy carrier. Green hydrogen is produced when renewable energy sources (i.e., wind or solar power) are used for the electrolysis of water [3]. While this creates opportunities to help reduce greenhouse gas emissions, a substantial amount of energy is consumed using this process. Ammonia is a promising energy carrier for long-term and large-scale energy storage due to its high hydrogen content, high energy density, facile storage/transportation, and low carbon emissions [4].

7.2 Battery Storage Systems

Batteries are a ubiquitous form of energy storage. We find them in vehicles, residences, in hybrid configurations, and in utility-scale electrical generation systems. When a battery capable of storing electricity is being charged, it is a system load; when it is discharging, it appears to the system to be a source of generation. The amount of energy that is released from the energy storage system is always less than what was initially supplied for charging. The ratio of usable energy released to the system, divided by the energy initially supplied to the battery, is called the storage medium's *turn-around efficiency*; the turn-around efficiencies of different batteries will vary. Losses usually take the form of heat. To increase the total storage capabilities and match system loads, batteries are grouped together in series or parallel wiring configurations.

While hydropower will remain one of the primary sources of renewable electricity generation, the use of wind power and solar energy has been increasing. Worldwide, total solar energy capacity is predicted to surpass natural gas-fired generation by 2026 and coal shortly thereafter [5]. Renewable energy is expected to almost double its rate of capacity additions in the U.S. from 2016–2021 compared to 2022–2027; capacity additions will more than double in the European Union during the same period [5]. This transition drives the application of energy storage systems. Battery energy storage adds stability to electrical networks using intermittent energy sources. It helps solve the intermittency problems by storing extra energy produced for later use.

7.2.1 Types of battery storage systems

Lead-acid is one of the oldest and most developed battery technologies. It is a low cost and popular storage choice for power quality, uninterruptible power supply systems, and some spinning reserve applications. Most of its components can be recycled. However, its application for energy management has been limited due to the short cycle life (3 to 6 years) of its components.

More appropriate battery systems for utility applications include sodium-sulfur (NaS), zinc-bromide (ZnBr), and vanadium-redox (VRB) batteries [6]. A NaS battery is a type of molten metal battery that has a high energy density, high efficiency of charge/discharge (89% to 92%), long cycle life, and is fabricated from inexpensive materials that can be recycled. Such features make these battery systems more appropriate for utility energy storage than lead-acid batteries. ZnBr is a type of flow battery that offers one of the highest cell voltages. However, these types of batteries use more expensive materials. Bromine is a highly toxic material if inhaled or absorbed, and special handling is required. VRB batteries offer a rapid response time and are often used for frequency regulation and load shifting. Their lower self-discharge rate makes them most appropriate in applications that require long-term battery storage and reduced maintenance. VRB energy storage systems are suited for buffering the irregular output of utility-scale wind and solar photovoltaic systems.

7.3 Increasing Storage Capacities

Energy storage capacities in the U.S. began to increase after 2018. In 2019, the U.S. added 430 MW of large-scale storage capacity, up from 311 MW in 2018 [7]. Afterwards, battery energy storage system (BESS) capacity in the U.S. began to grow rapidly [8]. Before 2020, the largest U.S. battery storage project was the 40 MW Vista Energy Storage in San Diego County, California [7]. Behind-the-meter energy storage capacities have been difficult to estimate, causing most estimates of total storage capacities to be understated.

The recent growth of energy storage systems is remarkable. At the end of 2019, 163 large-scale battery storage systems were operating in the U.S., a 28% increase from 2018 [9]. Many of these facilities were co-located with solar photovoltaic sites [9]. As of October 2022, a total of 7.8 GW of utility-scale battery storage was operating in the U.S.; by the end of 2023, battery storage capacity had increased to 16 GW [8].

The 250 MW Gateway Energy Storage System in California, which began operating in 2020, marked the beginning of large utility-scale battery storage installations in the U.S. [8]. At present, the 409 MW Manatee

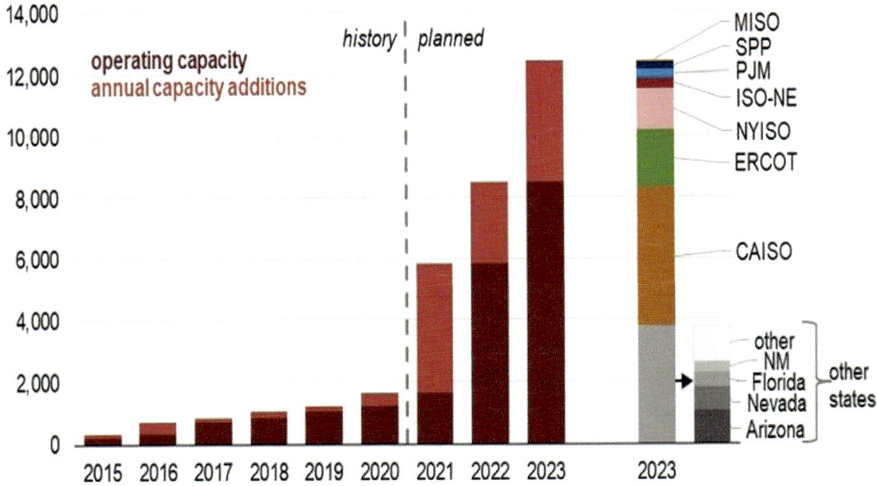

Figure 7.2 U.S. battery storage capacity in megawatts (2015–2023) [9].

Energy Storage project in Florida is the largest operating battery storage facility in the country. Developers have scheduled more than 23 large-scale BESS projects to be deployed by 2025; these range in storage capacity from 250 MW to 650 MW [8]. Large-scale projects (from 250 MW to 725 MW) are underway in Florida, Hawaii, New York, Oregon, and other states [7]. More utility-scale battery storage capacity additions are planned for the near future (see Figure 7.2) [9].

Power plant owners plan to add 62.8 gigawatts (GW) of new utility-scale electric-generating capacity in 2024 [10]. This addition would be 55% more added capacity than the 40.4 GW added in 2023 (the most since 2003) and points to a continued rise in industry activity. A portion of this added generation capacity uses renewables.

As energy storage technologies advance, there is a growing realization in the utility industry that electrical energy storage will be a key component of future transmission networks [6]. When variable output generation is being used, there is more incentive to consider BESS. Future energy storage systems will likely utilize hybrid configurations. They will be incorporated with renewable generation systems (e.g., solar farms and microgrids) to help resolve intermittency issues and improve resilience. Key drivers for energy storage development include: 1) helping match electrical supply with demand; 2) replacing inefficient auxiliary power production; 3) ensuring grid stability with diversified supply; and 4) facilitating distributed generation [11].

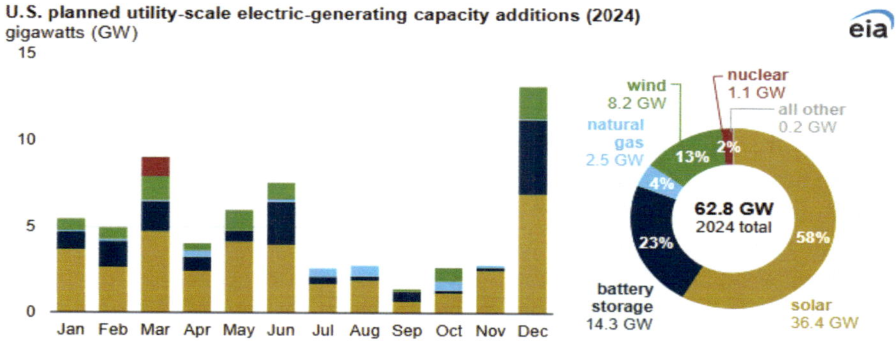

Figure 7.3 U.S. planned utility-scale electric-generating capacity additions [10].

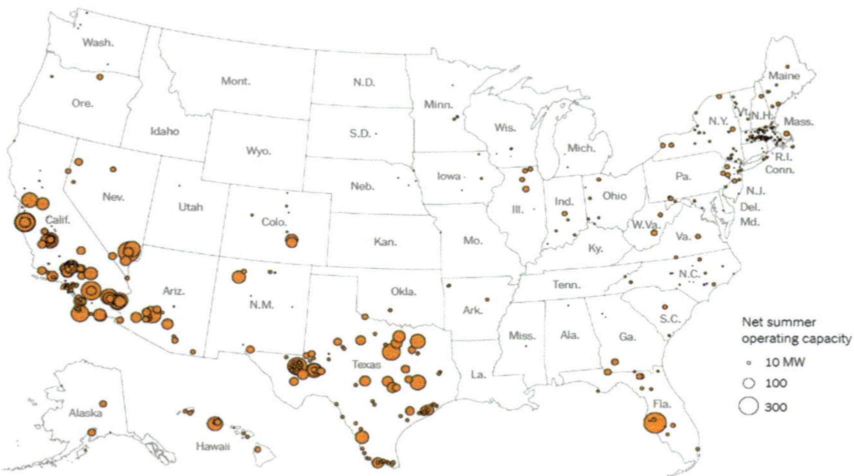

Figure 7.4 U.S. battery storage plants as of 2024 (U.S. Energy Information Administration).

While solar is expected to account for the largest share of new electric generation capacity in 2024, at 58%, battery storage is surprisingly anticipated to provide the second largest share, at 23% (see Figure 7.3) [10]. Led by the states of Texas and California, battery storage capacity in the U.S. is expected to double in 2024 (see Figure 7.4) [10]. The Inflation Reduction Act (IRA) has accelerated the development of energy storage by introducing investment tax credits (ITCs) for stand-alone storage [10]. Prior to the IRA, batteries qualified for federal tax credits only if they were co-located with solar [10].

Lithium-ion (L-I) batteries coupled with natural gas plants and solar farms will account for the bulk of planned additions [7]. These batteries offer high-performance, higher energy densities, are quick charging, and have low lifecycle maintenance costs. A challenge with the use of L-I batteries for certain utility-scale applications is that multiple charge-discharge cycles over the course of a single day may cause them to degrade, which shortens their operating life cycle. They are also unable to discharge at full power for more than four to six hours; this prevents them from stabilizing the electric grid for solar or wind energy applications during consecutive days of cloudy or calm weather conditions [12].

7.4 Comparing Battery Energy Storage Costs to Other Technologies

A question that is difficult to answer is what does it cost to install energy storage systems? The costs vary widely due to the type of energy being stored (e.g., thermal or electricity), the specific technologies used to store the energy, the length of time that the energy will be stored, storage capacity, geographic location, and ambient conditions. Actual cost data for installed energy storage projects are often not released by developers; when cost data are released, it is often not available until well after project completion. In the U.S., costs tend to be lower in states that have larger energy storage market activity (e.g., California, Illinois, and Texas) [13].

It is widely accepted that pumped hydropower storage is often the most cost-efficient type of energy storage. However, site conditions and geography must be considered, and tracking installed costs is based on limited data. Battery storage is often the preferred approach due to its versatility (i.e., batteries can be located almost anywhere).

For large-scale utility battery storage applications, the installed costs have stabilized at about \$1,500/kW. Average battery energy storage capital costs in 2019 were \$589/kWh; battery storage costs (\$s/kWh) fell 72% between 2015 and 2019 [9]. While somewhat counterintuitive, short duration (2-hour) energy storage of kWh for mid-size (60 MW) utility-scale projects is more costly than long-term storage; kW short-duration (2-hour) energy storage capacity is less costly than long-term storage [14].

Costs for current and projected future 4-hour L-I battery systems are found in Figure 7.5. It shows the cost projections for the power and energy components of the battery [15]. These components are combined to give a total system cost; the system cost (in \$/kWh) is the power component divided by the duration plus the energy component [15]. Table 7.1 provides the overnight

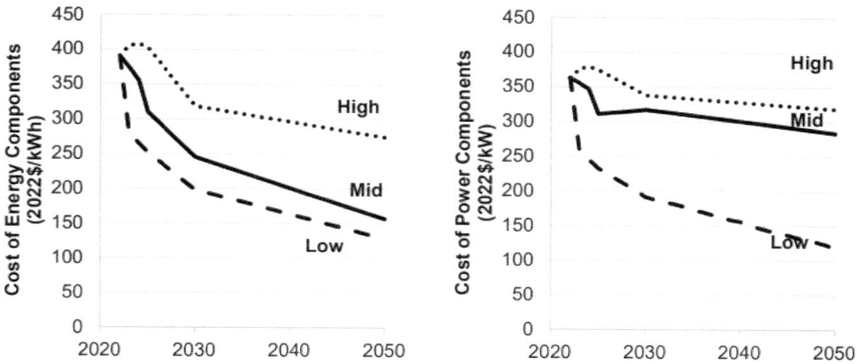

Figure 7.5 Cost projections for energy (left) and power (right) components of L-I systems [15].

costs of many new central station generation technologies that are being used by electric utilities in the U.S. Though the capacities of battery energy storage systems vary, the installation costs ($1,316/kW) and maintenance costs ($25.96/kW-year) are lower than most other prevailing technologies (see Table 7.1). Lead times for the installation of battery banks are minimal.

The power and energy costs offered in Figure 7.5 can be used to estimate the capital costs for other durations. Figure 7.6 shows the cost projections for 2-, 4-, and 6-hour duration batteries (using the mid only) [15]. Long- and short- duration batteries typically have a lower capital cost than mid-duration batteries [15]. Figure 7.6 (left) demonstrates why it is important to cite the duration whenever providing a capital cost in $/kWh or $/kW [15].

The alternative to not providing energy storage capacity can be costly, as it allows excess energy to be wasted. In the United Kingdom (UK), a lack of storage meant that 1.35 TWh of wind energy production between October 2022 and January 2023 was lost [16]. This wasted enough wind-generated electricity to power 1.2 million homes every day (based on average household usage in the UK) [16]. This was estimated to cause the UK to import £60 billion of natural gas to meet the high electric demand during the four-month period [16].

7.5 Conclusion

Energy storage technologies are being reinvented and becoming more powerful as their capacities increase. Applications for energy storage are becoming

Table 7.1 U.S. cost and characteristics of new central station generation technologies [17].

Technology	First available year[a]	Size (MW)	Lead time (years)	Base overnight cost[b] (2021$/kW)	Technology optimism factor[c]	Total overnight cost[d,e] (2021$/kW)	Variable O&M[f] (2021 $/MWh)	Fixed O&M (2021$ kW-y)	Heat rate[g] (Btu/kWh)
Ultra-supercritical coal (USC)	2025	650	4	$4,074	1.00	$4,074	$4.71	$42.49	8,638
USC with 30% carbon capture and sequestration (CCS)	2025	650	4	$5,045	1.01	$5,096	$7.41	$56.84	9,751
USC with 90% CCS	2025	650	4	$6,495	1.02	$6,625	$11.49	$62.34	12,507
Combined-cycle— single-shaft	2024	418	3	$1,201	1.00	$1,201	$2.67	$14.76	6,431
Combined-cycle— multi-shaft	2024	1,083	3	$1,062	1.00	$1,062	$1.96	$12.77	6,370
Combined-cycle with 90% CCS	2024	377	3	$2,736	1.04	$2,845	$6.11	$28.89	7,124
Internal combustion engine	2023	21	2	$2,018	1.00	$2,018	$5.96	$36.81	8,295
Combustion turbine—aeroderivative[h]	2023	105	2	$1,294	1.00	$1,294	$4.92	$17.06	9,124
Combustion turbine—industrial frame	2023	237	2	$785	1.00	$785	$4.71	$7.33	9,905
Fuel cells	2024	10	3	$6,639	1.09	$7,224	$0.62	$32.23	6,469
Nuclear—light water reactor	2027	2,156	6	$6,695	1.05	$7,030	$2.48	$127.35	10,443
Nuclear—small modular reactor	2028	600	6	$6,861	1.10	$7,547	$3.14	$99.46	10,443
Distributed generation—base	2024	2	3	$1,731	1.00	$1,731	$9.01	$20.27	8,923
Distributed generation—peak	2023	1	2	$2,079	1.00	$2,079	$9.01	$20.27	9,907
Battery storage	2022	50	1	$1,316	1.00	$1,316	$0.00	$25.96	NA
Biomass	2025	50	4	$4,524	1.00	$4,525	$5.06	$131.62	13,500
Geothermal[i,l]	2025	50	4	$3,076	1.00	$3,076	$1.21	$143.22	8,813
Conventional hydropower[j]	2025	100	4	$3,083	1.00	$3,083	$1.46	$43.78	NA
Wind[e]	2024	200	3	$1,718	1.00	$1,718	$0.00	$27.57	NA
Wind offshore[i]	2025	400	4	$4,833	1.25	$6,041	$0.00	$115.16	NA
Solar thermal[i]	2024	115	3	$7,895	1.00	$7,895	$0.00	$89.39	NA
Solar photovoltaic (PV) with tracking[e,i,k]	2023	150	2	$1,327	1.00	$1,327	$0.00	$15.97	NA
Solar PV with storage[i,k]	2023	150	2	$1,748	1.00	$1,748	$0.00	$33.67	NA

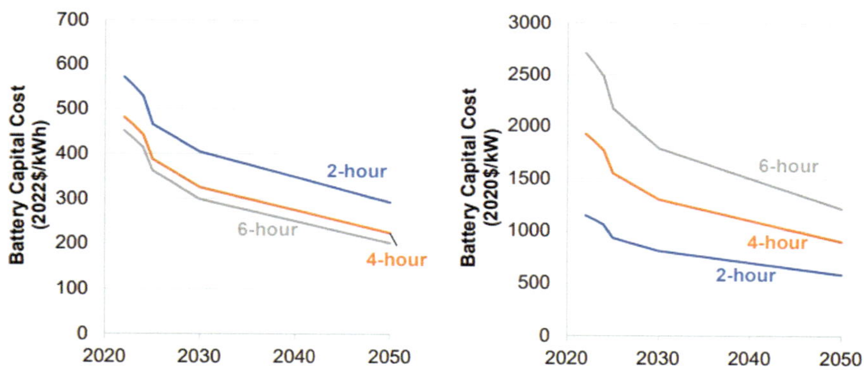

Figure 7.6 Cost projections for 2-, 4-, and 6-hour duration batteries using the mid cost projection [15].

more commonplace and specialized; their benefits are being widely acknowledged. There is a growing realization that these systems will be key technologies to support future electrical transmission networks, particularly those heavily dependent on intermittent energy resources [18]. The lack of energy storage can be costly because fossil fuels are often used to generate electricity to meet peak demand. The other direct benefits of energy storage systems include: 1) enabling a match between energy supplies and demand; 2) expediting the replacement of inefficient power production; 3) helping ensure system stability and security of supply; and 4) facilitating distributed generation [18]. If the use of fossil fuels to generate electricity is avoided, carbon emissions can be reduced.

Utility-scale energy storage units are a new electric power resource; their use has been growing in recent years, particularly in the U.S. [13]. Based on planning data compiled by the U.S. Energy Information Administration, an additional 10,000 megawatts of large-scale battery storage will likely be installed and connected to U.S. electric grids between 2021 and 2023—10 times the total amount of maximum generation capacity by all energy storage systems in 2019 [9]. For large-scale battery storage systems that were planned for installation from 2021 through 2023, most are either standalone systems or co-located with solar arrays [19]. The use of battery energy storage systems combined with solar applications is projected to outpace other types of capacity additions (see Figure 7.7).

The costs for installed energy storage systems, though challenging to compile, are generally declining. Many factors influence how the costs for energy storage systems will evolve [15]. These factors include market demand, supply chain expansions or constraints, material costs and

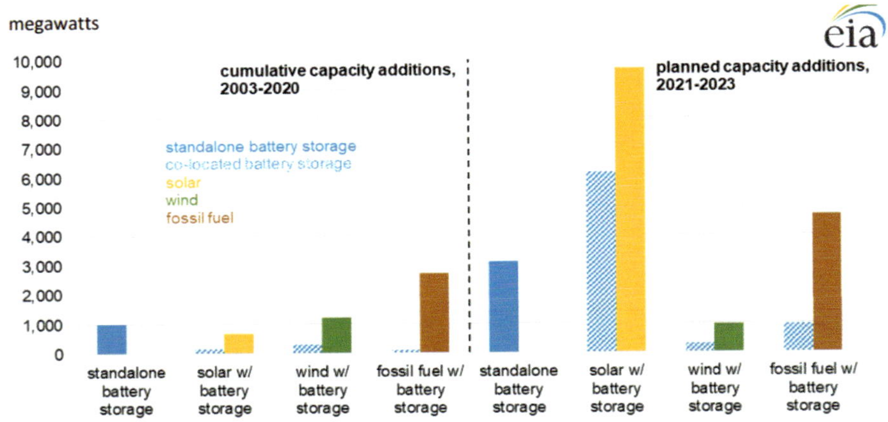

Figure 7.7 U.S. large-scale battery storage capacity additions [19].

availability, and interplay with other technologies such as electric vehicles [15]. Regardless, energy storage systems are often competitive in costs with other forms of energy generation. A key benefit is that they can be installed relatively quickly. As the market for energy storage systems continues to expand, we will likely see applications of hybrid generation configurations. When properly sized and designed, battery storage systems can reduce the amount of wasted electricity and smooth generation. Energy storage is being coupled with renewable energy generation to help resolve intermittency and reliability concerns.

We can design batteries for high-energy density, for multi-day discharge at full power, or for relatively inexpensive domestic supply chains [12]. This ability to match a battery's design to the specific application requirements is a new feature on the energy storage landscape [12]. Soon, we will see applications that demonstrate the need for multiple types of energy storage systems serving a single generation site, each fulfilling a specialized function. There will likely be new battery technologies coming available with greater capacities and longer lifecycles.

References

[1] U.S. Department of Energy. What is pumped storage hydropower? https://www.energy.gov/eere/water/pumped-storage-hydropower, accessed 17 May 2024.

[2] U.S. Department of Energy (2023). U.S. hydropower market report, 46–48. https://www.energy.gov/sites/default/files/2023-09/U.S.%20Hydropower%20Market%20Report%202023%20Edition.pdf, accessed 17 May 2024.

[3] Wiegland-Jackson, L. (2022, September). On the path to net zero: green hydrogen as a decarbonization solution. *International Journal of Strategic Energy and Environmental Planning*, 4(5), 8.

[4] World-Energy (2021, May 21). Ammonia is promising for energy storage. https://www.world-energy.org/article/17861.html, accessed 21 January 2023.

[5] International Energy Agency (2022). Renewables 2022, renewable electricity. https://www.iea.org/reports/renewables-2022/renewable-electricity, 12 December 2022. License: CC BY 4.0.

[6] Roosa, S. (2021). *Fundamentals of Microgrids: Development and Implementation*, 66–121. CRC Press: Boca Raton, Florida.

[7] Hering, G. (2020, March 4). In a 'groundbreaking year,' massive U.S. battery storage projects underway in 2020. https://www.spglobal.com/marketintelligence/en/news-insights/latest-news-headlines/in-a-groundbreaking-year-massive-us-battery-storage-projects-underway-in-2020-57373153, accessed 13 February 2023.

[8] U.S. Energy Information Administration (2022, December 8). U.S. battery storage capacity will increase by 2025. https://www.eia.gov/todayinenergy/detail.php?id=54939&utm_source=LinkedIn&utm_medium=EIAsocial&utm_id=Amplification, accessed 1 January 2023.

[9] U.S. Energy Information Administration (2021, August 6). Battery storage in the United States: an update on market trends. https://www.eia.gov/analysis/studies/electricity/batterystorage, accessed 21 January 2023.

[10] U.S. Energy Information Administration (2024, February 15). Solar and battery storage to make up 81% of new U.S. electric-generating capacity additions in 2024. https://www.eia.gov/todayinenergy/detail.php?id=61424, accessed 15 May 2024.

[11] Mista, A., Venkataramani, G., Gowrishankar, S., Ayyasam, E. and Ramalingam, V. (2018). Renewable energy based smart microgrids—a pathway to green port development. Chapter 18, in Roosa, S. (ed.) *International Solutions to Sustainable Energy, Policies and Applications*, 328–330. The Fairmont Press, Inc.: Lilburn, Georgia.

[12] Everett, L. (2022, March 2). The future of energy storage. Interview with George Crabtree, director of the Joint Center for Energy Storage Research. https://www.labmanager.com/ask-the-expert/the-future-of-energy-storage-27649#, accessed 7 February 2023.

[13] U.S. Energy Information Administration (2019, July 10). U.S. utility-scale battery storage power capacity to grow substantially by 2023. https://www.eia.gov/todayinenergy/detail.php?id=40072, accessed 13 February 2023.

[14] National Renewable Energy Laboratory (2021). Utility scale battery storage. https://atb.nrel.gov/electricity/2021/utility-scale_battery_storage, accessed 21 January 2023.

[15] Cole, W. and Karmaker, A. (2023). Cost projections for utility-scale battery storage: 2023 update. NREL. https://docs.nrel.gov/docs/fy23osti/85332.pdf, accessed 27 July 2025.

[16] Mavrokefalidis, D. (2023, February 7). Wasted wind energy could have powered 1.2m homes. https://www.energylivenews.com/2023/02/07/uk-blew-away-enough-wind-energy-to-power-1-2m-homes-every-day, accessed 8 February 2023.

[17] U.S. Energy Information Administration (2022, March). Cost and performance characteristics of new generating technologies. Annual energy outlook 2022. https://www.eia.gov/outlooks/aeo/assumptions/pdf/table_8.2.pdf, accessed 15 May 2024.

[18] Venkataramani, A., Gowrishankar, S. Ayyasam, E. and Ramalingam, V. (2018). Renewable energy based smart microgrids—a pathway to green port development, chapter 18, in Roosa, S. (ed.) *International Solutions to Sustainable Energy, Policies and Applications*, 330–333. The Fairmont Press, Inc.: Lilburn, Georgia.

[19] U.S. Energy Information Administration (2021, August 10). Battery storage in the United States: an update on market trends, 4. https://www.eia.gov/analysis/studies/electricity/batterystorage/pdf/battery_storage_2021.pdf, accessed 25 May 2024.

8

Matching Electricity Supply with Demand: Integrating Renewable Energy Systems into the U.S. Power Grid

Satish C. Boregowda and Suresh Chand

This chapter presents a conceptual approach to assess the impacts of integrating renewable energy systems into the conventional utility grid. It considers performance, economic, and reliability. There is a push for reliance on renewable energy systems due to concerns about climate change. Renewable portfolio standards expect that U.S. states will produce a certain percentage of electricity using renewable energy sources (e.g., solar, wind, and geothermal) by a certain future date. The U.S. Environmental Protection Agency (EPA) has mandated strict emission controls on greenhouse gases (GHGs) and harmful pollutants such as sulfur dioxide, nitric oxide, lead, and mercury. This poses major challenges for utility companies and their suppliers to meet increasing demand for electricity as renewable energy systems are integrated with conventional utility systems. The methodology begins with a theoretical example of a daily load (demand) curve of a utility firm. This study considers three possible scenarios: a) conventional power systems providing 100% electricity; b) a hybrid power system with renewable energy systems providing 25% of the electricity without battery storage systems; and c) a hybrid power system with renewable energy systems providing 25% of the electricity supplemented with battery storage systems. The impact of these scenarios on three techno-economic assessments, performance, economics, and reliability, is examined at a system level.

Conventional power systems include coal, natural gas, and nuclear generation units, while the renewable systems include solar photovoltaic (PV),

DOI: 10.1201/9788743811312-8

solar thermal, wind, geothermal, and battery energy storage units. In this study, all the conventional power units were retained as part of the hybrid system while the renewables were integrated to meet 25% of the electricity demand. The results demonstrate that the integration of renewable energy systems leads to a significant increase in the levelized cost of electricity and a modest increase in reliability. This conceptual study provides a futuristic framework to investigate the techno-economic complexity of the integration of renewable energy systems to match electricity supply with demand.

8.1 Introduction

During the past several decades, there has been a strong emphasis on the integration of renewable energy systems into the U.S. electricity grid, considering climate change concerns. This study examines the impacts of the integration of renewable energy systems on three techno-economic elements: performance, economics, and reliability. The intermittent electrical generation of certain renewable energy sources (e.g., solar and wind power) and the geographical limitations of geothermal energy pose operational problems when integrating renewable resources due to the widely variable demand for electricity. Using batteries for energy storage could overcome the intermittency of renewable energy sources and support grid stabilization. However, the smaller capacities of battery systems pose challenges to meet time-varying electrical demand. Their use leads to higher costs and reliability issues. Furthermore, there are problems with scale, economics, and reliability.

8.1.1 Approach to the study

Our study begins with an assessment of daily load curve data of a typical U.S. city (see Appendix A, Figure A8.1 and Table A8.1), using a model electric company with 17 conventional power generation systems [1]. The operating and dispatch schedules are provided in Table A8.2. The objective is to integrate renewable energy systems into the utility system to produce at least 25% of total electricity as required by a typical renewable portfolio standard [2]. The cost data for the conventional and renewable power generation units and for battery storage were obtained from the U.S. Energy Information Administration (EIA) [3].

The conventional power generation system is used as the base case. The backup system for integrating renewable energy systems is assumed to produce 25% of the daily electricity to form a hybrid system. For the purposes of

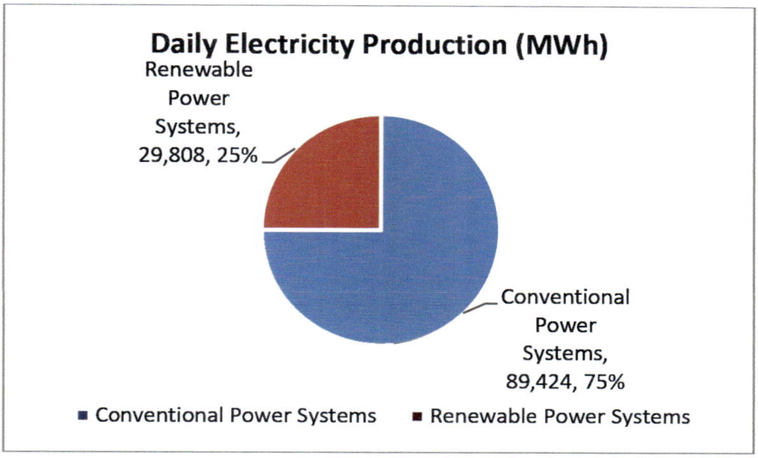

Figure 8.1 Hybrid system with renewable energy systems provides 25% of the total electricity generation.

this study, hybrid systems are defined as those that use both fossil fuel-fired generation and renewable generation systems.

Only one scenario for 25% renewable energy integration is provided in this study. The results indicate a significant difference between conventional and hybrid utility systems with renewable energy systems and battery energy storage systems (BESS) in terms of performance, economic, and reliability variables. It is shown that there is a modest increase in the reliability of the overall power grid due to the integration of renewable generation. There is increased reliability using battery storage systems, but this gain in reliability increases the levelized cost of electricity. The reliability and sourcing of BESS are concerns that need to be addressed in the future. A recent survey conducted by Parker et al. provides numerous opportunities and policy challenges facing the electric power industry [5]. Our study addresses one of the major techno-economic challenges faced by the utility industry—the integration of renewable energy systems into existing electric grids.

8.2 Methodology

The methodology for this study involves assessments of three targeted approaches to providing utility-scale electrical generation. The elements critical to the techno-economic evaluation of any electric grid (i.e., performance, economic, and reliability) are evaluated for three specific scenarios. These scenarios include a conventional fossil fuel-fired electrical generation system, a hybrid system with renewable energy generation, and a hybrid system with renewable generation combined with battery storage. We next investigate each of these essential elements.

8.2.1 Performance analysis

The study begins by assessing daily load curve data depicted in Appendix A, Figure A8.1 and Tables A8.1 and A8.2. These data represent a mid-sized utility company operating 17 conventional power generation units to meet the demand for electricity in its service territory. Using Table 1 data, we can calculate the key plant performance parameters: daily electricity generation (*DEG*), average load (*AL*), load factor (*LF*), capacity factor (*CF*), and reserve factor (*RF*) using Equations 8.1–8.5 [6].

$$DEG = \sum \text{Load (i)} \times \Delta\text{t}\ (i) \tag{8.1}$$

$$\text{Average load } (AL) = DEG\ /\ 24 \text{ hours} \tag{8.2}$$

$$\text{Load factor } (LF) = \text{Average load / Peak load} \tag{8.3}$$

$$\text{Capacity factor } (CF) = \text{Average load / Rated capacity} \tag{8.4}$$

$$\text{Reserve factor } (RF) = \text{Load factor / Capacity factor} \tag{8.5}$$

Sample calculations for the conventional generation system are provided here for the convenience of the reader. Using the data from Appendix A, Tables A8.1 and A8.2, we obtain the following metrics:

$$\begin{aligned} DEG &= (3{,}888 \text{ MW} \times 1 \text{ hour}) + (3{,}696 \text{ MW} \times 1 \text{ hr.}) + \ldots + \\ &\quad (4{,}800 \text{ MW} \times 1\text{hour}) \\ &= 119{,}232 \text{ MWh} \end{aligned}$$

$$\text{Average load } (AL) = 119{,}232 \text{ MWh / 24 hours} = 4{,}968 \text{ MW}$$

$$\text{Load factor } (LF) = 4{,}968 \text{ MW / } 6{,}000 \text{ MW} = 0.83$$

$$\text{Capacity factor } (CF) = 4{,}968 \text{ MW / } 6{,}825 \text{ MW} = 0.73$$

$$\text{Reserve factor } (RF) = 0.83\ /\ 0.73 = 1.14$$

If the rated capacity of the plant is equal to the peak load, then the capacity factor and load factor would be equal. The difference between the load factor and capacity factor is an indication of the reserve capacity represented by the reserve factor. The difference between the operating capacity and the load at different hours is the spinning reserve (see Appendix A, Figure A8.1). This is the buffer capacity that can be activated immediately to match electricity supply with demand when demand increases or supply is disrupted.

Without batteries to store electricity, there will be situations when electricity produced by renewables is more than the demand (not needed by the customers) and is not directed to the grid. This excess electricity produced by renewables is unusable. With battery storage capabilities, this excess electricity is available for later use. The stored battery power works as a source of electricity when the batteries are being discharged.

8.2.2 Economic evaluation

It is important to assess the economic performance of a utility system. We calculate the variables capital cost (*CCOST*), operating and maintenance cost (*OMCOST*), fuel cost (*FCOST*), annualized cost (*ACOST*), and the levelized cost of electricity (*LCOE*) using Equations 8.6–8.10:

$$CCOST = \sum \text{[Rated capacity unit } (i) \times \text{Capital cost unit } (i)] \quad (8.6)$$

$$OMCOST = \sum \text{[Rated capacity unit } (i) \times \text{O\&M cost unit } (i)] \quad (8.7)$$

$$FCOST = \sum \text{[Generation unit } (i) \text{ x Fuel cost unit } (i)] \quad (8.8)$$

$$ACOST = CCOST \times [\mathrm{I} \times (1 + \mathrm{I})^{\mathrm{N}}/(1+\mathrm{I})^{\mathrm{N}} - 1)] \quad (8.9)$$

$$LCOE = (ACOST + \mathrm{OMCOST} + FCOST)/DEG \quad (8.10)$$

The interest rate (*I*) and the length of the project (*N*) were assumed to be 7% and 20 years, respectively. The detailed cost data were selected from original EIA tabular data and presented in Appendix A, Tables A4–A6 for the generation units used in our study [3,4]. The operations and maintenance (O&M) costs were obtained by multiplying the unit O&M cost by their unit rated capacities. However, the fuel costs incurred by the conventional power systems depend on the power generated by each unit. The levelized cost of electricity in $/kWh is calculated using Equation 8.10.

8.2.3 Reliability assessment

The objective of a reliability analysis is to derive suitable measures, criteria, and indices of reliable and dependable performance based on component outage data and configuration [7]. For this study, reliability is defined as the probability that the system functions to generate the full power, P(S). In other words, the performance measure of reliability is 1 when 100% of the demand is met, and 0 otherwise.

We assume that the energy system consists of several power blocks. Each block consists of one or more units. We let $P(B_d)$ denote the probability that Block *d* operates successfully. The reliability factors for different blocks ($P(B_d)$, $d = 1, 2, 3\ldots$) are approximated based on the historical data and the availability of fuel source for the units in Block *d* [8].

For the case when the energy system consists of *k* power blocks and power is needed from each of these to meet the entire demand, then the reliability is a multiplication function of the probability of each of *k* energy blocks functioning. This is equivalent to a series system with *k* blocks in series, as in Equation 8.11.

$$P(S) = P(B_1) \times P(B_2) \times P(B_3) \times \ldots . \times P(B_k) \tag{8.11}$$

For the case when the system consists of *k* blocks and power is needed from any one of these to meet the entire demand. Using Equation 8.12:

$$P(S) = 1 - [1 - P(B_1)] \times [1 - P(B_2)] \times \ldots . \times [1 - P(B_k)] \tag{8.12}$$

This is equivalent to a parallel system with *k* blocks in parallel [8].

For this study, the reliability assessment is performed for configurations including both series and series subsystems connected in parallel. These configurations approximately represent the physical arrangement of generation units.

The reliability assessment is performed for the following three scenarios outlined in Appendix A, Tables A8.4–A8.6: Scenario I, conventional utility system (17 units); Scenario II, hybrid utility system without battery storage (45 units); and Scenario III, hybrid utility system with battery storage (70 units).

The individual power generation units are combined into seven blocks: coal, nuclear, natural gas, solar, wind, geothermal, and battery, as shown in Appendix A, Table A8.7. For instance, the coal block comprises six plants, each with multiple boilers, steam turbines, condensers, and pumps to ensure that the electric demand is met during the times they are operational. The same is true with the nuclear block that comprises two plants, each with multiple reactors to ensure smooth operation when shut down for refueling. This logic applies to the remaining blocks, including natural gas, solar, wind, geothermal, and battery. None of the conventional power generation units are replaced while integrating solar, wind, geothermal, and battery units into the

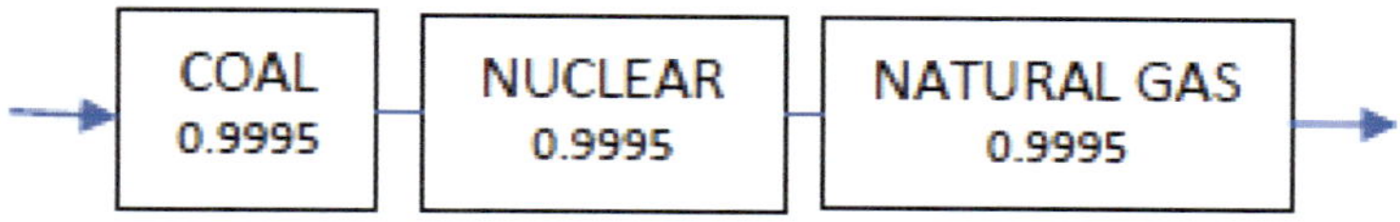

Figure 8.2 Reliability configuration of Scenario I, conventional power systems only.

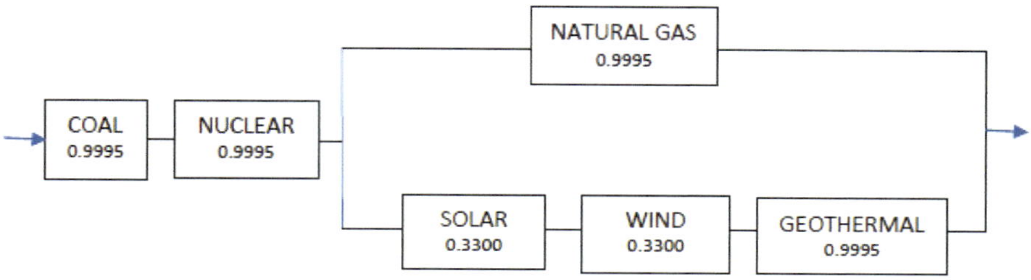

Figure 8.3 Reliability configuration of a Scenario II, hybrid system without batteries.

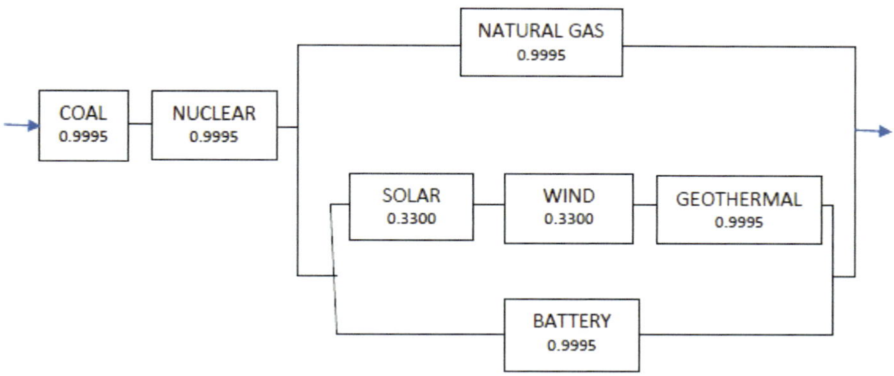

Figure 8.4 Reliability configuration of Scenario III, hybrid systems with battery energy storage.

system. This assumption ensures that the system maintains the same higher reliability as conventional systems, though the integration of renewable energy systems produces 25% of the total electricity or perhaps more.

Further investigation is needed when fossil-fuel units are replaced with renewable energy systems to produce 50%, 75%, and 100% of total electricity. The approximate operational reliabilities are presented in Appendix A, Table A8.7. These reliability numbers do not reflect actual values. They are used for benchmarking the scenarios for this study. We assume configurations in Figures 8.2–8.4 meet the electric demand by using different blocks.

8.3 Results and Discussion

We next analyze the variables pertaining to performance, economics, and reliability for three scenarios. The analysis starts with Scenario I, modeling a utility company operating 17 conventional power generation units. The cost data for conventional, renewable, and battery storage systems are provided in Appendix A, Tables A8.3–A8.6 and are adopted from EIA tabular data [3,4].

Scenario II involves the integration of renewable energy systems that generate 25% of the daily electricity generation per EIA guidelines. The utility system must produce 119,232 MWh of electricity per day to meet the demand for electricity. The objective is to produce 25% of the daily required electricity with the renewable energy systems: 119,232 MWh per day × 25% = 29,808 MWh per day. We assumed the use of all five renewable energy systems (solar thermal, solar PV, wind offshore, wind onshore, and geothermal) in equal portions for the sake of demonstration. It is expected that the average load will be used to select the rated capacities of renewable energy systems: 1,242 MW × 24 hours/day = 29,808 MWh/day.

It was assumed that the average load of 1,242 MW is met in equal proportion by solar thermal, solar PV, offshore wind, onshore wind, and geothermal generation. This means that an individual load of 248.5 MW (1,242 MW/5 units = 248.5 MW/unit) would be met by each of the five renewable energy systems.

Capacity factors from the Fowler Ridge Wind Farm (27.8%), the Ivanpah Solar Thermal Power Facility (24.1%), and the Blythe Mesa Solar PV Project (30.2%) were used to evaluate the needed capacities for installation of renewable energy systems [9–11]. On average, the capacity factor of 0.30 was chosen for the selection and sizing of solar thermal, solar PV, onshore wind, and offshore wind units as selected from U.S. Department of Energy tabular data [3]. The needed capacities of the solar and wind are selected as follows:

$$
\begin{aligned}
(\text{Rated capacity})_{\text{Solar, Wind}} &= \text{Average load / Capacity factor} \\
&= 248.5\ \text{MW}/0.30 = 828\ \text{MW}
\end{aligned}
$$

For geothermal units, the capacity factor of 0.82 ($= 215 \times 10^9 / 30^6 \times 8{,}760$ hours) was selected based on the actual performance of the Mammoth Geothermal Plant located in Nevada/California region [12].

$$
\begin{aligned}
(\text{Rated capacity})_{\text{Geothermal}} &= \text{Average load / Capacity factor} \\
&= 248.5\ \text{MW}/0.82 = 303\ \text{MW}
\end{aligned}
$$

In Scenario III, battery energy storage systems are added to support the renewable energy systems (considered in Scenario II) to overcome the intermittency of the renewable sources of wind and solar. For this scenario, the battery units needed to be capable of storing up to 1,242 MW to provide 25% of daily electricity to meet the renewable portfolio requirements. Approximately 25 battery units of 50 MW (approximately 1,242 MW/50 MW) were selected [4]. However, the conventional power systems remain intact to act as redundant systems in the event of failure of batteries and insufficient supplies.

Appendix Tables A8.4–A8.6 represent three configurations starting with a total of 17 units for conventional systems, 45 units with renewable energy systems, and 70 units with renewable energy systems plus battery storage. Appendix Table A8.7 provides estimates of operational reliability factors of units in both conventional and hybrid systems. These reliability factors are adjusted based on fuel source availability. The results and discussion, including the performance analysis, economic evaluation, and reliability assessment of the three scenarios, are presented in Figures 8.5–8.12.

8.3.1 Performance analysis

The daily electricity generation (*DEG*), average load (*AL*), and the load factor (*LF*) are the same for all three scenarios because the electrical demand is the same. The capacity factor (*CF*) measures the utilization of the plant capacity. A plant with a capacity factor of 100% means it is producing power continuously at its rated capacity. The reserve factor (*RF*) is a measure of buffer capacity greater than peak demand; it increases with the addition of the capacities of renewable energy systems and BESS (see Figure 8.5). The overall increase in capacity and number of assets is shown in Figures 8.6 and 8.7, respectively.

8.3.2 Economic evaluation

Figure 8.8 shows the capital costs associated with each of the three studied configurations. The conventional configuration (Scenario I) is the least costly; the renewable configurations with or without battery storage are more expensive (Scenarios II and III). The levelized cost of electricity (LCOE) is calculated based on the assumed interest rate of 7% and a project life of 20 years. More detailed analysis of the LCOE for different interest rates and project terms is warranted. The LCOE increases from $0.0611/kWh to $0.1030/kWh (see Figure 8.9). For a residence with an average monthly electricity consumption of 1,000 kWh, the monthly electricity costs increase from $61.05 to $103.01 per month (see Figure 8.10), or by approximately 65%.

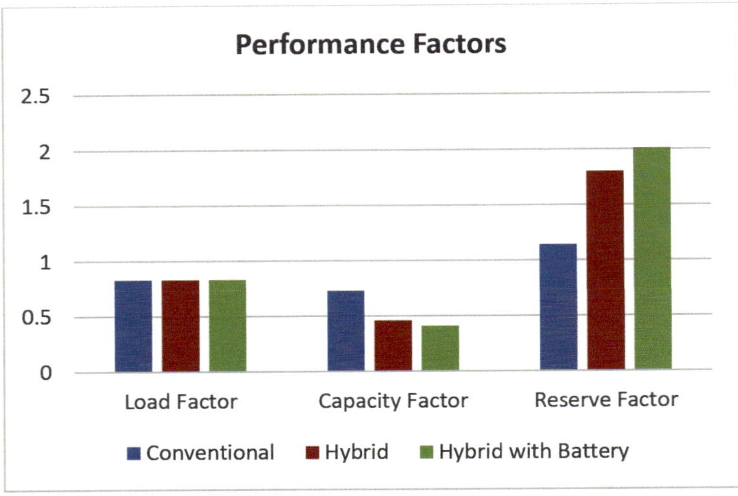

Figure 8.5 Overview of performance factors.

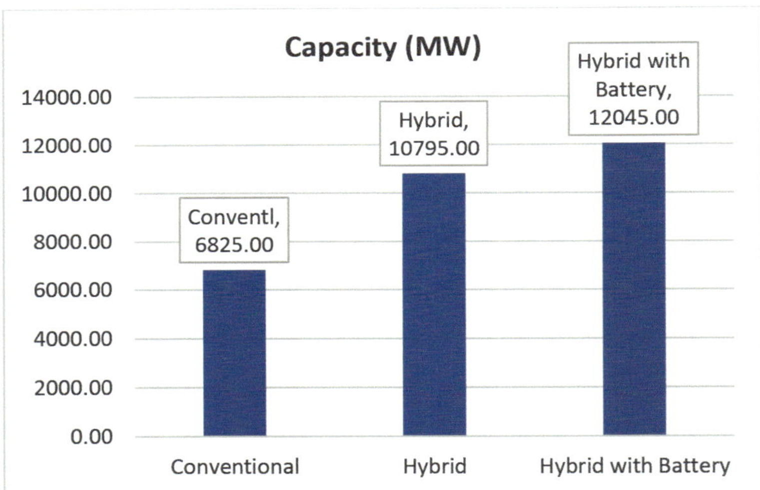

Figure 8.6 Increase in rated capacity.

8.3.3 Reliability assessment

Using the operational reliability factors in Table A8.7, the results of the reliability assessment are detailed in Figures 8.11 and 8.12. The equipment reliability factor of 0.9995 was assumed for all the units. Note that the value of 0.9995 implies unplanned downtime of about four hours annually [(1.0 – 0.9995) × 8,760 hours/year = 4.38 hours/year], which was reasonable.

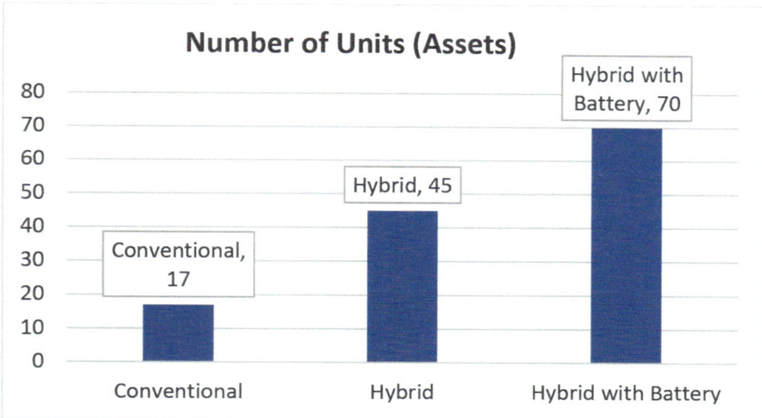

Figure 8.7 Increase in number of assets (units).

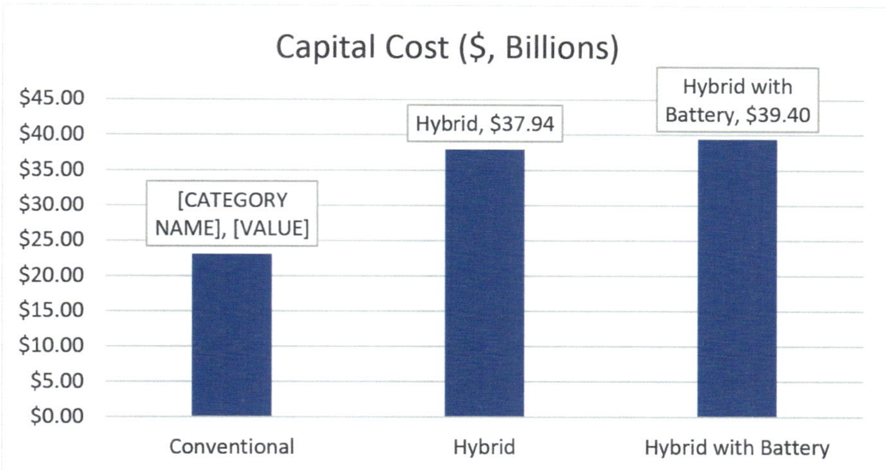

Figure 8.8 Capital cost ($, billion).

The operational reliability of the renewable energy systems would be much lower than the fossil-fuel and nuclear plants due to the generation intermittency of solar and wind energy resources.

The results show that the reliability of the conventional power system in Scenario I is 99.85%. This implies that the likelihood of a conventional power system generating power is 99.85% and the anticipated power outages total 13.13 hours annually [= (1 – 0.9985) × 8,760 hours/year]. With the integration of renewable energy systems, the reliability for Scenario II increases slightly to 99.86% and the expected power outage times decrease to

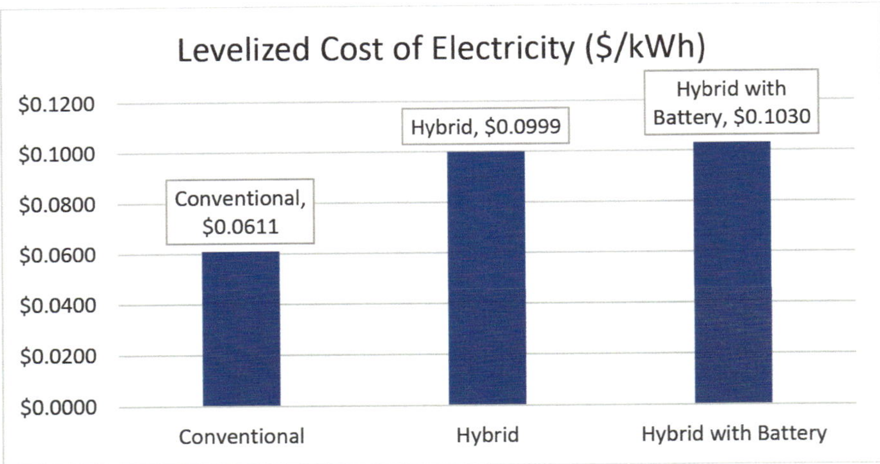

Figure 8.9 Levelized cost of electricity (I = 7%; N = 20 years).

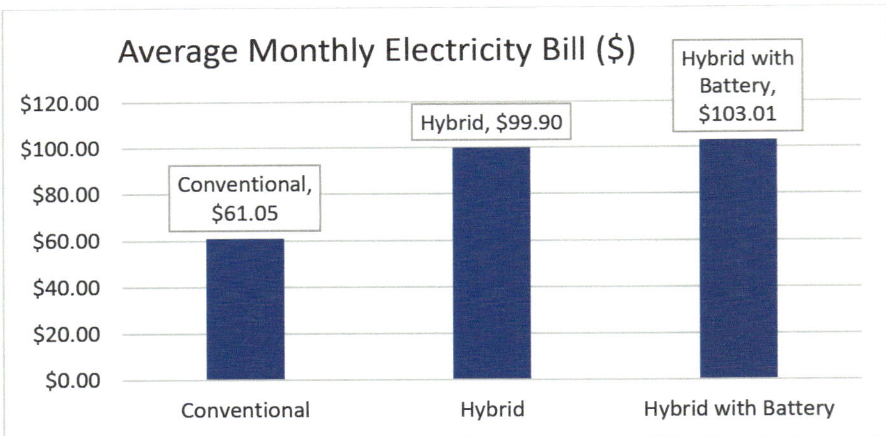

Figure 8.10 Average monthly home electricity costs ($).

12.66 hours yearly. In Scenario III, the addition of a BESS increases the operational reliability of the entire system to 99.89% while the expected power outage impact drops to 8.76 hours annually.

8.4 Conclusion

The main goal of this study was to understand the problem of renewable energy systems integration using a simple load curve without shutting down

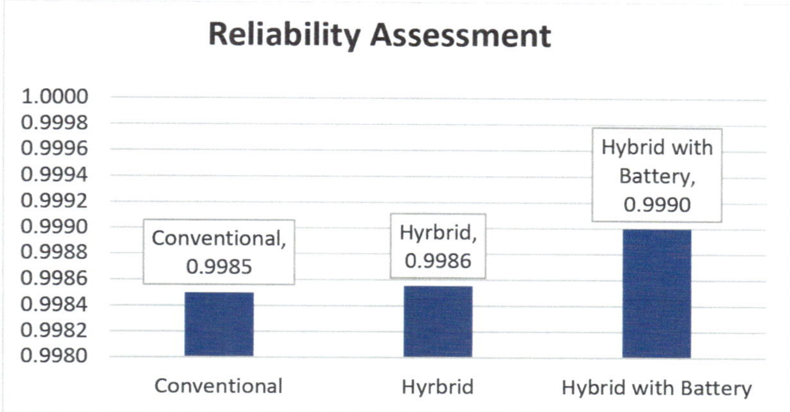

Figure 8.11 Reliability assessment of three scenarios: Conventional, hybrid, and hybrid with batteries.

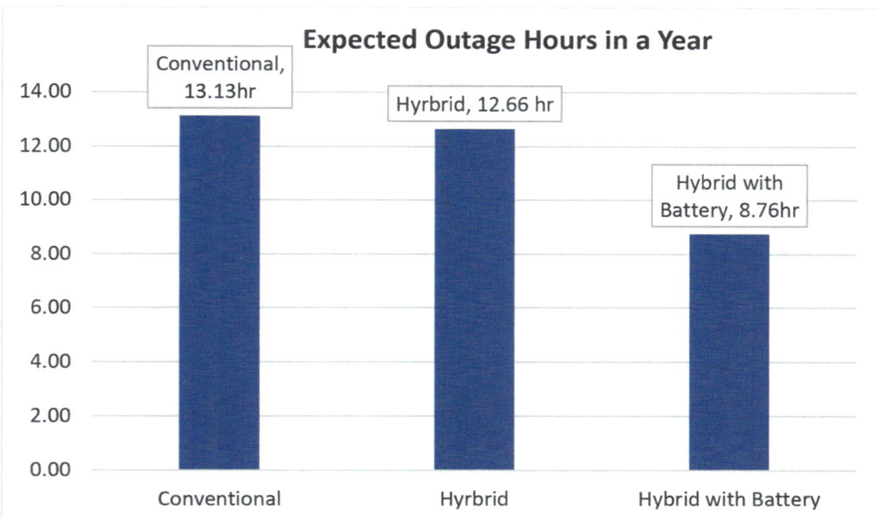

Figure 8.12 Reliability assessment in terms of expected outage hours.

the existing power plants. An objective of our study was to assess ways to produce 25% of the daily electricity using renewable energy systems. To achieve these ends, our study involved the assessment of a conceptual utility system comprised of 17 conventional power generation units. The impacts of three possible scenarios on performance, economics, and reliability were examined: Scenario I, a conventional power generation system; Scenario II, a hybrid renewable system without batteries; and Scenario III, a hybrid system with batteries for energy storage.

The findings indicate that the overall system capacity, the capital cost, the levelized cost of electricity, and the average monthly electricity costs increase with the integration of renewable energy and energy storage systems. The reliability of the system does not improve significantly because none of the conventional power generation units were replaced by the renewable energy units. All the conventional units were used as a backup, leading to a modest improvement in the reliability of the system. Furthermore, the addition of renewable energy systems reduces GHG emissions without adding additional fuel costs.

To further advance our knowledge and support the energy transition to integrate renewable energy and battery storage systems, more detailed investigations using real-time operational data and advanced simulation software are needed.

Acknowledgments

The authors thank professors Eckhard Groll and James Jones from the Purdue University School of Mechanical Engineering for supporting this research project.

References

[1] Li, K. and Priddy, P. (1985). *Power Plant System Design.* Wiley and Sons.

[2] National Conference of State Legislatures (2021, August 13). State renewable portfolio standards and goals. https://www.ncsl.org/research/energy/renewable-portfolio-standards.aspx, accessed 11 May 2022.

[3] U.S. Energy Information Administration (2013, April). Updated capital cost estimates for utility scale electricity generating plants. https://www.eia.gov/outlooks/capitalcost/pdf/updated_capcost.pdf, accessed 2 August 2025.

[4] U.S. Energy Information Administration (2022, March). Cost and performance characteristics of new generating technologies, annual energy outlook 2022. https://www.eia.gov/outlooks/aeo/assumptions/pdf/table_8.2.pdf, accessed 1 August 2025.

[5] Parker, G., Erciyes, B. and Kazan, O. (2019). Electric power industry: operational and public policy challenges and opportunities. *Production and Operations Management*, 28 (11), 2,673–2,919.

[6] Nag, P. (2015). *Power Plant Engineering,* 4th edition, McGraw-Hill.

[7] Rao, S. (2015). *Reliability Engineering,* 1st edition, Pearson Higher Education.

[8] Al-Shaalan, A. (2019, My 13). Reliability evaluation of power systems. DOI: https://www.intechopen.com/chapters/66913, accessed 2 August 2025.
[9] Fowler Ridge Wind Farm. https://en.wikipedia.org/wiki/Fowler_Ridge_Wind_Farm, accessed 27 May 2022.
[10] Ivanpah Solar Power Facility. https://en.wikipedia.org/wiki/Ivanpah_Solar_Power_Facility, accessed 27 May 2022.
[11] Blythe Mesa Solar Power Project. https://en.wikipedia.org/wiki/Blythe_Mesa_Solar_Power_Project, accessed 27 May 2022.
[12] Mammoth Geothermal Complex. https://en.wikipedia.org/wiki/Mammoth_Geothermal_Complex, accessed 2 August 2025.

Appendix A

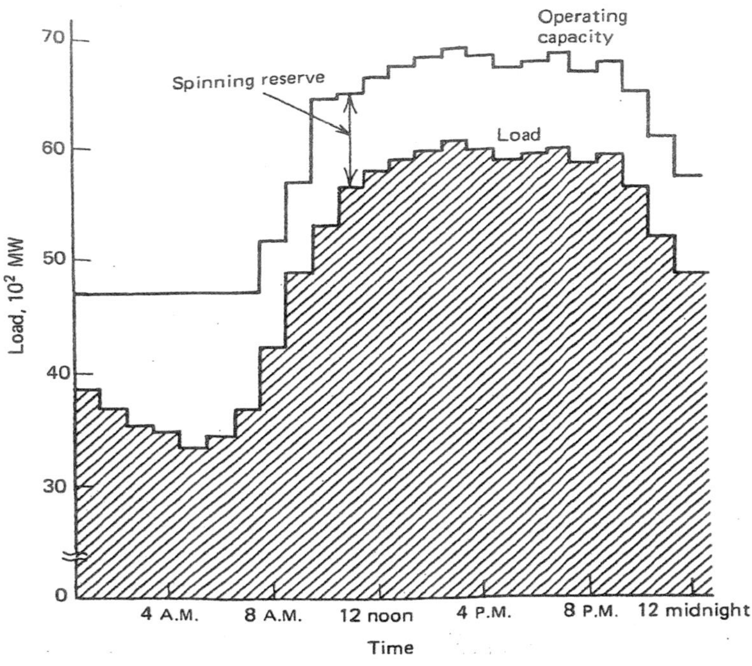

Figure A8.1 Daily load curve of a mid-sized U.S. city [1].

Table A8.1 Daily load data.

	1	2	3	4	5	6	7	8	9	10	11	12
Load (MW)	3,888	3,696	3,552	3,504	3,360	3,456	3,696	4,224	4,848	5,280	5,616	5,760
Hours	13	14	15	16	17	18	19	20	21	22	23	24
Load (MW)	5,856	5,904	6,000	5,904	5,808	5,856	5,904	5,760	5,856	5,568	5,136	4,800

Table A8.2 Daily operating schedule of the utility system.

Utility System Load Distribution in a 24-Hour Period																				
		Capacity in	Spinning	Unit Number																
	Load	Operation	Reserve	1	2	3	4	5	6	7	8	9	10	11	12	13	14	15	16	17
Hours	(MW)	(MW)	(MW)	Load Distribution (MW)																
100	3888	4700	812	483	513	574	719			800	800									
200	3696	4700	1004	416	450	534	696			800	800									
300	3552	4700	1148	341	418	508	685			800	800									
400	3564	4700	1196	312	410	500	682			800	800									
500	3360	4700	1340	240	382	472	665			800	800									
600	3456	4700	1244	283	402	492	678			800	800									
700	3696	4700	1004	416	450	534	696			800	800									
800	4224	5150	926	566	568	624	749		117	800	800									
900	4848	5650	802	687	664	695	800	63	339	800	800									
1000	5280	6400	1120	737	711	742	800	303	388	800	800	s	s							
1100	5616	6450	834	800	750	750	800	466	450	800	800	s	s							s
1200	5760	6575	815	800	750	750	800	500	450	800	800	s	110				s		s	s
1300	5856	6675	819	800	750	750	800	500	450	800	800	s	206			s	s	s	s	s
1400	5904	6750	846	800	750	750	800	500	450	800	800	s	254	s		s	s	s	s	s
1500	6000	6825	825	800	750	750	800	500	450	800	800	77	273	s	s	s	s	s	s	s
1600	5904	6750	846	800	750	750	800	500	450	800	800	s	254	s		s	s	s	s	s
1700	5808	6625	817	800	750	750	800	500	450	800	800	s	158				s	s	s	s
1800	5856	6675	819	800	750	750	800	500	450	800	800	s	206			s	s	s	s	s
1900	5904	6750	846	800	750	750	800	500	450	800	800	s	254	s		s	s	s	s	s
2000	5760	6575	815	800	750	750	800	500	450	800	800	s	110				s	s	s	s
2100	5856	6675	819	800	750	750	800	500	450	800	800	s	206			s	s	s	s	s
2200	5568	6400	832	800	750	750	800	418	450	800	800	s	s							
2300	5136	6000	864	716	692	722	800	239	376	800	800		s							
2400	4800	5650	850	680	658	689	800	50	323	800	800									

Note: "s" denotes the spinning reserve.

Table A8.3 Cost data for conventional power generation units of the utility system.

Unit number	Unit name	Fuel type (*)	Daily generation (MWh)	Rated capacity (MW)	Capital cost ($/kW)	O&M cost ($/kW-year)	Fuel cost ($/MWh)
1	Fbase1	C	15,417	800	$3,246	$37.80	$4.47
2	Fbase2	C	15,318	750	$3,246	$37.80	$4.47
3	Fbase3	C	16,080	750	$3,246	$37.80	$4.47
4	Fbase4	C	18,361	800	$3,246	$37.80	$4.47
5	Fcycle1	C	6,539	500	$3,246	$37.80	$4.47
6	Fcycle2	C	6,943	450	$3,246	$37.80	$4.47
7	Nuclear1	N	19,200	800	$5,530	$93.28	$2.14
8	Nuclear2	N	19,200	800	$5,530	$93.28	$2.14
9	Ccycle1	G	77	400	$917	$13.17	$3.60
10	Ccycle2	G	2,031	350	$917	$13.17	$3.60
11	Gas1	G	10	75	$973	$7.34	$15.45
12	Gas2	G	10	75	$973	$7.34	$15.45
13	Gas3	G	10	50	$973	$7.34	$15.45
14	Gas4	G	10	75	$973	$7.34	$15.45
15	Gas5	G	10	50	$973	$7.34	$15.45
16	Gas6	G	10	50	$973	$7.34	$15.45
17	Gas7	G	6	50	$973	$7.34	$15.45
Total			119,232	6825			

*C = Coal; N = Nuclear; G = Natural Gas

Source: http://www.eia.gov/forecasts/capitalcost/pdf/updated_capcost.pdf

Table A8.4 Conventional power systems.

Systems	Quantity	Installed capacity (MW)	Capital cost ($/kW)	O&M cost ($/kW-year)	Fuel cost ($/MWh)	Daily generation (MWh)
Coal (1-4)	4	3,100	$3,246	$37.80	$4.47	119,232 (100%)
Coal cycle (5-6)	2	950	$3,246	$37.80	$4.47	
Nuclear (7-8)	2	1,600	$5,530	$93.28	$2.14	
Natural gas cycle (9-10)	2	750	$917	$13.17	$3.60	
Gas turbine (11-17)	7	425	$973	$7.34	$15.45	
Total	17	6,825				119,232

Table A8.5 Hybrid power systems without battery storage.

Systems	Quantity	Installed capacity (MW)	Capital cost ($/kW)	O&M Cost ($/kW-year)	Fuel cost ($/MWh)	Daily generation (MWh)
Coal (1–4)	4	3,100	$3,246	$37.80	$4.47	89,424 (75%)
Coal cycle (5–6)	2	950	$3,246	$37.80	$4.47	
Nuclear (7–8)	2	1,600	$5,530	$93.28	$2.14	
Natural gas cycle (9–10)	2	750	$917	$13.17	$3.60	
Gas turbine (11–17)	7	425	$973	$7.34	$15.45	
Solar thermal (18–25)	8	920	$7,116	$85.82	$0.00	29,808 (25%)
Solar PV (26–31)	6	900	$1,248	$15.33	$0.00	
Wind onshore (32–36)	5	1,000	$1,846	$26.47	$0.00	
Wind offshore (37–38)	2	800	$5,453	$110.56	$0.00	
Geothermal binary (39–55)	7	350	$2,772	$137.50	$0.00	
Total	45	10,795				119,232

Table A8.6 Hybrid power systems with battery storage.

Systems	Quantity	Installed capacity (MW)	Capital cost ($/kW)	O&M cost ($/kW-year)	Fuel cost ($/MWh)	Daily generation (MWh)
Coal (1–4)	4	3,100	$3,246	$37.80	$4.47	89,424 (75%)
Coal Cycle (5–6)	2	950	$3,246	$37.80	$4.47	
Nuclear (7–8)	2	1,600	$5,530	$93.28	$2.14	
Natural gas cycle (9–10)	2	750	$917	$13.17	$3.60	
Gas Turbine (11–17)	7	425	$973	$7.34	$15.45	
Solar thermal (18–25)	8	920	$7,116	$85.82	$0.00	29,808 (25%)
Solar PV (26–31)	6	900	$1,248	$15.33	$0.00	
Wind onshore (32–36)	5	1,000	$1,846	$26.47	$0.00	
Wind offshore (37–38)	2	800	$5,453	$110.56	$0.00	
Geothermal binary (39–45)	7	350	$2,772	$137.50	$0.00	
Battery storage (46–70)	25	1,250	$1,165	$24.93	$0.00	
Total	70	12,045				119,232

Table A8.7 Approximate reliability factors of power generation units.

Blocks	Unit(s)	Equipment reliability factor	Fuel availability	Operational reliability
Coal	1-6	0.9995	1.0000	0.9995
Nuclear	7-8	0.9995	1.0000	0.9995
Natural gas	9-17	0.9995	1.0000	0.9995
Solar	18-31	0.9995	0.3300	0.3300
Wind	32-38	0.9995	0.3300	0.3300
Geothermal	39-45	0.9995	1.0000	0.9995
Battery	45-70	0.9995	1.0000	0.9995

9

Electricity Generation Capacity Planning with Nuclear, Renewables, and Batteries: Defining Pathways to Energy Security and Decarbonization

Satish C. Boregowda, Suresh Chand, and Stephen Roosa

This chapter presents a capacity planning approach to assess the impact of integrating nuclear energy systems on techno-economics and CO_2 emissions. While the U.S. Environmental Protection Agency (EPA) has imposed strict rules and standards to reduce pollution from fossil fuel-fired power plants, energy security and reliability are pervasive issues in the light of the climate crisis. Issues related to sustainability need to be addressed. The complexity of the problem at the interface of climate change and energy security poses major challenges to utility companies and electric grid operators who provide cost-effective and reliable electricity to their customers.

The main goal of our study is to better understand the integration of nuclear, renewables, and battery storage units with the help of a model utility system. To this end, this study utilizes a daily load curve from a model utility system to demonstrate electricity generation capacity planning and portfolio selection. It investigates three possible capacity planning scenarios: Scenario I comprised of coal, nuclear, and natural gas units; Scenario II comprised of nuclear and natural gas units; and Scenario III comprised of nuclear, natural gas, renewable, and battery energy storage units. Using U.S. Energy Information Administration data, the impact of capacity planning scenarios on techno-economic elements such as plant performance factors, economics, reliability, and CO_2 emissions is examined. The results indicate a significant variation in performance factors, the levelized cost of electricity (LCOE), and

DOI: 10.1201/9788743811312-9

CO_2 emissions, while the reliability remains the same across all three scenarios. With nuclear energy, renewables, and battery energy storage replacing coal, a reduction in CO_2 emissions of 99% is observed. In summary, this study provides a framework to assess the techno-economic complexity of electricity generation capacity planning with nuclear energy, renewables, and battery storage units.

9.1 Introduction

There has been an ongoing debate pertaining to the commissioning of nuclear power plants to produce carbon-free electricity. In the U.S., the Environmental Protection Agency (EPA) standards have imposed strict regulations to reduce pollution from fossil-fuel-fired power plants. These standards and high operational costs have been the principal reasons for the decommissioning of numerous coal-fired power plants across the U.S. [1]. It has become important to find alternatives to produce base load electricity, plus enhance the stability and security of electric grids. The International Energy Agency (IEA) defines energy security as the uninterrupted availability of energy sources at an affordable price with intricate relationships to national security and natural resources [2]. However, the quest to decarbonize energy infrastructure and produce affordable, clean, and reliable electricity poses challenges for industries and nation-states.

While natural gas (NG) plants produce less CO_2/kWh of electricity generated than coal- or oil-fired generating stations, they are more suited for producing power for intermediate and peak loads in conjunction with renewable energy systems. The rapid emergence of artificial intelligence (AI) data centers, electric vehicle (EV) charging infrastructure, and the growing need for air conditioning systems in buildings are straining local power grids and increasing electrical demand. Policies are in flux. There are now regional regulations focused on achieving greenhouse gas emissions (GHG) reduction goals.

For nuclear power, there are safety concerns in the minds of the public and policymakers after the partial meltdown at the Three-Mile Island plant in the U.S. (1979) and the nuclear disasters at the Chernobyl plant in Ukraine (1986) and Japan's Fukushima Daichi plant (2011) [3]. Despite concerns about nuclear waste disposal, the climate crisis has prompted many to reconsider nuclear energy as a viable option for producing electricity [4]. Given this complex situation, the U.S. government issued official statements supporting the generation of electricity using nuclear energy [5].

Table 9.1 Changes in energy use by resource.

Energy resources	2008 (quads)	2022 (quads)	% Change
Coal	22.42	9.91	–55.8%
Natural Gas	23.84	33.40	40.1%
Nuclear	8.45	8.05	–4.7%
Hydro	2.45	2.31	–5.7%
Biomass	3.81	4.88	28.1%
Wind	0.51	3.89	662.8%
Solar	0.09	1.87	1,977.8%
Geothermal	0.35	0.21	–40.0%
Petroleum	37.13	35.80	–3.6%

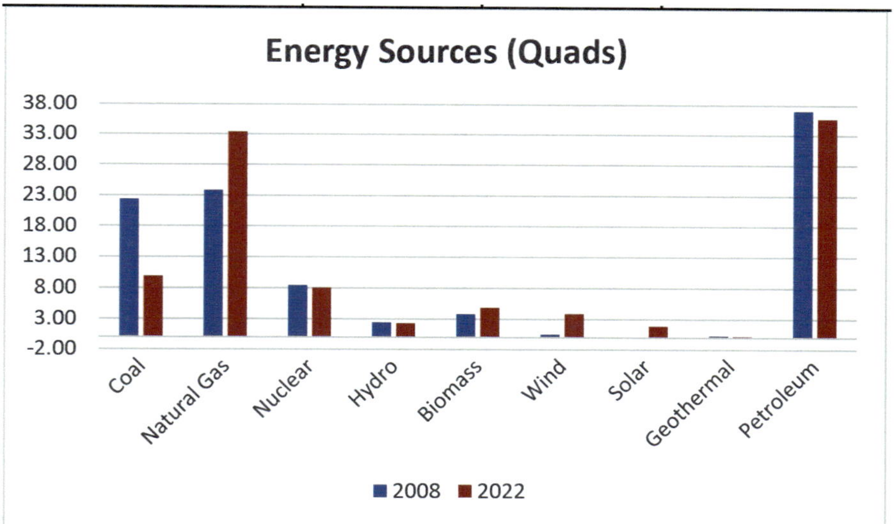

Figure 9.1 U.S. energy portfolio mix from 2008 to 2022.

A historical analysis of the U.S. portfolio of energy resources since 2008 shows that the energy mix used for electrical generation has evolved over the last 16 years. Data obtained from the Lawrence Livermore National Laboratory (LLNL) energy flow charts are reproduced in Table 9.1 and Figure 9.1 [6]. Using the 2008 base year, coal usage declined by 55.8% from 22.4 to 9.9 quads[1], while NG usage increased by 40.1% from 23.4 to 33.4 quads by 2022. We see substantial development of operating wind and

[1] quadrillion Btu = 293 × 10^9 kWh of electricity.

solar farms over the period, increasing by 677% and 1,977% respectively. While hydropower production declined by 5.7%, biomass usage has surged by 28.1% with the use of wood pellets in many retired coal-fired power plants and smaller biomass plants [7]. Although nuclear power production has declined by 4.7%, from 8.5 to 8.1 quads due to plant closures, it has the potential to reemerge as a viable option for producing electricity without the threat of increasing GHG emissions.

9.1.1 Approach to the study

The reassessment of nuclear energy to meet the climate goals involves both technical, operational, management, and financial challenges.[2] A recent survey conducted by Parker et al. indicated numerous such opportunities and policy challenges facing the electric power industry [10]. Our conceptual study addresses this complex issue by examining the impacts of nuclear integration along with renewables and battery storage units [11]. It further considers their impacts on CO_2 emissions, plant performance factors, economics, and grid reliability.

The study begins with the assessment of a daily load curve data (see Chapter 8, Appendix A, Figure A8.1 and Table A8.1) for a modeled electric utility company with 17 conventional power generation systems [12]. The operating and dispatch schedules are provided in Chapter 8, Appendix A, Table A8.2. Using the conventional power system shown in Appendix B, Table B9.1, three possible scenarios are considered (see Appendix B, Tables B9.2–B9.4). Appendix B, Table B9.5 provides the approximate reliability factor data; the cost data for the coal, nuclear, natural gas (NG), renewables, and battery energy storage system (BESS) units were obtained from the U.S. Department of Energy (DoE) and the Energy Information Administration (see Appendix B, Table B9.6) [13]. The energy resources used for the three selected scenarios are summarized in Table 9.2.

Our analysis presents three capacity planning scenarios:

- Scenario I comprised coal, nuclear, and natural gas units.
- Scenario II comprised nuclear and natural gas units.

---

[2] Two new nuclear reactors are now operating in Waynesboro, Georgia. Plant Vogtle units 3 and 4 have a capacity of 1,100 MW each and are expected to generate more than 17 million MWh of electricity, enough to power an estimated 1.6 million average American homes. The plants hold the distinction of producing the most expensive electricity of any plant of any type ever constructed [8,9,20].

Table 9.2 Three possible scenarios.

Scenario I	Scenario II	Scenario III	
Coal	Nuclear	Nuclear	
Nuclear			
Natural gas	Natural gas	Natural gas	Solar PV
			Wind onshore
			Wind offshore
			Battery storage

- Scenario III comprised nuclear, natural gas, renewable, and battery storage units.

With the addition of renewable and battery units in Scenario III, the capacity factor is reduced, which indicates excess reserve capacity. The levelized cost of electricity (LCOE) increases while grid reliability remains constant. The results indicate that Scenario I, using coal-fired units, produces 30 million metric tons of CO_2 per year. Scenario II with nuclear units instead of six coal-fired units produces much lower CO_2 emissions of 350,000 tons per year, a 98.8% reduction from Scenario I. Using renewable energy resources and BESS complements the NG-fired plants and further reduces CO_2 emissions to about 245,000 metric tons in Scenario III. The three scenarios demonstrate the fact that nuclear energy provides an effective pathway to achieving decarbonization and climate goals while maintaining energy security due to higher generation capacity.

9.2 Methodology

The methodology for this analysis involves an initial assessment of Scenario I followed by the nuclear-based Scenario II, and subsequently Scenario III with a mix of nuclear, renewables, and battery energy storage units. The performance, economics, reliability, and CO_2 emissions are evaluated for each of the three scenarios. We next investigate four key elements that are important for the techno-economic evaluation of power grids: performance, economics, reliability, and CO_2 emissions.

9.2.1 Performance analysis

The study begins with the assessment of a daily load curve data depicted in Chapter 8, Appendix A, Figure A8.1, and Tables B8.1 and B8.2. These data model a mid-sized utility company operating 17 conventional power generation units to meet consumer electricity demand. Using the Chapter 8, Appendix A, Table A8.1 data, we can calculate the plant performance parameters: daily electricity generation (*DEG*), average load (*AL*), load factor (*LF*), capacity factor (*CF*), and reserve factor (*RF*) using Equations 9.1–9.5 [14]:

$$DEG = \sum \text{Load}\ (i) * \Delta t\ (i) \tag{9.1}$$

$$\text{Average load}\ (AL) = DEG\ /\ 24\ \text{hours} \tag{9.2}$$

$$\text{Load factor}\ (LF) = \text{Average load / Peak load} \tag{9.3}$$

$$\text{Capacity factor}\ (CF) = \text{Average load / Rated capacity} \tag{9.4}$$

$$\text{Reserve factor}\ (RF) = \text{Load factor / Capacity factor} \tag{9.5}$$

Sample calculations for Scenario I are provided. Using the data from Chapter 8, Appendix A, Tables A8.1 and A8.2, we have the following metrics:

$$\begin{aligned} DEG &= [(3{,}888\ \text{MW} \times 1\ \text{hour}) + (3{,}696\ \text{MW} - 1\ \text{hour}) + \ldots \\ &\quad + (4{,}800\ \text{MW} - 1\ \text{hour}) \\ &= 119{,}232\ \text{MWh} \end{aligned}$$

$$\text{Average load}\ (AL) = 119{,}232\ \text{MWh} / 24\ \text{hours} = 4{,}968\ \text{MW}$$

$$\text{Load factor}\ (LF) = 4{,}968\ \text{MW} / 6{,}000\ \text{MW} = 0.83$$

$$\text{Capacity factor}\ (CF) = 4{,}968\ \text{MW} / 6{,}825\ \text{MW} = 0.73$$

$$\text{Reserve factor}\ (RF) = 0.83 / 0.73 = 1.14$$

If the rated capacity of the plant is equal to the peak load, then the capacity factor and load factor would be equal. The difference between the load factor and capacity factor is an indication of the reserve capacity. The difference between the operating capacity and the load at different hours is the spinning reserve (see Chapter 8, Appendix A, Figure A8.1). This is the buffer capacity that can be switched on immediately to match supply with demand when an increase in demand and/or supply disruption occurs.

Coal-fired plants are replaced by nuclear plants for Scenario II (see Appendix B, Table B9.3). The capacities of the nuclear units are assumed to be the same as those of coal-fired units to maintain the same capacity factors. Renewable energy resources and battery storage capabilities are integrated to create Scenario III, as shown in Table 9.2. With a capacity factor of 0.3, the renewable resource capacities are determined to match that of natural gas. This is because NG-fired units act as backup for solar and wind generation along with the batteries. The total capacity of 24 battery storage units is assumed to be equal to that of the NG-fired units (50 MW × 24 = 1,200 MW), which matches the NG electrical generation capacity of 1,175 MW.

9.2.2 Economic evaluation

It is important to assess the economic performance of an electric utility system. For this analysis, we calculate the variables capital cost (*CCOST*), operating and maintenance cost (*OMCOST*), fuel cost (*FCOST*), annualized cost (*ACOST*), and the levelized cost of electricity (*LCOE*) using Equations 9.5–9.10:

$$CCOST = \sum \text{[Rated capacity unit } (i) \text{ x Capital cost unit } (i)] \tag{9.6}$$

$$OMCOST = \sum \text{[Rated capacity unit } (i) \text{ x O\&M cost unit } (i)] \tag{9.7}$$

$$FCOST = \sum \text{[Generation unit } (i) \times \text{Fuel cost unit } (i)] \tag{9.8}$$

$$ACOST = CCOST \times [I \times (1 + \mathrm{I})^N/(1+\mathrm{I})^N - 1)] \tag{9.9}$$

$$LCOE = (ACOST + OMCOST + FCOST) / DEG \tag{9.10}$$

The interest rate (*I*) and the length of the project (*N*) were assumed to be 7% and 20 years for the sake of the analysis. Daily electric generation (*DEG*) is the average daily power output in kWh. The detailed cost data were selected from the EIA (see Appendix B, Table B9.6) and are presented for the generation units used in this study (see Appendix B, Tables B9.2–B9.4). The operations and maintenance (O&M) costs were obtained by multiplying the unit O&M cost with their unit rated capacities. However, the fuel costs incurred by the fossil fuel power systems depend on the power generated by each unit. The *LCOE* in units of $/kWh is calculated using Equation 9.10.

9.2.3 Reliability assessment

The objective of a reliability analysis is to derive suitable measures, criteria, and indices of reliable and dependable performance based on component outage data and system configuration [15]. For this study, reliability is defined as the probability that the system functions to generate full power [P(S)]. In other words, the performance measure of reliability is 1 when 100% of the demand is met, and 0 otherwise.

We assume that the energy system consists of several power blocks, where each block consists of one or more units. We let $P(B_d)$ denote the probability that Bock d operates successfully. The reliability factors for different blocks ($P(B_d)$, $d = 1, 2, 3...$) are approximated based on historic data and the availability of the fuel source for units in Block d [15–16].

For the case when the energy system consists of k power blocks and power is needed from each of these to meet the entire demand. The reliability is a multiplication function of the probability of each functioning k energy block. This is equivalent to a series with k blocks in series.

$$P(S) = P(B_1) \times P(B_2) \times P(B_3) \times .... \times P(B_k) \quad (9.11)$$

For the case when the system consists of k blocks and power is needed from any one of these to meet the entire demand, then…

$$P(S) = 1 - [1 - P(B_1)] \times [1 - P(B_2)] \times .... \times [1 - P(B_k)] \quad (9.12)$$

This is equivalent to a parallel system with k blocks in parallel [15–16].

For this study, the reliability assessment is performed for configurations including both series and series subsystems connected in parallel. These configurations approximately represent the physical arrangement of generation units. The reliability assessment for the following three scenarios is outlined in Appendix B, Tables B9.2–B9.4:

- Scenario I with coal, nuclear, and natural gas (17 units)
- Scenario II with nuclear and natural gas (17 units)
- Scenario III with nuclear, natural gas, renewables, and battery energy storage (61 units)

The individual power generation units are combined into seven blocks: coal, nuclear, natural gas, solar PV, onshore and offshore wind power, and battery energy storage (see Appendix B, Table B9.5). For instance, the coal block comprises six plants, each with multiple boilers, steam turbines, condensers, and pumps to ensure that they meet the electric demand when operational. The same is true with the nuclear block that comprises six nuclear plants, each with multiple reactors to ensure smooth operation during shutdown for refueling and maintenance. This same logic applies to the remaining blocks, including NG, solar, onshore and offshore wind power, and BESS. The approximate operational reliabilities are presented in Appendix B, Table B9.5. These reliability numbers are not intended to reflect actual values but are used for benchmarking the scenarios in this study. We assume the configurations in Figures 9.2–9.4 meet the demand by using different blocks.

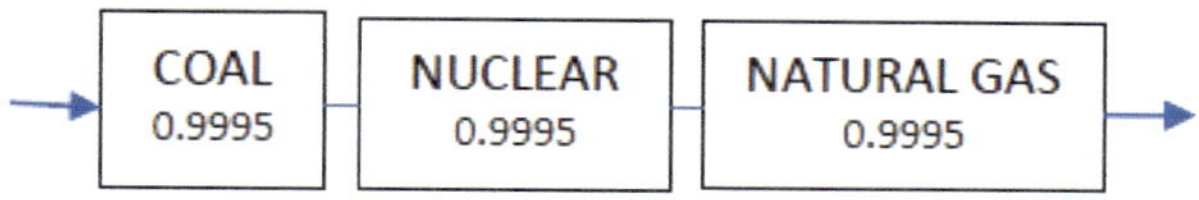

Figure 9.2 Reliability configuration of Scenario I with coal + nuclear + natural gas.

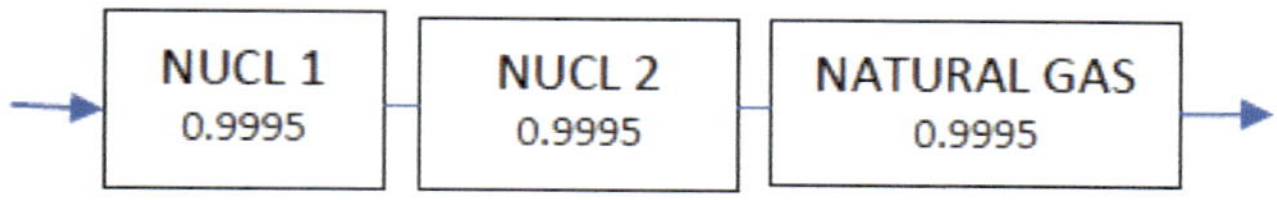

Figure 9.3 Reliability configuration of Scenario II with nuclear + natural gas without coal.

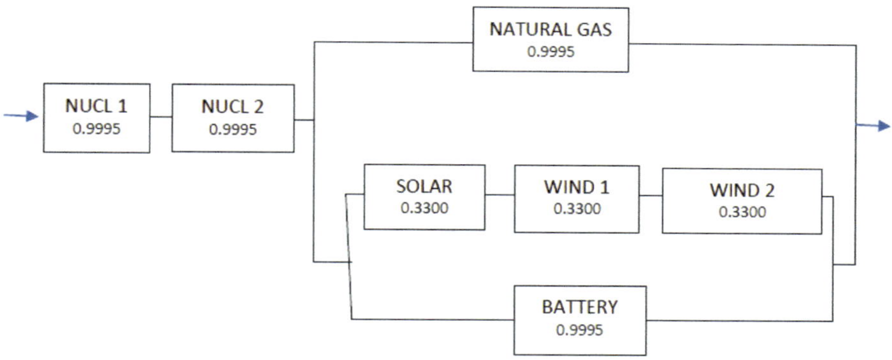

Figure 9.4 Reliability configuration of Scenario III with nuclear + natural gas + renewables + battery storage without coal.

9.2.4 CO_2 emissions

With these arrangements, we assume that Scenario III with both nuclear and renewables, along with battery energy storage, reduces the CO_2 emissions by about 50% from Scenario II; the overall capacity factor and reserve capacity of the system are increased. It is assumed that renewables will produce adequate electricity to offset the electricity produced by NG by approximately 30%. With the addition of batteries, it is assumed that renewables will provide an additional 20% of the electricity, for a total of 50% of the electricity that would have been produced by NG-fired units in Scenario II. This is designed to reduce the CO_2 emissions of Scenario II by 50% which achieves potential decarbonization goals.

9.3 Results and Discussion

In this section, we analyze the variables related to performance, economics, reliability, and CO_2 emissions for three scenarios, starting with the coal-based scenario. The cost data for conventional, renewable, and battery energy storage systems are provided in Appendix B, Tables B9.1–B9.5, which are adopted from the U.S. EIA (see Appendix B, Table B9.6). Scenario II involves a utility generation system with nuclear units replacing all six coal units. For Scenario III, we integrated three widely used renewable systems (solar PV, onshore and offshore wind power, and BESS units) in equal portions for the sake of demonstration. It is expected that the capacity of the NG-fired units, totaling 1,175 MW (750 MW + 425 MW) will be used as the basis to select the rated capacities of renewable energy systems for Scenario III. In doing so, the renewable systems and NG-fired units are coupled to complement each other with the objective of producing carbon-free electricity. We assume a hypothetical situation in which Scenario III produces electricity using nuclear power and renewable resources supported by battery storage, which provides near-zero CO_2 emissions.

This means that an individual electric load of 391.33 MW (1,175 MW/3 = 391.33 MW) will be met by each one the three renewable energy systems. Capacity factors used to evaluate the needed capacities for installation of the renewable energy systems are based on actual projects (27.8% from the Fowler Ridge Wind Farm and 30.2% from the Blythe Mesa Solar PV Project) [16–18]. On average, the capacity factor of 0.30 was chosen for the selection and sizing of solar PV, onshore wind, and offshore wind units whose rated capacities and data are obtained from the U.S. DOE (see Appendix B, Table B9.6). The needed capacities of the solar and wind are selected as follows:

$$\begin{aligned}\text{Rated capacity}_{\text{Solar PV}} &= \text{Average load / Capacity factor}\\ &= 391.33\text{ MW}/0.30 = 1{,}304\text{ MW}\end{aligned}$$

$$\begin{aligned}\text{Rated capacity}_{\text{Wind Onshore}} &= \text{Average load / Capacity factor}\\ &= 391.33\text{ MW}/0.30 = 1{,}304\text{ MW}\end{aligned}$$

$$\begin{aligned}\text{Rated capacity}_{\text{Wind Offshore}} &= \text{Average load / Capacity factor}\\ &= 391.33\text{ MW}/0.30 = 1{,}304\text{ MW}\end{aligned}$$

Based on these estimates, we required at least 1,350 MW of rated capacities of solar PV, onshore wind, and offshore wind to complement NG (see Appendix B, Table B9.4). The selected battery storage units are assumed to meet the electric demand equal to the capacity of the NG-fired units, which is 1,175 MW. We selected 24 50 MW battery storage units to meet the 1,175 MW (24 × 50 MW = 1,200 MW) demand that otherwise would have been met by the NG units. The idea behind the selection of renewable

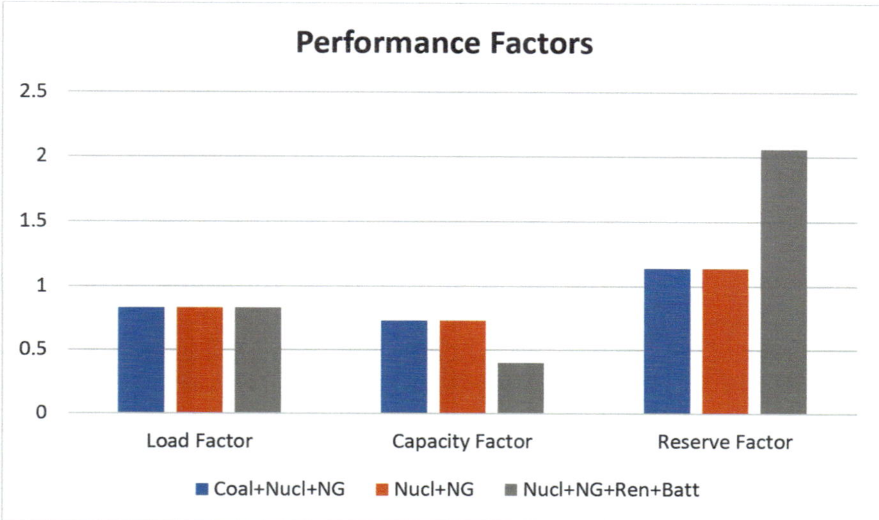

Figure 9.5 Overview of performance factors.

energy systems is to match the capacity of NG units. By doing so, the NG units would act as backup for the renewables while meeting intermediate and peak loads. In Scenario III with utility system III, we have assumed that renewables, along with supplemental battery energy storage, would meet the load of 1,175 MW, thus reducing the CO_2 emissions to nearly zero.

Appendix B, Tables B2–B4 represent three configurations starting with a total of 17 units for conventional systems, 17 units with nuclear systems, and 61 units with renewable energy systems plus battery storage. Appendix B, Table B9.5 provides estimates of operational reliability factors of units in both conventional and hybrid systems. These reliability factors are adjusted based on the availability of the fuel source. The results and discussion, including performance analysis, economic evaluation, reliability assessment, and CO_2 emissions of the three different configurations, are presented in Figures 9.5 through 9.13.

9.3.1 Performance analysis

The daily electricity generation (*DEG*), average load (*AL*), and the load factor (*LF*) remain the same for all three scenarios because the electric demand is the same. The capacity factor (*CF*) measures the utilization of the plant capacity. A plant with a capacity factor of 100% can produce electricity

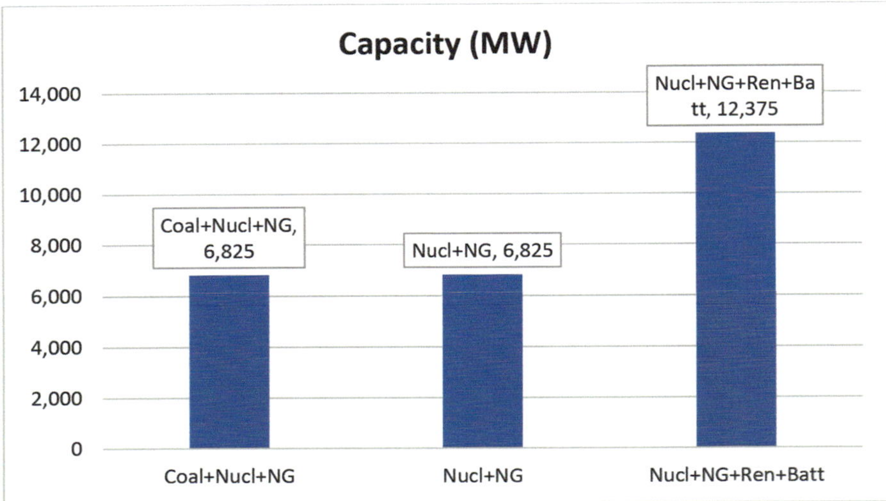

Figure 9.6 Increase in rated capacity.

continuously at its rated capacity. The capacity factors vary with increases in rated capacity. The reserve factor (*RF*) is a measure of buffer capacity that exceeds peak demand; *RF* increases with the addition of renewable energy and BESS capacities, resulting in increased reserve capacity (see Figure 9.5). The overall increase in capacity and number of assets is shown in Figures 9.6 and 9.7, respectively.

9.3.2 Economic evaluation

The capital cost of the alternatives increases with the addition of generation assets (see Figure 9.8). The *LCOE* is calculated based on the interest rate of 7% and the project life of 20 years for our system-level study. However, this warrants a detailed analysis of the *LCOE* for different interest rates and project lengths. The *LCOE* increases from \$0.08/kWh to \$0.11/kWh to \$0.15/kWh as shown in Figure 9.9. For an average monthly residential electricity consumption of 1,000 kWh, the monthly cost increases from \$82 to \$111 to \$153 per month, as shown in Figure 9.10.

9.3.3 Reliability assessment

Using the operational reliability factors in Appendix B, Table B9.5, the results of the reliability assessments are estimated and shown in Figures 11 and 12. The equipment reliability factor of 0.9995 was assumed for all units. Note

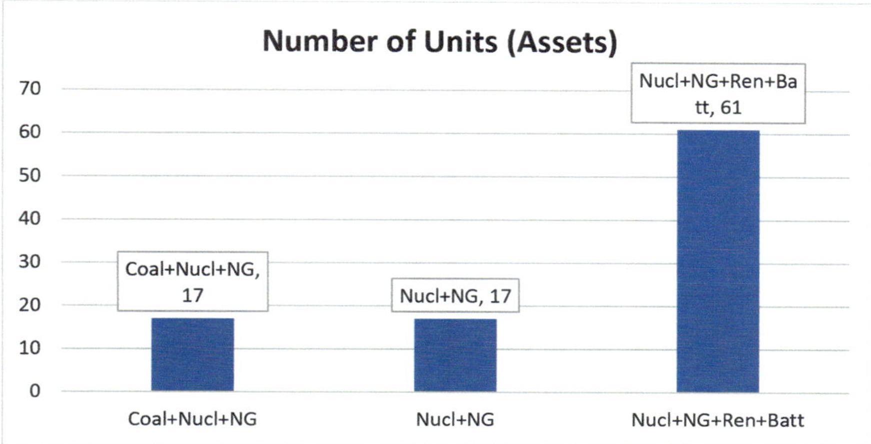

Figure 9.7 Increase in number of assets (units).

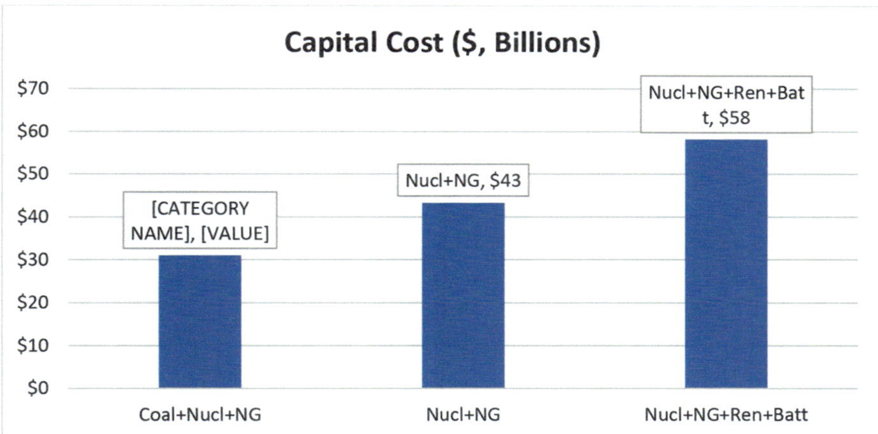

Figure 9.8 Capital costs (billion $).

that the value of 0.9995 implies a reasonable unplanned downtime of about 4 hours annually [(1.0 – .9995) × 8,760 = 4.38 hours/year]. The operational reliability of renewable energy systems is much lower than fossil-fuel and nuclear plants due to the intermittent generation of solar and wind resources.

The results show that the reliability of the coal-based Scenario I is 0.9985 (99.85%). This implies that the likelihood of a conventional power system generating power is 99.85%; the expected power outage time is 13.13 hours [= (1 – 0.9985) × 8,760 hours] annually. With nuclear power replacing coal,

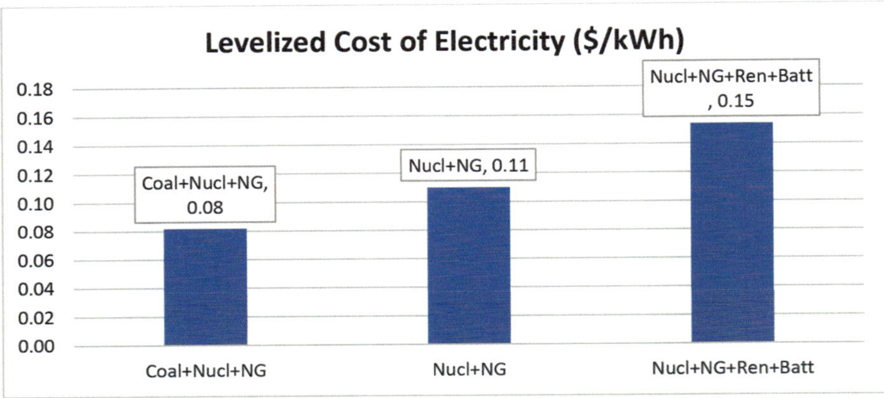

Figure 9.9 Levelized cost of electricity ($I = 7\%$; $N = 20$ years).

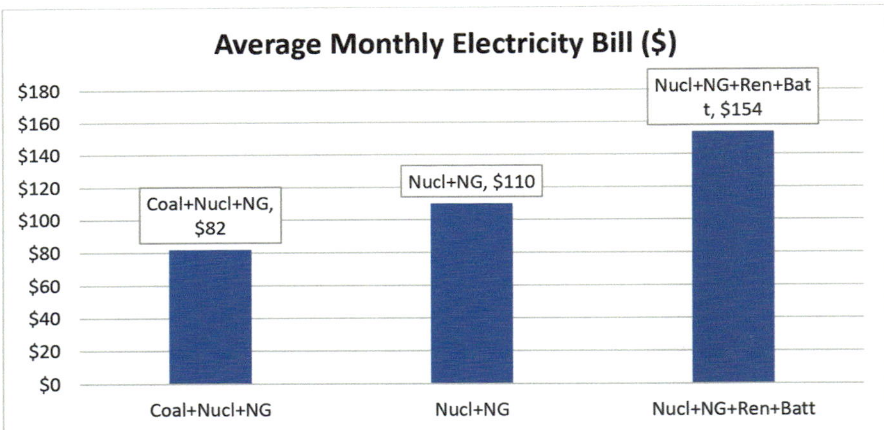

Figure 9.10 Average monthly home electricity bill ($).

the reliability in Scenario II remains at 0.9985 with the power outage time of 13.13 hours. In Scenario III, the addition of renewable energy systems and BESS units increases the operational reliability of the entire system to 0.9989 (99.89%) while the expected power outage time drops to 8.76 hours annually.

9.3.4 CO_2 emissions

Using the U.S. EIA standards of 2.30 lbs./kWh for coal and 0.97 lbs./kWh for natural gas, we calculated the CO_2 emissions for our three scenarios (see Table 9.3) [19]. With the integration of nuclear energy systems in Scenario II

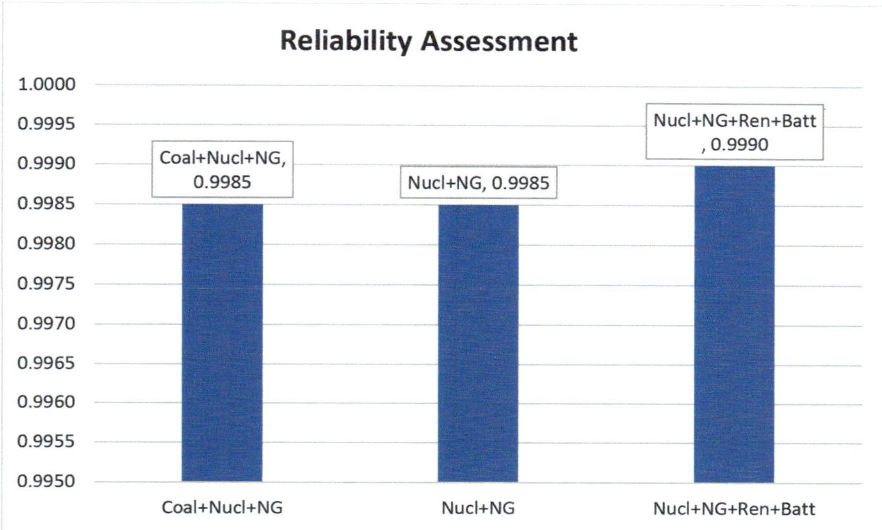

Figure 9.11 Reliability factors of three scenarios.

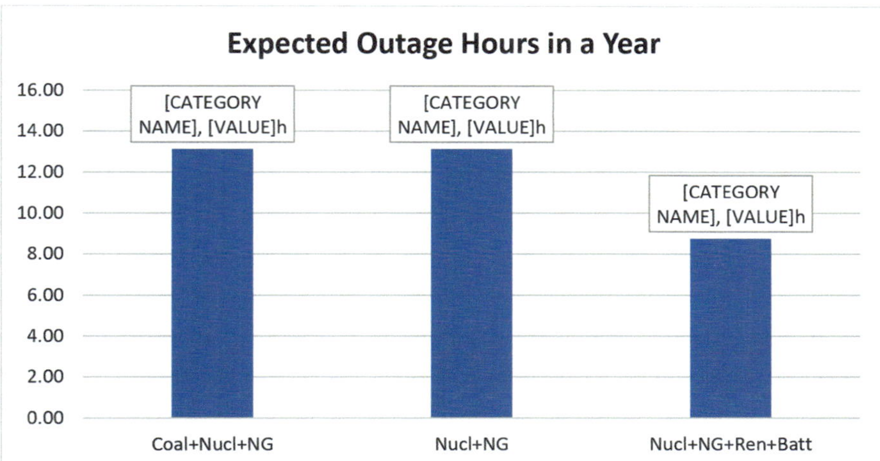

Figure 9.12 Expected annual outage hours for three scenarios.

for producing base load electricity, the CO_2 emissions can be reduced by 98.8% from 30.3 million metric tons of CO_2/year to approximately 0.35 metric tons/year (see Figure 9.13). Replacing coal with nuclear energy contributes to this decline in CO_2 emissions. With the integration of renewable energy systems and battery energy storage in Scenario III, CO_2 emissions are expected to decrease by about 50% to approximately 174,615 metric tons annually.

Table 9.3 Annual CO_2 production in metric tons for three scenarios.

Systems	CO_2 (lbs./kWh)	Daily Generation (MWh)	CO_2 million metric tons per year Scenario I	Scenario II	Scenario III
Coal	2.3	78,658	29,960,704	0	0
Natural gas	0.97	2174	349,231	349,231	174,615
Nuclear	0	38,400	0	0	0
Solar PV	0	0	0	0	0
Wind onshore	0	0	0	0	0
Wind offshore	0	0	0	0	0
Total		119,232	30,309,935	349,231	174,615

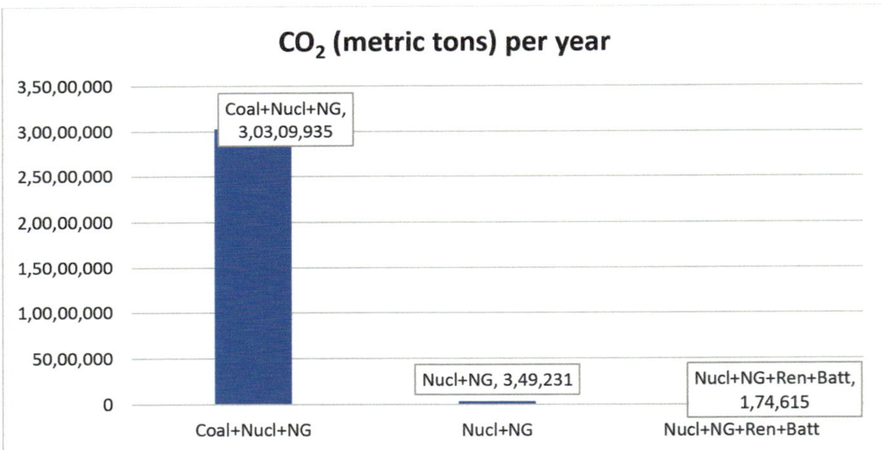

Figure 9.13 Impact of nuclear energy on CO_2 emissions.

It is assumed that renewables will produce enough electricity to approximately offset the electricity produced by natural gas by 30%. With the addition of BESS, it is assumed that renewables will provide another 20% of the needed electricity, a total of 50% of the electricity that would have been produced by the natural gas units in Scenario II. This reduces the CO_2 emissions of Scenario II by 50% leading to the potential of meeting or exceeding possible decarbonization goals. This is a probabilistic scenario that requires further investigation with real-time data obtained from a demonstration project for verification.

9.4 Conclusion

The main goal of our study was to better understand the integration of nuclear, renewables, and battery energy storage units with the help of a model utility system. To achieve this objective, this study involved assessing a conceptual utility system comprised of 17 conventional power generation units. The impact of three possible scenarios: Scenario I with coal, nuclear, and natural gas units; Scenario II with nuclear and natural gas units; and Scenario III with nuclear, natural gas, solar PV, onshore wind, offshore wind, and battery storage was examined.

It is clear from the results that the overall system capacity, the capital cost, levelized cost of electricity, and the average monthly electricity costs increase with the integration of nuclear, renewable energy systems, and batteries in Scenario III. The reliability of the system does not change significantly because nuclear units have reliability factors like coal and natural gas units. Replacing coal units with nuclear leads to a 99% reduction in CO_2 emissions, as evidenced in Scenario II. Furthermore, the addition of renewable energy systems and battery storage units in Scenario III leads to further reductions in CO_2 emissions, provided the renewables and natural gas units complement each other to meet at least 50% of the intermediate and peak demands.

Achieving sustainable development goals involves trade-offs. Nuclear energy, along with renewables and a BESS, can provide viable options for achieving decarbonization goals within the energy security framework, but challenges remain. Utility-scale nuclear energy projects carry associated financial risks that are mitigated when construction and financing costs are predictable—which is rarely the case [20]. There are challenges in implementing these solutions. Optimizing the potential electrical generation from the installation of a solar photovoltaic (PV) system provides intermittent generation because the production of electricity from a solar PV array varies due to factors such as the site location, weather conditions, and orientation [21]. Electrical generation from wind power systems adds to intermittency and requires management. Utility-scale BESS are declining in cost. Many electric utilities see them as a viable option for their operations. The challenges associated with variable forms of generation are being overcome, and we are seeing technological advancements that are overcoming their barriers.

Acknowledgments

The authors thank Professor Eckhard Groll and Professor James Jones from Purdue University School of Mechanical Engineering for supporting this research project.

References

[1] Biden-Harris administration finalizes suite of standards to reduce pollution from fossil fuel fired power plants. https://www.epa.gov/newsreleases/biden-harris-administration-finalizes-suite-standards-reduce-pollution-fossil-fuel, accessed 3 July 2024.

[2] Jacoby, K. (2009). Energy security: conceptualization of the International Energy Agency (IEA), in Brauch, H., et al. *Facing Global Environmental Change. Hexagon Series on Human and Environmental Security and Peace*, 4. Springer, Berlin, Heidelberg. https://doi.org/10.1007/978-3-540-68488-6_23, accessed 2 July 2024.

[3] Friedman, S. (2011). Three Mile Island, Chernobyl, and Fukushima: an analysis of traditional and new media coverage of nuclear accidents and radiation. *Bulletin of the Atomic Scientists*, 67(5), 55-65. https://doi.org/10.1177/0096340211421587.

[4] Munster, R. and Sylvest, C. (2015). Pro-nuclear environmentalism: Should we learn to stop worrying and love nuclear energy? *Technology and Culture*, 56 (4), 789–811. https://doi.org/10.1353/tech.2015.0107.

[5] Fact sheet: Biden-Harris administration announces new steps to bolster domestic nuclear industry and advance America's clean energy future. https://www.whitehouse.gov/briefing-room/statements-releases/2024/05/29/fact-sheet-biden-harris-administration-announces-new-steps-to-bolster-domestic-nuclear-industry-and-advance-americas-clean-energy- future/#:~:text=DOE's%20Advanced% 20Reactor%20Demonstration%20Program,BWX%20Technologies%2C%20and%20Southern%20Company, accessed 2 July 2024.

[6] Lawrence Livermore National Laboratory. Energy flow charts. https://flowcharts.llnl.gov/commodities/energy, accessed 2 July 2024.

[7] Tran, H., Juno, E. and Arunachalam, S. (2023). Emissions of wood palletization and bioenergy use in the United States. *Renewable Energy*, 219, Part 2, 119536, ISSN 0960 1481. https://doi.org/10.1016/j.renene.2023.119536.

[8] Associated Press (2023, August 17). Utilities begin loading radioactive fuel into a second new reactor at Georgia nuclear plant. https://www.usnews.com/news/business/articles/2023-08-17/utilities-begin-loading-radioactive-fuel-into-second-new-reactor-at-georgia-nuclear-plant, accessed 14 July 2023.

[9] Energy.gov (2019, April 18). 5 things you should know about Plant Vogtle. https://www.energy.gov/ne/articles/5-things-you-should-know-about-plant-vogtle, accessed 14 July 2023.

[10] Parker, G., Erciyes, B. and Kazan, O. (2019). Electric power industry: operational and public policy challenges and opportunities. *Production and Operations Management*, 28(11), 2,673–2,919.

[11] Boregowda, S. and Chand, S. (2022). Matching electricity supply with demand: integration of renewable energy systems into the U.S. power grid. *International Journal of Strategic Energy and Environmental Planning,* 4(6), 18–38.

[12] Li, K. and Priddy, P. (1985). *Power Plant System Design.* Wiley and Sons.

[13] U.S. Energy Information Administration. https://www.eia.gov/outlooks/aeo/assumptions/pdf/elec_cost_perf.pdf, accessed 25 June 2024.

[14] Nag, P. (2014). *Power Plant Engineering,* 4th edition. McGraw-Hill.

[15] Rao, S. (2015). *Reliability Engineering*, 1st edition. Pearson Higher Education.

[16] Al-Shaalan, A. Reliability evaluation of power systems. DOI: https://www.intechopen.com/chapters/66913, accessed July 3, 2024.

[17] Fowler Ridge wind farm. https://en.wikipedia.org/wiki/Fowler_Ridge_Wind_Farm accessed July 3, 2024.

[18] Blythe Mesa solar power project. https://en.wikipedia.org/wiki/Blythe_Mesa_Solar_Power_Project accessed July 3, 2024.

[19] U.S. Energy Information Administration. How much carbon dioxide is produced per kilowatt hour of U.S. electricity generation? https://www.eia.gov/tools/faqs/faq.php?id=74&t=11, accessed July 3, 2024.

[20] Roosa, S. (2024, September). Nuclear energy—Lessons from Georgia's new power plant. *Alternative Energy and Distributed Generation Journal*, 6(3), 5–12.

[21] Roosa, S. (2024, September). The variable output of solar PV arrays. *International Journal of Strategic Energy and Environmental Planning,* 6(5), 5–12.

Appendix B

Table B9.1 Cost data for conventional power generation units of the utility system.

Unit number	Unit name	Fuel type (*)	Daily generation (MWh)	Rated capacity (MW)	Capital cost ($/kW)	O&M cost ($/ kW-year)	Fuel cost ($/MWh)
1	Fbase1	C	15,417	800	$4,507	$45.68	$5.06
2	Fbase2	C	15,318	750	$4,507	$45.68	$5.06
3	Fbase3	C	16,080	750	$4,507	$45.68	$5.06
4	Fbase4	C	18,361	800	$4,507	$45.68	$5.06
5	Fcycle1	C	6,539	500	$4,507	$45.68	$5.06
6	Fcycle2	C	6,943	450	$4,507	$45.68	$5.06
7	Nuclear1	N	19,200	800	$7,406	$136.91	$2.67
8	Nuclear2	N	19,200	800	$7,406	$136.91	$2.67
9	Ccycle1	G	77	400	$1,176	$13.73	$2.10
10	Ccycle2	G	2,031	350	$1,176	$13.73	$2.10
11	Gas1	G	10	75	$867	$7.88	$5.06
12	Gas2	G	10	75	$867	$7.88	$5.06
13	Gas3	G	10	50	$867	$7.88	$5.06
14	Gas4	G	10	75	$867	$7.88	$5.06
15	Gas5	G	10	50	$867	$7.88	$5.06
16	Gas6	G	10	50	$867	$7.88	$5.06
17	Gas7	G	6	50	$867	$7.88	$5.06
Total			119,232	6,825			

*C = Coal; N = Nuclear; G = Natural Gas
Source: https://www.eia.gov/outlooks/aeo/assumptions/pdf/elec_cost_perf.pdf.

Table B9.2 Scenario I, coal, nuclear, and natural gas.

Systems	Quantity	Installed capacity (MW)	Capital cost ($/kW)	O&M cost ($/ kW-year)	Fuel cost ($/ MWh))	Daily generation (MWh)
Coal (1–4)	4	3,100	$4,507	$45.68	$5.06	119,232
Coal cycle (5–6)	2	950	$4,507	$45.68	$5.06	
Nuclear (7–8)	2	1,600	$7,406	$136.91	$2.67	
Natural Gas cycle (9–10)	2	750	$1,176	$13.73	$2.10	
Gas turbine (11–17)	7	425	$867	$7.88	$5.06	
Total	17	6,825				119,232

Table B9.3 Scenario II, using nuclear and natural gas.

Systems	Quantity	Installed capacity (MW)	Capital cost ($/kW)	O&M cost ($/ kW-yr.)	Fuel cost ($/ MWh)	Daily generation (MWh)
LWR (1–4)	4	3,100	$7,406	$136.91	$2.67	119,232
SMR (5–6)	2	950	$7,590	$106.92	$3.38	
LWR (7–8)	2	1,600	$7,406	$136.91	$2.67	
Natural Gas cycle (9–10)	2	750	$1,176	$13.73	$2.10	
Gas turbine (11–17)	7	425	$867	$7.88	$5.06	
Total	17	6,825				119,232

Abbreviations: LWR → Light Water Reactor
SMR → Small Modular Reactor

Table B9.4 Scenario III, using nuclear, natural gas, renewables, and battery storage.

Systems	Quantity	Installed capacity (MW)	Capital cost ($/kW)	O&M cost ($/ kW-year)	Fuel cost ($/ MWh)	Daily generation (MWh)
LWR (1-4)	4	3,100	$7,406	$136.91	$2.67	119,232
SMR (5-6)	2	950	$7,590	$106.92	$3.38	
Nuclear (7-8)	2	1,600	$7,406	$136.91	$2.67	
Natural Gas Cycle (9-10)	2	750	$1,176	$13.73	$2.10	
Gas Turbine (11-17)	7	425	$867	$7.88	$5.06	
Solar PV (18-27)	9	1,350	$1,448	$17.16	$0.00	
Wind Onshore (28-34)	7	1,400	$2,098	$29.64	$0.00	
Wind Offshore (35-37)	4	1,600	$5,338	$123.81	$0.00	
Battery Storage (38-61)	24	1,200	$1,270	$45.78	$0.00	
Total	61	12,375				119,232

Table B9.5 Approximate reliability factors of power generation units.

Blocks	Equipment reliability factor	Fuel availability	Operational reliability
Coal	0.9995	1.0000	0.9995
Nuclear	0.9995	1.0000	0.9995
Natural Gas	0.9995	1.0000	0.9995
Wind onshore	0.9995	0.3000	0.3000
Wind offshore	0.9995	0.3000	0.3000
Solar PV	0.9995	0.3000	0.3000
Battery	0.9995	1.0000	0.9995

Table B9.6 2023 cost estimates of coal, nuclear, natural gas, renewables, and battery storage units [13].

Technology	First available year[a]	Size (MW)	Lead time (years)	Base overnight cost[b] (2022$/kW)
Ultra-supercritical coal (USC)	2026	650	4	4,507
USC with 30% carbon capture and sequestration (CCS)	2026	650	4	5,577
USC with 90% CCS	2026	650	4	7,176
Combined-cycle – single-shaft	2025	418	3	1,330
Combined-cycle – multi-shaft	2025	1,083	3	1,176
Combined-cycle with 90% CCS	2025	377	3	3,019
Internal combustion engine	2024	21	2	2,240
Combustion turbine – aeroderivative[h]	2024	105	2	1,428
Combustion turbine – industrial frame	2024	237	2	867
Fuel cells	2025	10	3	6,771
Nuclear – light water reactor	2028	2,156	6	7,406
Nuclear – small modular reactor	2028	600	6	7,590
Distributed generation – base	2025	2	3	1,915
Distributed generation – peak	2024	1	2	2,300
Battery storage	2023	50	1	1,270
Biomass	2026	50	4	4,996
Geothermal[i,j]	2026	50	4	3,403
Conventional hydropower	2026	100	4	3,421
Wind	2025	200	3	2,098
Wind offshore[i]	2026	400	4	5,338
Solar thermal[i]	2025	115	3	8,732
Solar photovoltaic (PV) with tracking[e, i, k]	2024	150	2	1,448
Solar PV with storage[i, k]	2024	150	2	1,808

Data source: Sargent & Lundy, *Cost and Performance Estimates for New Utility-Scale Electric Power Generating Technologies*, December 2019; Hydroelectric: Oak Ridge National Lab, *An Assessment of Energy Potential at Non-Powered Dams in the United States*, 2012; and Idaho National Engineering and Environmental Laboratory, *Estimation of Economic Parameters of U.S. Hydropower Resources*, 2003;

Technological optimism factor	Total overnight cost[d,e] (2022$/kW)	Variable O&M[f] (2022$/MWh)	Fixed O&M (2022$/kW-y)	Heat rate[g] (Btu/kWh)
1.00	4,507	5.06	45.68	8,638
1.01	5,633	7.97	61.11	9,751
1.02	7,319	12.35	67.02	12,507
1.00	1,330	2.87	15.87	6,431
1.00	1,176	2.10	13.73	6,370
1.04	3,140	6.57	31.06	7,124
1.00	2,240	6.40	39.57	8,295
1.00	1,428	5.29	18.35	9,124
1.00	867	5.06	7.88	9,905
1.08	7,291	0.66	34.65	6,469
1.05	7,777	2.67	136.91	10,447
1.10	8,349	3.38	106.92	10,447
1.00	1,915	9.69	21.79	8,912
1.00	2,300	9.69	21.79	9,894
1.00	1,270	0.00	45.76	NA
1.00	4,998	5.44	141.50	13,500
1.00	3,403	1.31	153.98	8,881
1.00	3,421	1.57	47.06	NA
1.00	2,098	0.00	29.64	NA
1.25	6,672	0.00	123.81	NA
1.00	8,732	0.00	96.10	NA
1.00	1,448	0.00	17.16	NA
1.00	1,808	0.00	32.42	NA

Geothermal: National Renewable Energy Laboratory, *Updated U.S. Geothermal Supply Curve*, 2010.
Note: MW = megawatt, kW = kilowatt, MWh = megawatthour, kWy = kilowattyear, kWh = kilowatthour; Btu = British thermal unit.

[a] The first year that a new unit could become operational.

[b] Base cost includes project contingency costs.

10

Distributed Energy Resources and Demand Response: The Tension Between Reliability and Flexibility

Nicolette Sowa

Electric grids are trending toward the use of more renewable energy resources, which creates more intermittent electrical loads. Between 2010 and 2020, the growth of renewable energy resources outpaced all other forms of energy generation in the U.S., increasing by 42% [1]. Distributed energy resources (DERs) are on track to grow exponentially nationwide, with capacity expected to increase by 217 GW from 2024 to 2028 [2]. Renewable portfolio standards (RPS) and clean energy standards (CES) established in 31 states plus the District of Columbia are critical drivers of DER growth, with half of renewable energy generation in the U.S. attributed to states' renewable energy requirements [3].

Demand response is a valuable tool to mitigate unpredictable loads from renewables and utilize DERs to effectively shift loads. Demand response programs incentivize customers to reduce their energy usage during peak times of load and stress on the grid. However, there is an inherent tension that exists in asking DER resources to participate in demand response programs because they include customers with variable energy use needs that are often determined by external factors (heat, drought, unexpected business trends, etc.). DERs, such as solar panels, add the most value by responding to real-time flexibility to periods of peak energy consumption, while traditional demand response programs require load reduction commitments months in advance to ensure grid reliability.

DOI: 10.1201/9788743811312-10

In the evolving energy landscape, demand response will integrate aggregated DERs at a rapid pace, fueled by Federal Energy Regulatory Commission (FERC) Order 2222, initially issued in September 2020 [4]. Demand response programs must become more responsive to real-time signals while providing more accurate load reduction forecasts [5]. Energy providers, such as utilities, must be able to rely on demand response resources with certainty to address electrical system peak loads while also responding in real-time to usage spikes. This requires the flexibility to achieve as much load reduction as possible at any given moment. This chapter explores how demand response can expand and stack programs to resolve the tension between reliability and flexibility with an approach that includes: 1) program options; 2) policies that appropriately value DERs; 3) implementation and utilization of technology solutions; and 4) a focus on customer experience.

10.1 Introduction

Electric grid resources are trending toward greater use of variable renewable energy, which creates intermittent loads on electric grids. In the U.S., California increased its utility-scale wind and solar output by 29% between 2023 and 2024; New York has a goal of receiving 70% of its energy from renewables by 2030 [6,7]. DERs are on track to grow exponentially, fueled by clean energy standards (CES) and renewable portfolio standards (RPS) adopted by 31 states and the District of Columbia as of 2022 [8]. Electrical generation from renewable energy sources in the U.S. is poised to exceed natural gas-fired power generation by the mid-2030s, largely due to industry growth, economic viability, and furthered by states' CES and RPS goals [4,9]. These standards will help advance the implementation of DERs in the U.S. by requiring roughly 300 terawatt-hours (TWh) of added clean electricity to be brought online by 2030 and an additional 800 TWh by 2050 [10].

Global energy demand is expected to grow at least 25% between 2022 and 2040, with an increase of more than 50% in electricity demand largely due to electrification efforts, fast-growing technologies such as artificial intelligence (AI), and a greater need for cooling systems as Earth's average temperature continues to rise [8,11]. A record USD $1.8 trillion was invested globally in clean energy and distributed energy resources in 2023 [12]. Over 50% of the needed clean energy is likely to be generated by renewables (primarily solar photovoltaic and wind) by 2050 [13]. Over 500 GW of new renewable energy capacity is expected to be added globally from 2024 to 2030 [14].

Newer DER technologies, such as electric vehicles (EVs), battery energy storage systems, and smart devices, present both challenges and opportunities for utilities and energy providers. For example, the increased adoption of EVs strains the electric grid as more customers charge their vehicles from home. However, EVs can also provide ancillary services such as reserve control that contribute to system reliability as the use of renewable generation increases [5]. Integrated DER resources provide both intermittent loads that must be shaped and resources that can shape intermittent loads. Control—either direct or indirect—of these resources is crucial to maintaining grid stability.

Demand response is a change in the power consumption of electric utility customers to better match the demand for power with the supplies available. It involves supporting efforts to balance the grid by providing incentives to shift or shed electricity demand in wholesale and ancillary power markets [15]. Demand response programs are valuable tools used by utilities and energy providers to mitigate unpredictable loads from renewables and utilize DERs to effectively shape loads. By offering mechanisms by which events or behaviors that respond to load—actual or forecasted—are incentivized, the demand for electricity with the supply available on the grid can be optimized. However, there is an inherent tension that exists in asking DER owners to participate in demand response programs. Energy customers have variable loads that are often determined by external factors (heat, drought, unexpected business trends, etc.). They are often hesitant to commit to firm load reductions. The success of demand response relies on the actions of people and the use of technologies, not just power plants, to help meet the variable demand for electricity [16].

10.2 The Integration of Renewables and DERs

The *duck-curve* is an all-too-familiar graphic visualization of the challenges posed by renewables to grid operators (see Figure 1). The *belly* of the duck demonstrates the ever-increasing generation provided by solar resources, while the *neck* represents the ramp that occurs when the sun starts to set, and demand simultaneously increases [17]. Distributed battery storage systems, for example, will allow excess solar produced during the *belly* of the duck to be stored and then dispatched later in the day to reduce the steep ramp of the *neck*.

The duck shape is somewhat specific to California's experience due to factors such as weather patterns, policies, and programs. However, the lessons learned are universal:

- At times, intermittent energy resources will result in over-generation.

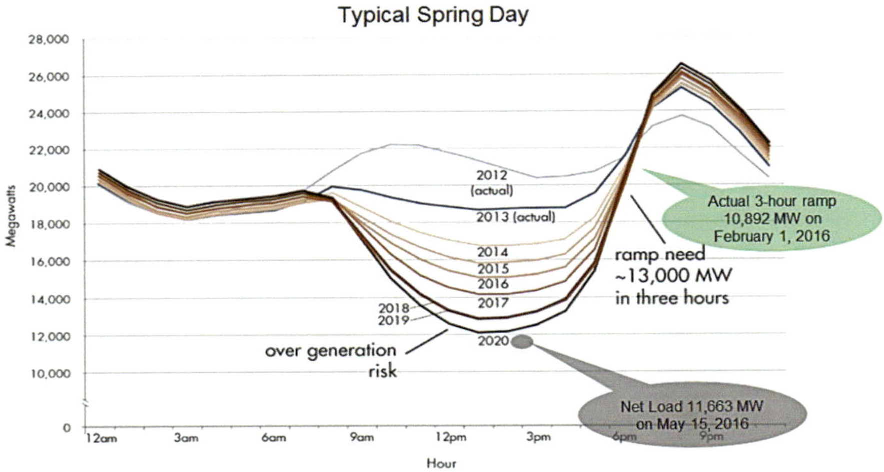

Figure 10.1 Duck curve showing steep ramping needs and overgeneration risk [14].

- Consumer behaviors can drastically affect the shape of a load curve.
- The share of renewable energy at any given time may be generally predictable, but may deviate from forecasts.

The economic and operational challenges of fluctuations in energy supply, imperfect forecasting, and uncertainty in resource adequacy, such as demand response, mean it is important to embrace technology and flexibility. They are the keys to electric grid reliability in the new energy landscape [18]. Integrated DERs add complexity to the management of demand, but that complexity also provides the answer to the challenges of variable and intermittent loads. A balance can be struck at saturation points; however, this poses burdens on the electric grid (see Figure 10.2). Therefore, it is important to create opportunities to engage DERs in advanced load management. For example, DERs can be enrolled directly or aggregated through third parties in a variety of demand response options ranging from event-based direct load control to long-term behavioral incentives that shift load. Demand response programs can be designed for specific customers, technologies, and to satisfy localized peak demand.

10.3 Power Moves: Demand Response

Demand response is more than just a mechanism to reduce peak load. The modern concept of demand response introduced by Lawrence Berkeley

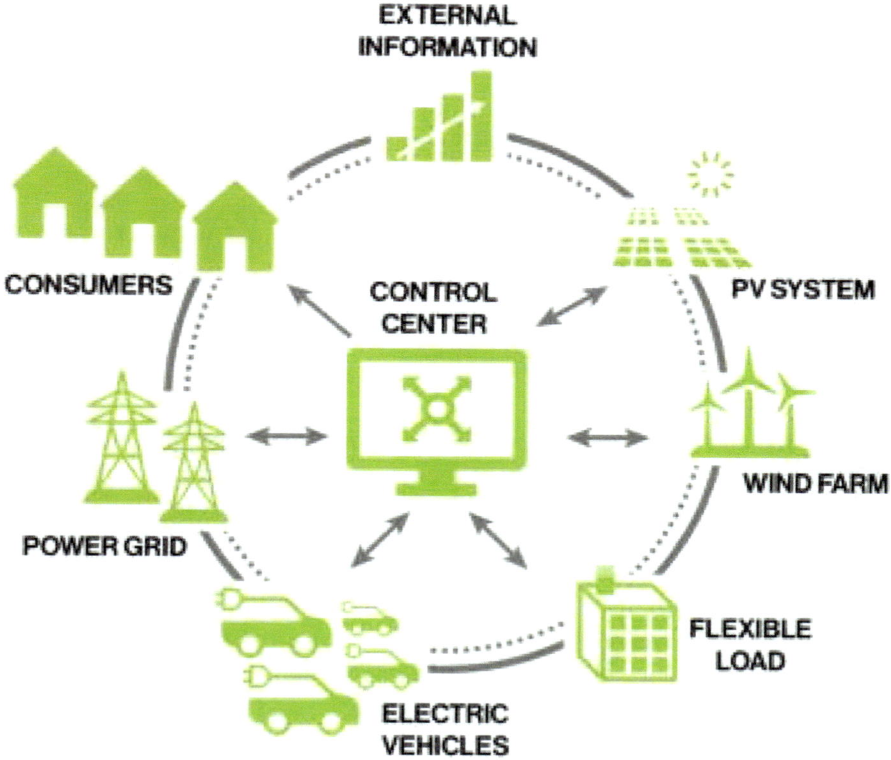

Figure 10.2 The integration of renewables and DERs [19].

National Laboratory in 2017, provides shape, shift, shed, and shimmy, while capturing the dynamic and diverse benefits demand response offers for both long- and short-term energy solutions [20]. DERs have a unique ability to help demand response satisfy its potential. To level up and maximize DER participation, demand response programs must become more responsive to real-time signals while providing more accurate load reduction forecasts [5]. Energy providers need to be able to rely on demand response resources with certainty while addressing unexpected spikes in peak load. This process requires the flexibility to achieve as much load reduction as possible at any given moment.

Demand response has the potential to expand and stack programs to resolve the tension between reliability and flexibility with an approach that includes: 1) program options, 2) policies that appropriately value DERs, 3) implementation and utilization of technology solutions, and 4) a focus on customer experience. Each facet of this approach overlaps with and supports the others, generating a feedback loop capable of responding at a pace that will match the continued evolution of distributed resources.

10.3.1 Program options

The basis of program options is the recognition that demand response resources are also customers. Some customers may participate for one-time rebates, others for consistent incentives, and others to simply contribute to a clean energy future. Though they bring various behaviors and types of loads that can be challenging to control, they are all valuable in responding to demand events and trends.

At the core of demand response programs are the concepts of dispatchable (event) vs. non dispatchable (time-of-use) loads with the independent system operator (ISO) on the supply-side vs. the load serving entity (LSE) that triggers load modifications [16]. This is a great place to start. However, to avoid getting trapped within the traditional structures of these concepts, and miss the opportunity to harness the power of DERs, the next step is to align program options with the four primary benefits of demand response: 1) peak energy demand shaving, 2) flattening increasingly steep system ramps, 3) firming intermittent energy resources, and 4) relieving network congestion stress [16].

With these benefits as a guide, program times and incentives can be evaluated that work for DER customers while benefiting the grid operators. For example, asking a customer with solar generation capabilities to respond to a peak-load event in the early evening is not consistent with its intended uses to either power the home or charge batteries. Incentives must be tiered to compensate customers according to factors beyond load reductions. These include event response time or load shifting that more permanently contribute to grid stability, both of which DERs are well-equipped to provide. While a plethora of options can appeal to a wider range of customers, it is notable that prioritizing simplicity and usefulness in program design will help drive engagement and add the most value. Hurdles to enrollment and earning incentives should be minimized to ensure program success.

DER participants who have more consistent loads and consistently available load reduction options can be part of long-term resource planning, while customers who can provide short-term notice of load reduction capabilities and short-term load reduction can help prevent outages with real-time response during peak events. Demand response programs are far more cost-effective than building, or even black-starting, a peak-load power station.

Creating a range of programs that align with DER capabilities and providing valuable incentives for participation can reduce the friction between reliability and intermittent generation.

10.3.2 Policies

In the evolving energy landscape, demand response will rapidly integrate with aggregated DERs. This growth is supported by FERC Order 2222, which builds upon Order Nos. 741 and 841, and allows DERs to be aggregated and participate as resources in markets [4,21]. State-level policies that allow third-party aggregation of DERs and provide direct incentives or rebates for DER implementation and/or demand response enrollment increase the available grid capacity. When valued appropriately, with transparent values assigned to the span of DER load types and use cases, owners of demand response resources receive incentives that increase their profitability and support the growth of businesses advancing a variety of clean energy initiatives. Energy providers, demand response participants, customers, the economy, and the environment will benefit. Integrating DERs entails policies that encourage the deployment of demand response resources to better manage electricity demand and harmonize intermittent energy resources [8].

As DERs become more widely available, policies that emphasize regional coordination need to support integration into demand response. Regional refers to both a region larger than the ISO and the localized balancing of resources. Coordinating more closely with the broader regions served by the energy provider—including states receiving imports or providing exports—provides opportunities to mitigate large-scale emergencies, such as natural disasters or extreme climate events. Alternatively, empowering local communities to manage DERs helps alleviate transmission congestion, creates grid resiliency, and supports optimal community permitting and siting regulations.

Like program options, future policy design must account for the variety of DER customer- and load-types available and the added value. Including demand response resources in long-term procurement and integrated resource plans is important, but it meets limited needs and fails to incentivize customers to reduce load further during peak events or flatten intermittent ramping. This flexibility becomes increasingly critical as tomorrow's electric grids become progressively dominated by variable power generation such as wind and solar PV, and electric grid planning and resource adequacy calculations become more nuanced [5,15].

Appropriate policies can provide relief to the tension between reliability and flexibility by advancing initiatives that do not limit the participation of DERs in demand response programs. It is important to design policies and programs that recognize the full spectrum of value available within DERs that improve grid reliability. Policies must be coupled with technologies to ensure that the flexible grid services DERs provide—such as exports and real-time responses—can be realized expeditiously.

10.3.3 Technology

DERs require automated demand response and advanced communications technologies. While utilities and energy providers do not necessarily need to provide the integration technology for DERs to participate in demand response programs, independent system operators and utilities must invest in enterprise platforms that will provide the widest range of availability in terms of community and technology partnerships [19]. In other words, the challenge is not how to predict the future of technology advancements, but how to construct a foundation that adapts to them. This challenge spurs innovation and presents business opportunities for technology providers that benefit customers. Energy providers become the orchestrators of a network of DER platforms that seamlessly integrate workable technologies into an integrated energy system and broader infrastructure platform [19].

Energy storage solutions have also become paramount to successfully balancing the fluctuating nature of variable renewable electricity and providing flexibility to the electric grid [23]. In 2024, in California, over 3,400 GWh of utility-scale solar and wind energy was added to the electric grid—an increase of 29% compared to 2023 [6]. Coupled with solar and wind power, energy storage enables the incorporation and expansion of cost-effective sources of electricity generation [23]. Flexible energy technologies and platforms enable intermittent resources to provide more flexible energy services. The utilization of the rapid response capabilities of energy storage systems will provide greater reliability [12].

Several components are important to ensure long-term success as technological solutions continue to evolve. Examples include advanced communications for real-time energy management, robust data management tools, and user-friendly operational portals that support ease of enrollment and participation. Network model management supports advanced bi-directional data communication and analysis while allowing long-term and real-time monitoring and signaling of grid conditions [22]. With these and other technology tools, utility companies and energy providers can scale from control to collaboration, encouraging flexible program options and participation by tracking where and how devices connect within the system—and how grid performance is impacted [22].

10.3.4 Customer experience

DERs are owned, generated, and consumed by customers. The customer experience is paramount to engagement in demand response. The way a program engages customers determines the success of its impact on the electric grid. Whether the customers are residential or commercial, direct-enrolled or enrolled through a third-party aggregator, the goal is to make enrollment and participation easy and the incentives rewarding. Customization of the customer experience can vary widely. Commercial customers in New York City have very different load profiles and energy needs than agricultural customers in rural Oregon, even if they utilize similar energy storage systems.

Energy as a service requires a fundamental shift from considering energy users as utility ratepayers to the concept of energy managers and partners [24]. Value can be added in the form of new programs, policies, and technologies, which create choices and flexibility for customers [24]. As demand response evolves along with energy as a service and DER integration, the focus on the customer experience will result in people-driven solutions that can help program participants adapt to the diverse changes affecting the energy landscape and realize the full potential of clean energy [16].

With technology and policy advancements, energy customers have more information and options available to them than ever before. They are not just consumers of electricity with technological solutions that enable engagement—they are active participants in managing the energy they consume. With the right combination of information and tools, consumers can become powerful partners with energy providers in balancing real-time electrical system reliability with future planning and asset management.

10.4 Conclusions

As the energy transition accelerates, the convergence of distributed energy resources (DERs), real-time demand response, and customer-driven flexibility is redefining how grid reliability is achieved. Utilities and energy providers grapple with incorporating an unparalleled influx of intermittent energy resources while simultaneously preparing for the retirement of thousands of megawatts of conventional power stations [16]. The new energy landscape requires active participation from customers with both slow- and real-time distributed energy resources. It offers transparent incentives for the value provided to the electric grid in times of peak load as a long-term solution to shift loads.

DERs now represent not just sources of load or generation, but flexible, intelligent assets that can be optimized through policies, technologies, and customer engagement. Modern demand response must move beyond legacy models and operate as a dynamic, integrated system that can respond in real-time to evolving grid needs. Creating participation pathways that consider program design, refining policy frameworks to value DER's role in peak load management, deploying scalable and interoperable technology tools, and prioritizing the customer experience are evolving processes. The flexibility energy providers build into these pathways will create the reliability required to optimize DERs.

The inherent variability of DERs provides opportunities to create flexible, customized solutions that will benefit customers and better manage loads [5]. FERC Order 2222, maturing technologies like AI-optimized load control, and the proliferation of electric vehicles and batteries have opened the door to bidirectional energy relationships. Customers are becoming prosumers—proactive participants in grid stability. As intermittent renewable energy becomes dominant, reliability will no longer be defined by dispatchable peak power plants alone but by the ability to leverage DERs to flex, respond, and adapt.

Ultimately, resolving the tension between reliability and flexibility depends on how well we align incentives, technologies, and human behaviors. The grid of the future is not a single infrastructure—it is a collective, real-time partnership between utilities, customers, and technologies. Demand response, when designed with inclusivity and responsiveness in mind, will be the linchpin that transforms DER complexity into an asset—delivering not just stability, but resilience and equity in a clean energy future.

References

[1] Center for Climate and Energy Solutions (2023, August 27). Renewable energy. https://www.c2es.org/content/renewable-energy/, accessed 12 August 2025.

[2] Martucci, B. (2024, July 3). U.S. to add 217 GW of distributed energy resource capacity through 2028, Wood Mackenzie projects. UtilityDive. https://www.utilitydive.com/news/wood-mackenzie sees-217-gw-new-distributed-energy-resources-2028/720581, accessed 15 August 2025.

[3] Energy Information Administration (2022, February 1). Five states updated or adopted new clean energy standards in 2021.

[4] Federal Energy Regulatory Commission (2020, October 21). 18 CFR Part 35. Docket no. RM18-9- 000; Order no. 2222. Participation of distributed energy resource aggregations in markets operated by regional transmission organizations and independent system operators.
[5] Trabish, K. (2020, August 24). Ensuring DER inclusion in capacity markets may require a rethink of resource adequacy. https://www.utilitydive.com/news/ensuring-der-inclusion-in-capacity-markets may-require-a-rethink-of-resourc/583590/, accessed 13 December 2020.
[6] Energy Information Administration (2025, May 28). Solar and wind power curtailments are increasing in California. https://www.eia.gov/todayinenergy/detail.php?id=65364#: %2C%20CAISO%20curtailed%203.4,up%20in%20the%20evening%20hours, accessed 13 August 2025.
[7] New York State Energy Research and Development Authority. Renewable energy. https://www.nyserda.ny.gov/Impact-Renewable Energy#:~:text=Leading%20New%20York's%20Clean%20Energy%20Transition&text=%2C%20 which%20sets%20nation%2D-leading%20goals,emission%20electric%20grid%20by%202040, accessed 13 August 2025.
[8] World Nuclear Association (2025, January 6). World energy needs and nuclear power. https://world-nuclear.org/information-library/current-and-future-generation/world-energy-needs-and nuclear- power#:~:text=The%20world%20will%20need%20significantly,half%20between%202022%20and%202040, accessed 13 August 2025.
[9] RPCM Team (2025, May 15). Rethinking energy demand through 2050: review of the EIA annual energy outlook 2025. RBAC Inc. https://rbac.com/rethinking-energy-demand-through-2050-review-of-the-eia-annual-energy-outlook-%20in%20natural,sees%20a%20%2D1.7%25%20CAGR, accessed 13 August 2025.
[10] Barbose, G. (2023, June). U.S. state renewables portfolio and clean electricity standards: 2023 status update. Berkeley Lab. https://emp.lbl.gov/publications/us-state-renewables portfolio-clean, accessed 13 August 2025.
[11] Wiatros-Motyka, M., Fulghum, N. and Jones, D. (2024, May). Global electricity review 2024. https://ember-energy.org/app/uploads/2024/05/Report-Global-Electricity-Review-2024.pdf, accessed 13 August 2025.
[12] International Energy Agency (2023). Net zero roadmap: a global pathway to keep the 1.5 °C goal in reach. https://www.iea.org/reports/net-zero-roadmap-a-global-pathway-to-keep-the-15-0c-goal-in reach, License: CC BY 4.0, accessed August 13, 2025.

[13] International Renewable Energy Agency (2017, October 19). Electricity storage and renewables: costs and markets to 2030. https://www.climateaction.org/white-papers/IRENA-electricity-storageand-renewables-costs-and-markets-to-2030, accessed 13 August 2025.
[14] International Energy Agency (2024, October 9). Massive global growth of renewables to 2030 is set to match entire power capacity of major economies today, moving world closer to tripling goal. https://www.iea.org/news/massive-global-growth-of-renewables-to-2030-is-set-to-match-entirepower-capacity-of-major-economies-today-moving-world-closer-to-tripling-goal, accessed August 13, 2025.
[15] International Energy Agency (2023, July 11). Demand response. https://www.iea.org/energy-system/energy-efficiency-and-demand/demand-response, accessed 14 August 2025.
[16] Panfil, M. and Fine, J. (2015). Putting demand response to work for California. Environmental Defense Fund. https://www.edf.org/sites/default/files/demand-response-california.pdf, accessed 13 December 2020.
[17] California ISO. Fast facts. https://www.caiso.com/documents/flexibleresourceshelprenewables_ fastfacts.pdf, accessed 13 December 2020.
[18] Sustainability Directory (2025, April 10). Why is intermittency a challenge for renewables? https://energy.sustainability-directory.com/question/why-is-intermittency-a-challenge-for renewables/#:~:-text=The%20economics%20of%20intermittency%20also,costs%20and%20 lower%20electricity%20prices, accessed 13 August 2025.
[19] Asmus, P., Metz, A., Gunjan, P., Labastida, R., Shepard, S. and Wood, E. (2020). Guidehouse Insights. Integrated DER: orchestrating the grid's last mile. https://guidehouse.com/- /media/www/site/insights/energy/2020/gh_integrated-der_2020.pdf, accessed 13 August 2025.
[20] Alstone, P., Potter, J., Piette, A. et al. (2017, March). 2025 California demand response potential study: charting California's demand response future. Final report on phase 2 results. https://buildings.lbl.gov/publications/2025-california-demand-response, accessed 13 August 2025.
[21] Federal Energy Regulatory Commission (2020, June 19). Demand response. https://www.ferc.gov/industries-data/electric/power-sales-and-markets/demand-response, accessed 13 December 2020.
[22] Jacquemin, S. (2025, July 24). Network model management: collaboration in and out of the control room. Factor this power engineering. https://www.renewableenergyworld.com/power-grid/smart-grids/

network-model-management- collaboration-in-and-out-of-the-control-room, accessed 14 August 2025.

[23] American Clean Power (2021, December 1). Clean energy storage facts. https://cleanpower.org/facts/clean-energy-storage/#:~:text=Energy%20 storage%20is%20the%20only, later%20when%20it%20is%20needed, accessed 14 August 2025.

[24] Leksono, E., Mardiatna, I., Prihatmanto, A., Budiyono, A., Joelianto, E. and Sunaryo, A. (2023). Energy as a service (EaaS): a paradigm shift in energy management. *The Journal of Instrumentation, Automation & Systems*, 10(3).

11

Initial Capital Cost Estimate for Clean Hydrogen Electricity Produced by a 500 MWh Baseload Plant

Adam Gribb, Marianne Ramos, and Samantha Gribb

In 2023, the U.S. federal government awarded $7 billion to develop regional clean hydrogen (CH) hubs; several U.S. states began allowing investor-owned utilities to operate CH projects. For example, the U.S. state of Colorado has laws that define CH projects as those ventures using specific types of hydrogen-producing equipment. These include electrolyzers, dedicated renewable energy resources for electrolysis, upgraded turbine-generators at an electric generating station that uses hydrogen as an energy source, plus purchasing clean hydrogen from third parties. This chapter develops an initial (systemwide) capital cost estimate for a 500 MWh CH-combusting facility that continually supplies electricity to the grid (i.e., baseload generation plant).

The method used in the estimate is based on an initial energy balance of a new 500 MW electric generation system. The energy source for electrolyzers comes from renewable energy generation comprised of 50% wind and 50% solar generation, adequately sized transmission, and storage components to provide continuous generation. An appropriately sized storage system is assumed to have 10 units—one unit in operation, six units charged and ready, two units charging, and one spare unit. Given these assumptions, a preliminary capital cost estimate was developed based on recently published wind, solar, transmission, and energy storage costs.

DOI: 10.1201/9788743811312-11

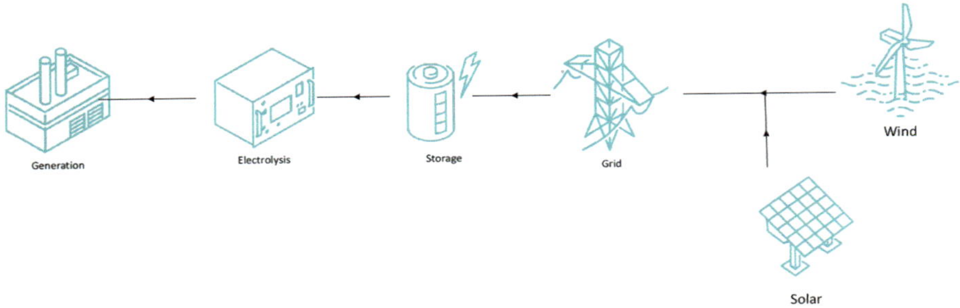

Figure 11.1 Green hydrogen system.[1]

11.1 Introduction

Spinning wind turbines and sunlight can excite electrons into generating electrical energy. At a local interconnection point, the electricity starts flowing onto the grid. In doing so, some of it is transformed into heat energy and lost due to voltage changes and overcoming the resistance in transmission lines. As the electricity flows into a battery energy storage system (BESS), it continues to lose more energy as it is converted into chemical energy. During electrolysis, the electrons plunge into water, cleaving hydrogen (H_2) atoms from water molecules and transforming their electrical energy into chemical energy. These H_2 atoms form molecules, which are collected and shipped to an electric generating facility.

At the generator (e.g., combined cycle with multiple shafts), the hydrogen atoms ignite in the air and release nearly all their chemical energy as heat. Most of this heat boils water, making steam, and increasing the kinetic energy of the combustion gases. The steam and combustion gases provide the pressure to drive the turbines, converting some of their kinetic energy into mechanical energy. These spinning turbines excite new electrons to make electricity by converting mechanical energy back into electrical energy. This arrangement is represented in Figure 11.1.

The goals of this chapter are:

- Define the scope of a theoretical electrical system that converts renewable energy into green hydrogen, which can be used to fuel baseload electric generation to serve consumer loads.
- Identify the major system components of the theoretical system, including generation, transmission, and energy storage.

[1] Icons are from Enapter Energy Icon set. https://handbook.enapter.com/icons.html.

- Develop cost estimates to install the theoretical system. Assume only wind and sunlight are used as the energy resources to generate electricity to create green hydrogen.

11.2 Methodology

To achieve the stated objectives, five analysis phases were developed. Phase I defines the system scope, its components, and identifies each component's standard efficiencies. Phase II identifies and discusses informational limitations for each of the identified components. Phase III documents the required definitions and assumptions that are incorporated into the description of each of the system's identified components. Phase IV determines the assumptions for energy requirements and standard costs associated with each element in the theoretical electrical system. Phase V provides an example of an initial cost estimate for a base case and selected alternative cases.

11.2.1 Phase I: Scope definition

Define the scope of a theoretical electrical system, each of the system's components, and each component's standard efficiencies.

Table 11.1 Flow of energy for system components and efficiencies for clean hydrogen generation.

Component	Energy	Efficiency	Loss	Capacity factor[2]
Renewable energy generation,wind	Kinetic, mechanical, and electrical			42% [1]
Renewable energy generation,solar	Heat and electrical			28% [2]
High voltage transmission	Electrical	95% [3]	5%	
Industrial storage	Chemical, electrical.	80% [4]	20%	
Electrolysis to produce CH	Electrical, chemical.	64% [5]	36%	
Dispatchable electric generation burning CH	Chemical, heat, kinetic, mechanical, and electric.	60% [6]	40%	

[2] Because there are times when renewable generation will be offline, capacity factor is defined as the percentage of time when the resource is available.

11.2.2 Phase II: Identify the available information for each system component

The authors' research determined that much of the referenced efficiency and costing information for several of the CH system components was of limited value. Many of the energy transformation efficiencies were qualified as *rules of thumb* rather than based on recent scientific findings. They lacked reliable documentation, and the references provided relied on similar information sources. Another concern was the conflicting and optimistic cost estimates for turnkey generation systems; such costs are questionable. Finally, procurement and installation costs for renewable generation and BESS equipment will likely be lower in the future than they are today, based on the cost savings reaped from innovation and economies of scale.

To provide the most accurate and up-to-date information, the authors selected data from the U.S. Energy Information Administration (EIA) and National Renewable Energy Laboratory (NREL) when possible. The authors used such sources to address costing issues, adjusted costs for inflation, and averaged these values into the consensus costs. Other sources that were used include the U.S. Geological Survey, Xcel Energy, and physics coursework from Stanford University.

11.2.3 Phase III: Definitions and assumptions associated with each component

Colorado has unambiguous legal definitions for *green hydrogen, eligible energy resources*, and *renewable energy resources*. We believed it was prudent to use these more precise legal definitions to determine what types of initial renewable energy would be used to create green hydrogen as a fuel source for dispatchable base electricity generation. The following are quoted directly from the Colorado Revised Statutes:

- Green hydrogen is "...hydrogen derived from a clean energy resource... that uses water as the source of the hydrogen."[3]
- Eligible energy resources are "...recycled energy, renewable energy resources, and renewable energy storage. Fossil and nuclear fuels and their derivatives are not eligible energy resources." [4]

[3] CRS §40-3.2-108.

[4] CRS §40-2-125.

Renewable energy resources are "...solar, wind, geothermal, biomass that is greenhouse gas (GHG) neutral, new hydroelectricity with a nameplate rating of 10 megawatts (MW) or less, and hydroelectricity in existence on 1 January 2005, with a nameplate rating of 30 MW or less ...".[5] Given these definitions, the following assumptions were adopted to provide definition of the process and its components:

- Baseload generation is designed to be dispatchable and operates continuously as required to meet base demand.
- Natural gas-fired and H_2-fired generation equipment have the same cost ($/kWh) and efficiencies.
- The 30-year average inflation rate (excluding 2021 and 2022) for non-residential buildings is 4.2% [7].
- An electrolyzer made in Europe or the U.S. costs $2,500/kW [8].
- Annual renewable energy generation is 50% wind and 50% solar.
- Accuracy of planning estimate—no excess renewable energy paid for by ratepayers is available for off-system sales.
- A typical BESS for industrial energy storage is 80% efficient with a 20% to 80% operating range. The equipment scoping requirement is 10 days of energy storage (two winter storm Uri): one unit is discharging, two units are charging, five units are always ready, two units are prepared in case of weather conditions, and one unit is a true spare [9].

11.2.4 Phase IV: Assumed energy requirements and standard cost for each system component

The first law of thermodynamics states that *energy cannot be created or destroyed.* When energy is transformed from one state to another (e.g., from wind, which is mechanical rotation, to electricity, which is electrons flowing through wires or from electrical current to energy stored in H_2 gas), a percentage of its power is lost due to system inefficiencies or transformed into heat. This loss in resultant power is often substantial. For example, combined cycle natural gas (NG) and H_2 generating facilities with multiple shafts are

[5] CRS §40-2-124.

Table 11.2 System losses and requirements associated with generating 500 MWh from a green hydrogen fuel source (in MWh).

	Green hydrogen	Electrolysis	Storage	Transmission
Name plate (MWh)	500	833	1,302	1,628
Efficiency	60%	64%	80%	95%
MWh in	833	1,302	1,628	1,713

	Wind	Solar
MWh out	857	857
Capacity factor	42%	28%
Name plate (MWh)	2,040	3,027

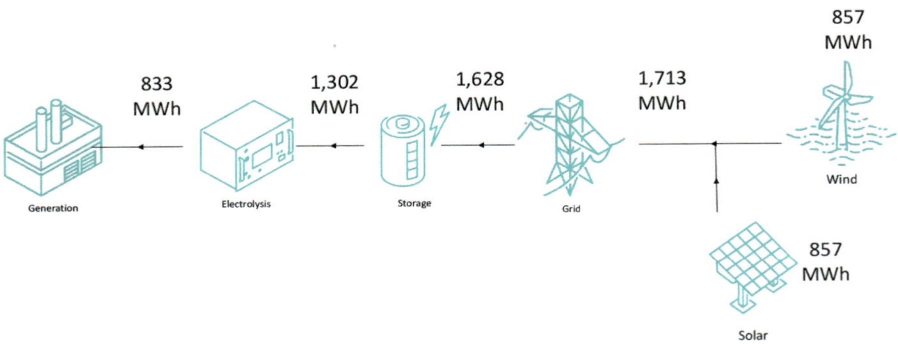

Figure 11.2 System requirements to generate 500 MWh from a green hydrogen fuel source.

approximately 60% efficient. This means that to obtain 500 MWh of electric energy from a generating unit, 1.6 times more energy (i.e., 833 MWh) from H_2 will have to go into the generating facility.

$$500 \text{ MWh} / 60\% = 833 \text{ MWh hydrogen}$$

Each component identified in the theoretical electrical generating system has some efficiency losses as the energy transforms. The same method as previously discussed can be used to calculate the energy losses associated with each system component. These calculations are summarized in Table 11.2 and Figure 11.2. The calculations shown in Table 11.2 and the configuration in Figure 11.2 indicate what is needed to supply enough H_2 fuel to support a 500 MWh electricity demand on a 24-hour/day basis. It requires the installation of 857 MWh of wind generation, 857 MWh of solar generation, 1,713 MWh of transmission capacity, 1,628 MWh of industrial battery

Table 11.3 Systemwide efficiencies (percentage).

	Green hydrogen	Electrolysis	Industrial storage	Transmission
Name plate (MWh)	1.00	1.67	2.60	3.26
Efficiency	60%	64%	80%	95%
MWh in	1.67	2.60	3.26	3.43
	Wind	**Solar**		
MWh out	1.71	1.71		
Capacity factor	42%	28%		
Name plate (MWh)	4.08	6.05		
			Efficiency based on input	29.2%
			Efficiency based on renewable nameplate	9.87%

storage, 1,302 MWh of electrolyzers, and 833 MWh of dispatchable H_2-fired electrical generation.

Another important consideration is the estimate of electric system efficiency. To calculate overall system efficiency, the input energy required by each system component was divided by the 500 MWh system output. This calculation was performed in two ways based on the initial required input or on renewable nameplate information. The summary for the efficiency of each component as calculated by the two methods is found in Table 11.3.

11.2.5 Phase V: Example of an initial cost estimate

Once component sizing is determined, the values from Phase IV can be used to ascertain initial scoping cost estimates for procurement and construction for each of the theoretical electrical system components. Scoping cost estimates provide an order of magnitude estimate based on the actual costs of similar installations that have been recently completed. These cost estimates are created to help gauge project feasibility. If initial scoping estimates indicate that a project is economically viable, more accurate cost estimates can be developed. These might be based on factors such as the exact location of the proposed sites, known electric demand requirements, equipment availability, and defined local supply chains.

Table 11.4 Generation and system component costs using different reference sources ($/kWh).

Source		**EIA**	**EIA**	**NREL**
Year		2021	2023	2022
Combined cycle (MS)	$/kWh		$1,176	
CC with 90% capture	$/kWh		$3,140	
Electrolysis	$/kWh	$1,428		
Wind	$/kWh	$1,428	$2,098	$1,540
Solar	$/kWh	$1,561		$1,373
Storage	$/kWh	$1,796	$1,270	
Transmission[6]	$/kWh	$500		
Nuclear - light water	$/kWh		$7,777	
Nuclear - small modular	$/kWh		$8,349	

Since no large scale H_2-fired baseload generating units have recently been constructed, we used three sources for the cost estimates: 1) EIA's construction cost data for electric generators installed in 2021 [10]; 2) EIA's cost and performance characteristics of new generating technologies, Annual Energy Outlook 2023 [11]; and 3) NREL's 2023 annual technology baseline [12].[6] Construction costs, on a $/kWh basis, were created using different cost estimating references (i.e., EIA and NREL) for several of studied alternatives (see Table 11.4).

To arrive at a consensus cost, we corrected for inflation and then averaged each component's calculated values in Table 11.4. A 30-year average inflation rate (excluding 2021 and 2022) of 4.2% for nonresidential construction was assumed. To simplify very large, quantified figures, all values were converted from $/kWh to $/MWh. The estimated construction costs, on a $/MWh basis, for all studied cases are summarized in Table 11.5.

The cost of each system component was determined by multiplying the component's energy sizing in MWh (see Table 11.2) by the price for each energy component shown in Table 11.5 ($/MWH). As an example, the estimated cost of a 500 MWh combined cycle with multiple shafts is:

$$500 \text{ MWh x } \$3{,}270{,}000/\text{MWh} = \$1{,}613 \text{ million}$$

To provide a more robust understanding of the system, its components, and costs, the following alternatives were studied:

[6] The most recent historical data in the 2023 ATB is from 2022.

Table 11.5 Estimated 2024 generation construction cost ($/MWh).

Source		EIA	EIA	NREL	Average
Year		2021	2023	2022	
Combined cycle (MS)	$/MWh		$1,230,000		$1,230,000
CC with 90% capture	$/MWh		$3,270,000		$3,270,000
Electrolysis	$/MWh	$1,330,000			$1,330,000
Wind	$/MWh	$1,600,000	$2,190.000	$1,670,000	$1,820,000
Solar	$/MWh	$1,770,000		$1,490,000	$1,630,000
Storage	$/MWh	$2,030,000	$1,320,000		$1,680,000
Transmission[7]	$/MWh	$566,000			$566,000
Nuclear – Light water[8]	$/MWh		$8,100,000		$8,400,000
Nuclear – Small modular[8]	$/MWh		$8,700,000		

Alternative 1: 500 MWh of electricity using green H_2 from renewable energy via electrolysis, with 80% efficiency storage (base case).

Alternative 2: The base case with storage efficiency increased to 86% efficiency.

Alternative 3: The base case with no required energy storage.

Alternative 4: The base case with energy storage reduced to 5 days.

Alternative 5: A case based on renewables supplying energy storage only, no electrolysis and no generation.

Alternative 6: A case with 3 days of energy storage and 10% extra hydrogen to store extra H_2 as a buffer to during times of low renewable generation (3 days plus extra H_2).[9]

Alternative 7: A case based on nuclear generation.

Alternative 8: A case with NG generation including 90% carbon capture.

Alternative 9: A case based on current NG generation.

[7] The transmission costs are based on Colorado's Power Pathway, submitted to the Colorado Public Utilities Commission by Xcel Energy on March 3, 2021.

[8] Nuclear, light water and nuclear, small modular are averaged into one cost.

[9] Hydrogen storage was not included in this cost estimate.

Table 11.6 Component and total system cost of 500 MW of green hydrogen as base generation.

	Green hydrogen	Electrolysis	Storage	Transmission
Name plate (MWh)	500	833	1,302	1,628
Efficiency	60%	64%	80%	95%
MWh in	833	1,302	1,628	1,713
Cost ($MM)	$613	$1,100	$1,090,000	$921
	Wind	**Solar**		
MWh out	857	857		
Capacity factor	42%	28%		
Name plate (MWh)	2,040	3,027		
Cost ($MM)	$1,560	$1,400		
Total ($MM)				$1,100,000

Example of industrial storage use in Alternatives: From Table 11.2, the system will require 1,302 MWh going from energy storage to electrolysis. The round-trip efficiency for industrial energy storage is 80%. Additionally, the current operating range for the BESS is 20% to 80%. Thus, the required name plate capacity for 1,302 MWh of storage is:

1,302 MWh × (1/80% efficiency) × [1/(80% – 20%) BESS = 2,713 MWh

Therefore, 10 days of energy storage requires a 651,000 MWh nameplate capacity.

11.3 Results

The estimated cost of a system that can provide 500 MWh of electricity from green H_2, as previously defined (i.e., as base generation), is $1.10 trillion. The individual component breakout and total costs for this base case analysis are found in Table 11.6.

As noted in Table 11.3, the system efficiencies for the base case alternative are 29.2% and 9.87% based on system input and renewable nameplate capacity, respectively. The same processes were used in each of the alternative cases. Results from these alternative calculations are summarized in Table 11.7.

11.4 Conclusions

A preliminary cost estimate for a baseload, hydrogen-fired, 500 MWh electric generating plant was developed. The system costs over $1 trillion and has an

Table 11.7 Cost and efficiencies for base and alternative cases.

Case	Storage (in units)	Total alternative cost ($ MM)	Percent of base (in %)	Alternative efficiency Input (%)	Name plate (in %)
Alternative 1: Base	10	$1,100,000	---	29.2%	9.87%
Alternative 2: 86% efficient	10	$1,020,000	92.7%	31.4%	10.6%
Alternative 3: No storage	0	$4,830	0.44%	36.5%	12.3%
Alternative 4: 5 days	5	$491,000	44.6%	32.8%	11.1%
Alternative 5: Storage only	5	$211,000	19.2%	76.0%	25.7%
Alternative 6: 3 days storage plus extra H2	3	$326,000	29.6%	32.8%	11.1%
Alternative 7: Nuclear generation	N/A	$4,200	0.38%	35.0%	35.0%
Alternative 8: NG generation with 90% Carbon capture	N/A	$1,640	0.15%	59.0%	59.0%
Alternative 9: Current NG	N/A	$613	0.06%	60.0%	60.0%

estimated overall operating efficiency of 29.2%. The cascade of energy losses across the studied system compounds these issues. The base case system loses over 70% of the energy entering the electric grid. Over 90% of the renewable nameplate capacity is required to meet the definition of green hydrogen.

There are other less costly ways to provide generation. An NG plant that collects 90% of its own CO_2 emissions costs $1.64 billion with an estimated long-term operating efficiency of 60%. Compared to the clean hydrogen plant, using NG with 90% carbon capture represents just 0.15% of the initial cost and is twice as efficient under long-term operation. Similarly, a nuclear generator with no direct CO_2 equivalent emission would have 0.38% of the initial installation cost of a H_2-fired system and have roughly twice the long-term efficiency.

A new 2,200 MW nuclear power plant, the first in the U.S. in over 30 years, is now operating in Waynesboro, Georgia (Plant Vogtle reactors 3 and 4) [13]. Despite delays, labor issues, cost overruns, and inflation increasing the project cost to over $31 billion, this cost is much less than the estimate for the baseload hydrogen plant [14]. Large hydropower plants are much more efficient and much less costly. China's Three Gorges Dam project, the world's largest electrical generating station of any type, cost the equivalent of $32 billion; it has a generating capacity of 22,500 MW, over 10 times that of Plant Vogtle's units 3 and 4 combined, and provides electrical generation using a renewable energy resource [13].

Given these preliminary estimates, the only reasonable conclusion that can be made is that green H_2-fired electric generation is too expensive and should not be viewed as a viable alternative to solving issues related to fossil fuel-fired GHG emissions. While technically possible, the use of green hydrogen to produce baseload electrical generation is not economically feasible. Reasons for this lack of viability include the current cost of industrial energy storage and low overall system inefficiencies. While green H_2 baseload generation may not be a reasonable option today, it could become more viable in the future.

References

[1] U.S. Geological Survey. How many homes can an average wind turbine power?https://www.usgs.gov/faqs/how-many-homes-can-average-wind-turbine-power.

[2] Tableau Software (2021). Literature. https://public.tableau.com/views/2021Literature/rojections?:embed=y&Technology=UtilityPV&:embed=y&:showVizHome=n&:bootstrapWhenNotified=y&:apiID=handler4 , accessed June 20, 2024.

[3] AC transmission line losses. http://large.stanford.edu/courses/2010/ph240/harting1.

[4] Energy Information Administration. Utility-scale batteries and pumped storage return about 80% of the electricity they store. https://www.eia.gov/todayinenergy/detail.php?id=46756, accessed July 7, 2022.

[5] Technology brief: analysis of current-day commercial electrolyzers. n.d.

[6] betweenpixels (2015, December 25). Natural gas power plant efficiency. Walton Gas (blog). https://www.waltongas.com/natural-gas-power-plant-efficiency.

[7] Edzarenski (2024, January 17). Construction inflation 2024. Construction Analytics (blog). https://edzarenski.com/2024/01/17/construction-inflation-2024.

[8] Parkes, R. (2024, March 4). Cost of electrolyzers for green hydrogen production is rising instead of falling: BNEF. Hydrogen news and intelligence. *Hydrogen Insight.* https://www.hydrogeninsight.com/electrolysers/cost-of-electrolysers-for-green-hydrogen-production-is-rising-instead-of-falling-bnef/2-1-1607220.

[9] Environment Texas Research and Policy Center (2022, January 31). The Texas freeze: timeline of events. https://environmentamerica.org/texas/center/articles/the-texas-freeze-timeline-of-events.

[10] Electricity—construction cost data for electric generators. https://www.eia.gov/electricity/generatorcosts/index.php.

[11] Cost and performance characteristics of new generating technologies. Annual energy outlook 2023, March 2023.

[12] Tableau Software (2023). Literature. https://public.tableau.com/views/2023LiteratureProjections?:embed=y&:toolbar=no&-Technology=UtilityPV&Metric%20ilter=CAPEX&:embed=y&:-showVizHome=n&:bootstrapWhenNotified=y&:apiID=handler4, accessed June 20, 2024.

[13] Roosa, S. (2024, September). Nuclear energy—lessons from Georgia's new power plant. *Alternative Energy and Distributed Generation Journal*, 6(3), 5–12.

[14] Thibault, M. (2023, August 3). $30B Bechtel-built nuclear plant starts operation in Georgia. https://www.constructiondive.com/news/bechtel-nuclear-power-plant-startsoperation-georgia/689738, accessed 29 October 2023.

12

Reinventing the Electric Grid

Stephen A. Roosa

It is time to reinvent our electric grids, transforming them into infrastructure that is more usable, durable, and efficient. Most of today's electric grids are based on early 20th-century models. They are configured to meet the requirements of an antiquated and centralized electricity distribution system. They are designed to deliver electricity in a one-way direction, which provides generation, transmission, and distribution to customers. The power grids of the future need to be more responsive, resilient, and capable of providing uninterrupted and expanded services. Improving transmission and distribution system efficiencies requires upgrading technologies and modifying system configurations. Tomorrow's electric grids need to be able to quickly incorporate the added generation from renewable and nuclear energy resources. How can this be accomplished? Selectively upgrading electrical distribution systems, adding energy storage systems, and using smart microgrids are ways to improve the operation of electric grids. Other possibilities include reconfiguring electrical systems into peer-to-peer networks and virtual power plants. This chapter explores opportunities to structure electrical energy delivery systems to better meet the needs of tomorrow's electrical consumers.

12.1 Introduction

Electricity is provided to markets using a one-way process of generation, transmission, and distribution. The interconnected system we call the electric grid is designed for this purpose. The configuration of today's electric grids originated from antiquated distribution models. Failure of a central production plant or any of its direct links to customers creates disruptions and

DOI: 10.1201/9788743811312-12

the potential for cascading power outages. Surprisingly, the U.S. has more electrical power outages than any other country in the world. The increased use of automated and digital systems makes these outages problematic. Today's power grids need to be more responsive to consumer needs. They need greater resiliency to ward off the sophisticated physical and cyberattacks that cause power interruptions and failures. The national electric grids must evolve to become more efficient, and their transmission systems must be capable of providing uninterrupted services. Despite the need to improve electric grid infrastructure, lowering the cost of development, operations, and maintenance remains the primary priority of public utility companies. More efficient electrical infrastructure reduces system losses, advances the energy transition, and supports decarbonization.

There are no easy solutions. Utility-scale electrical generation facilities are based on centralized models of generation and distribution. Improvements to the electric grid often occur after catastrophic failures that follow major weather events that damage electrical infrastructure. These tend to be hurried replacements of damaged components with upgraded equipment, but using similar technologies. In the rush to restore critical services, rarely is time available to redesign the system. Rather than being reactive to disruptive events, electric grid infrastructure can be designed to be more resilient. Grid infrastructure must be hardened to resist physical damage and cyberattacks while expanding customer services. To meet such requirements, electric grids need to be reinvented. This includes upgraded generation, distribution, and transmission systems. There are ways to transition from central distribution configurations to distributive generation while simultaneously improving system efficiencies and meeting environmental objectives. Existing electrical transmission and distribution systems can be designed to carry greater capacities. Though costly, critical power lines can be placed underground. The model for the future involves developing smarter electric grids with interconnected microgrids, using distributed renewable and nuclear generation, and offering virtual electricity supply systems.

12.2 What Are Smart Grids?

Smart grids are modern electricity distribution systems that are designed to monitor, protect, and automatically optimize the operation of interconnected elements, including generation equipment, high-voltage distribution, automation systems, and energy storage [1]. Like the bulk power grid, smart microgrids generate, distribute, and regulate the flow of electricity to consumers, but do so locally using lower voltages [2]. They include small power generation

Table 12. 1 Characteristics of traditional and smart grids.

Traditional Grid Characteristics	Smart Grid Characteristics
Electric Machinery	Digital
One-way communication flow	Bidirectional communications
Centralized power generation	Distributed power generation
Centralized distribution network	Decentralized or distributed network
Small number of sensors, if any	Full grid sensor array
Manual system monitoring	Automatic monitoring
Manual recovery	Automatic/automated recovery
Failures and power outages	Adaptive and islanded
Few user options	Many user options

and distribution systems, energy storage, energy conversion devices, related load management, monitoring, and protection devices. They operate as autonomous systems that can provide local systems control, protection, and management. Table 12.1 compares certain characteristics of traditional grids to smart grids. A key difference from the consumer perspective is that traditional grids are less resilient and more prone to failures than smart grids.

12.2.1 Creating value using smart microgrids

Microgrids are electricity distribution systems that contain loads and distributed energy resources (DER) that may serve a local university campus, military base, industrial site, or a specific community [3]. Often used to supply remote villages or towns, their costs achieve parity due to the expense of running and maintaining power lines over long distances. Their configurations include sources of electrical generation, wiring and interconnections, a central control system, and loads. They may also use some type of energy storage and be interconnected with their host electric grid. A microgrid typically supports either decentralized or distributed network configurations. They are designed to meet various conditions and targeted user requirements for electrical power at lower voltages. Microgrids can generate electricity using fossil fuels, renewable energy systems, and potentially modular nuclear reactors [3]. Microgrids can be designed to supply alternating (AC) or direct (DC) current. In the U.S. almost all are AC, but many have systems that natively generate DC power (e.g., fuel cells, solar photovoltaic systems, etc.). Distribution and transmission losses are reduced because of the proximity of generation sources to the consumer loads.

Matching the electrical energy supplied by microgrids with loads can be difficult. The intermittency of power generation associated with renewable energy can create problems matching power generation with electrical demand. When grid-interconnected, this may contribute to instability [4]. This problem can be resolved using load management systems and software. Smart microgrids use direct digital control (DDC) systems to overcome these issues and enable their customers to connect to the supply networks in real time. Microgrids use a DDC central controller which, if so designed, can be integrated with peer-to-peer networks.

Smart microgrids are characterized by bidirectional flow of both electricity and information to create an automated and distributed energy delivery network [1]. Smart microgrids offer the benefit of distributed computing and communications to deliver real-time information and enable instantaneous balancing of electrical supply and demand at the specific level required for each discrete device [5]. A control system is a key feature of smart microgrids. Smart microgrid controls typically include a central computer system. It serves as a load dispatch location from which the distribution of generation is managed; it is where the monitoring and control of system stability and load frequency are performed. Algorithmic software platforms provide decision-making and predictive intelligence capabilities. Some elements of the microgrid control system are decentralized and capable of independent operation [5]. Electronic power conditioning and control of the production and distribution of electricity add value to successful smart microgrid applications.

As microgrids are being connected to smart grids, new theories have emerged redefining how they are deployed. These nascent microgrids have variously been called advanced microgrids, advanced remote microgrids, and smart microgrids [5]. There are many benefits associated with smart microgrids as they employ a wide range of technologies. They can improve system resiliency and reliability, reduce voltages, lower electrical generation costs, and minimize the need for transmission infrastructure. System efficiencies are improved as transmission losses are minimal. Smart microgrids are an ideal way to integrate renewable resources on the community level and allow for customer participation in the electricity enterprise [2].

12.3 Reimagining Nuclear Energy

The vision for regional clean energy grids often involves local generation of electricity using a renewable energy source, perhaps from wind, solar, or hydropower. Excess power from DERs could be stored for later use using battery energy storage, pumped hydropower, compressed air storage, or directed to an electrolysis facility to create green hydrogen.

For nuclear power stations, uranium is a viable fuel for electrical generation. Sourced from a mineral and using a reactor, it is neither a renewable energy resource nor a fossil fuel. Nuclear power is typically classified as a clean energy source when used to produce heat and generate electricity because greenhouse gases are not emitted. Like solar and wind power, the potential for nuclear power as a clean energy resource appears unlimited.

12.3.1 Conventional nuclear generation

When we think of nuclear plants, we focus on utility-scale generation facilities with fission reactors using enriched uranium-235. This technology offers seemingly infinite supplies of energy and, perhaps more importantly, provides base-load electrical power generation at high operating capacities [3].

With a fleet of about 440 operating reactors, nuclear energy accounts for almost 10% of the world's electrical energy production [3]. Nuclear reactors that generate electricity in the U.S. fall into two main categories: boiling water reactors (BWRs) and pressurized water reactors (PWRs) [6]. Both systems boil water to make steam (BWRs within the reactor and PWRs outside the reactor) [6]. The steam must be cooled after it is directed through a turbine to produce electricity, which necessitates its signature cooling towers.

Utility-scale nuclear reactors that supply power to electric grids have their limitations. Design, permitting, development, and prove-out may require a decade or more. Reactor delivery costs are difficult to estimate. Other viable concerns include refined uranium's price volatility, how the converted uranium is sourced (about 40% of the refined global supply comes from Russia), the ultimate disposition and storage of spent fuel rods, and public concerns about safety and the potential of meltdowns.

Another more common concern is that traditional nuclear plants withdraw and consume vast volumes of water. This is the primary reason why nuclear plants are located near large natural sources of fresh water. They use more water per unit of electricity produced than coal plants using similar cooling technologies because nuclear plants operate at lower temperatures and turbine efficiencies. Once-through nuclear plants withdraw 25,000 to 60,000 gallons (94,635 to 227,125 liters) of water per MW of electricity produced and consume 100 to 400 gallons (379 to 1,514 liters) per MW [7]. Recirculating systems withdraw 800 to 2,000 gallons (3,028 to 9,842 liters) of water per MW of electricity produced and consume 160 to 800 gallons (607 to 3,028 liters) per MW [7]. The availability of water can be a serious problem in regions where water is already scarce or during periods of drought. If water becomes unavailable, the reactors must be shut down.

12.3.2 Integral molten salt reactors

An alternative to conventional nuclear power generation is to produce electricity with modular nuclear systems. The smart microgrids of the future could be powered using small nuclear reactors (SNRs). Advanced SNRs are being developed in the U.S. and can be used to generate heat and electrical power. They vary in output from a few to hundreds of megawatts and use various types of liquid coolants. SNRs offer advantages that include their safety, small size, variable output capabilities, reduced capital investment, ability to be sited in locations not possible for larger nuclear plants, and the capability to incrementally add power [8]. They can operate either independently or connected to a host electric grid and can power a campus or small community in the event of grid failure [9]. SNRs can stock two years or more of their fuel requirements on-site; this allows them to maintain power during and after extreme weather events or other threats to the electric grid [9]. The U.S. Navy has about 100 small nuclear reactors that are used by 86 submarines and aircraft carriers [10]. They provide real-world models of the use of package nuclear reactors and have logged over 5,400 reactor years of accident-free operation [10].

The next generation of small reactors might use thorium as fuel and molten salt as a heat sink to provide electricity with no nuclear waste [11]. Integrating smart grid capabilities in distribution and load management can ensure that nuclear-powered microgrids would manage the ebb and flow of demand and production—at greater efficiencies and capacities than today's aging power generation infrastructure [11].

While there are major costs, regulatory ramifications, permitting, and construction delays that are associated with large nuclear reactors, these hurdles might be overcome with smaller versions of nuclear technology. While fusion reactor demonstration projects are conceivable, small nuclear reactors are more likely to become available in the near term. Integral molten salt reactor (IMSR) technology is an example of applying smaller-scale, modular technology. It is conceivable that assembly line and modular production of these reactors is possible.

IMSRs can operate using a variety of fuels, including spent nuclear wastes [12]. Molten salt can be used as both the fuel and the coolant. IMSRs can work in combination with renewable energy facilities such as solar and wind to produce continuous utility-grade, fossil-fuel-free energy with no carbon emissions [12]. IMSR plants can operate under ambient pressures, making them much safer than conventional nuclear plants. They are also not subject to the potential of radioactive gas explosions, and there is no risk of meltdown upon failure, which alleviates most public safety concerns [12].

With these advantages and the need to reduce greenhouse gas emissions on a world scale, is it time to consider package nuclear power plants for electrical generation? Possibly, yet there are high costs for the development of package nuclear energy systems and unresolved environmental concerns. Regulatory structures are not in place to enable the wide adoption of package nuclear systems. While several SNRs are being developed in the U.S., none have yet been deployed. There are hopes that assembly line production will lower manufacturing costs once modular designs are stationed and proven in field operations.

12.4 Efficient Transmission and Distribution Systems

Electricity must be consumed the instant it is produced. Adding energy storage is one way to reduce system losses. While unused electricity can be stored, it is difficult to do so in large quantities unless a mechanical energy storage system is used. Typical types of storage include flywheels, battery energy storage systems, pumped hydropower, or compressor air storage. Losses of electricity occur during the process of storing and recovering it after storage. Energy storage improves load balancing capabilities and helps optimize transmission efficiencies. Being among the least costly forms of bulk energy storage, pumped storage hydropower accounts for 96% of U.S. utility-scale energy storage capacity.

There are about 511,100 circuit miles (822,534 km) of high-voltage transmission lines in North America greater than 100 kilovolts [13]. An electrical transmission line is a long, specially designed conductor that carries bulk amounts of generated power at very high voltages from one distribution station to another [14].

12.4.1 Estimating transmission line efficiency

Transmission line efficiency is the ratio of power received to power sent, indicating how much power reaches its destination versus how much is sent. Losses occur due to line lengths, line losses, voltage drops, power factors, heat conversion in transformers, and other factors. Transmission efficiency is defined as the ratio of receiving end power, P_R, to the sending end power, P_S, expressed as a percentage. The efficiency can be calculated using Equation 12.1.

$$\%\ nT = (P_R / P_s) \times 100 = (V_R I_R \cos\theta_R /\ V_S I_S \cos\theta_R) \times 100 \qquad (12.1)$$

In Equation 12.1, $\cos\theta_s$ is the sending end power factor, $\cos\theta_R$ is the receiving end power factor, V_S is the sending end voltage per phase, and V_R is the receiving end voltage per phase.

Efficient transmission over long distances requires high voltages, normally overhead, which help reduce line losses. Overhead transmission wires depend on air for insulation, requiring that lines maintain minimum clearances. Voltage levels are changed using transformers. Transformers help maintain the efficiency of the electrical grid by stepping up voltages for long-distance transmission and stepping down voltages to distribute electricity to end users. They also generate heat, which results in system losses. Transformer efficiencies typically range from 95% to 99%; however, large power transformers can operate at efficiencies over 99%.

Transmitting electricity at high voltage reduces the fraction of energy lost to Joule heating, which varies by the conductor type, electrical current, and transmission distance. For example, a span of 100 miles (160 km) at 765 kV carrying 1,000 MW of power can have losses of 0.5% to 1.1%. A 345 kV line carrying the same load across the same distance has losses of 4.2% [15]. For a given amount of power, higher voltages reduce the current and thus the resistive losses. Raising the voltage by a factor of 10 reduces the current by a corresponding factor of 10 and the losses by a factor of 100, provided the same-sized conductors are used. Transmission and distribution losses from 2013 to 2019 in the U.S. were estimated to be approximately 5%. Transmission systems with higher power factors have lower losses. There are ways to reduce these losses.

12.4.2 Reconductoring

There is a need to add transmission capacity to enable the addition of renewables and nuclear power production. As of 2022, more than 10,000 power plants and energy storage projects were awaiting permission to connect to the U.S. electric grid—95% were zero-carbon resources [16]. Boosting transmission capacity with traditional approaches usually requires using grid-enhancing technologies to optimize the throughput of electricity, or more commonly, constructing new transmission lines. New power lines can take up to 10 years to plan, permit, and build. Most of this new electric generation is distributed but must be carried by the existing electric grids. Some utility companies are considering alternative means of increasing transmission capabilities. This is because obtaining approval for right-of-way for new transmission is time-consuming, requires regulatory approvals, may be met with local opposition, and the costs for the infrastructure are prohibitive.

Another way to increase power transmission capacity and efficiency is to replace the high-voltage power lines. *Reconductoring* is the replacement-in-place of existing transmission lines with higher-capacity lines. This retrofit process uses the existing transmission infrastructure and does not require the purchase of additional land. It can be performed more quickly with fewer regulatory hurdles.

Most power lines in use today have a steel core surrounded by aluminum strands. Replacing the steel with a lighter, stronger composite material such as carbon fiber reduces line losses, doubles the transmission capacity, and allows the transmission lines to operate with less sag under heavier electrical loads. Composites are less prone to expansion at higher temperatures than steel. Although these more technologically advanced transmission lines are more expensive than using steel, total reconductoring costs are less than half that of installing a new line. A reconductoring project in southeastern Texas upgraded 240 miles of transmission lines at a cost of $900,000 per mile, versus a 3,600-mile greenfield project that averaged $1.9 million per mile [16].

12.5 The Era of Virtual Power Plants

One of the major issues with the traditional electrical grid is the losses at the point of generation. The efficiency of different types of power plants based on heat rates varies; coal-fired generation is 32%, nuclear 33%, and natural gas 44% [17]. Solar panels are 18% to 25% efficient; modern wind turbine generators range in efficiency from 35% to 47% [17]. Hydropower offers the highest efficiencies. High-head hydropower projects operate at over 90% efficiency when converting moving water into electrical current; some of the generators at our largest dam sites approach efficiencies of 99% [17]. Unlike fossil fuel plants that rely on the Rankine cycle, many renewable energy technologies lack a combustion process and do not have high heat conversion loses. Considering the entire fleet of U.S. energy sources, more energy is lost in conversion processes than is turned into electricity.

Creating opportunities to use selectable and dispatchable power generation is possible when distributed generation systems are available. It is often challenging to configure these typically remote systems to take advantage of differences in generation efficiency.

12.5.1 What are virtual power plants?

A virtual power plant (VPP) is a network of decentralized, medium-scale power-generating systems, such as wind farms, solar parks, combined heat and power systems, and energy storage systems that supply electricity to end-users [18]. VPPs have been defined as cloud-based, internet-accessible distributed power plants that aggregate the capacities of energy resources for the purposes of enhancing power generation, exchanging or selling electricity on the market, or providing demand-side options for load reduction [19]. By pooling diverse electric generation assets and linking them through a central control center, the VPP can be designed to adapt in relation to the hot electric grid as a single controllable entity in a manner like microgrids [5]. Regardless, the various generation assets are remotely dispatched through a common control center, and though interconnected, they are independent in their operation and ownership [18].

A VPP uses advanced software and technologies to monitor and coordinate power generation assets and controllable loads, simulating the processes that would be expected from a conventional electrical power plant [5]. The bidirectional data exchange between the individual plants and the VPP enables the transmission of control commands and provides real-time data on the capacity and utilization of the networked units [18]. During periods of peak load, the VPP relieves the load on the grid by smartly distributing the generated electricity [18]. The power generation and consumption of the networked units become commodities that can be traded on an energy exchange [18]. If provisionally configured, it is possible to sell and purchase units of electricity using a personal smartphone.

12.5.2 Applications for virtual power plants

The real-world working applications for virtual power plants are diverse. Though counterintuitive, a virtual power plant may not necessarily involve the actual deployment of electrical generation sources [5]. Many VPPs in the U.S. tap utility demand response and critical peak pricing programs as resources. When aggregated, they mimic certain characteristics of a traditional power plant delivering peak capacity, energy, or grid reliability regulation services as needed by a utility or independent system operator [10]. In Europe, a VPP typically refers to aggregating supply-side resources, most often a diverse pool of renewable distributed energy generation and/or wholesale renewable energy sources [21].

More confounding, VPPs might also refer to the ability of commercial consumers in countries such as Denmark to purchase wholesale capacity via auction from baseload fossil fuel facilities for short periods of time [5]. Statkraft, a Norwegian company, has operated one of Europe's biggest power generation facilities as a VPP since 2011 [22]. Their VPP in Germany has a 12,000 MW capacity, enough to power five million homes [22].

VPPs often provide services to the host grid in exchange for preferred pricing. Flexible power equipment (e.g., motors and pumps) can be operated on optimized price schedules by using electricity when it is inexpensive and electric demand is low [18]. The VPP's central control system helps stabilize grid power before balancing is required [18]. If a grid imbalance is imminent, the signals from the operators are processed in the central control system and converted into control instructions for the qualified units to balance the grid by delivering frequency control reserves [18]. If an unanticipated high feed-in develops, generation assets can be shut down within seconds to avert critical grid situations [18]. When excess electrical supplies are available, they can be stored. According to Andreas Bader, vice president for sales for Statkraft, "We can connect batteries from Spain with wind farms in Germany, and that makes it scalable" [22].

Virtual power plants can be used to establish virtual microgrids. Microgrids can be configured to operate either as a VVP or mimic their capabilities and services. Distribution scale forecasting can help the host utility company predict when distributed generation might create distribution system imbalances; when dispatchable, localized power can be provided when generation prices are increasing [19]. When such events happen, VVPs can help provide peaking power. Because VPPs can be aggregated at various levels, they can be used for both electric grid reliability and operational optimization [19]. VVPs offer rapid and flexible deployment and can be quickly reconfigured or scaled back in response to changing needs [20]. As the networking capabilities of VPPs improve, they will likely become valuable resources that help improve electric grid reliability and resiliency.

12.6 Conclusion

There are problems with the way electricity is being distributed. The power grids of the future need to be more resilient and responsive to consumer needs. Reliability is a primary consumer concern, which functionally means that supplies must be adequate and available continuously [13]. There are technologies available to improve reliability and reduce the losses that occur in the generation, distribution, and transmission of electricity using centralized distribution configurations. This creates opportunities for distributed generation systems that can provide power locally and regionally. More efficient electrical infrastructure reduces system losses, advances the energy transition, and supports decarbonization.

Deployment costs for renewable energy systems have been declining, which makes their use in microgrids more feasible. Many renewable energy technologies are proven and scalable. Compared to conventional fossil fuel-fired generation, they have lower development costs, fewer permitting obstacles, do not pollute the air and water, and do not emit greenhouse gases. Advanced small nuclear reactors are being developed in the U.S. and can be used in tomorrow's smart microgrids to produce heat and generate electricity. However, the availability and comparatively lower cost of natural gas and renewables such as wind power, solar, and hydropower make most nuclear power plants a comparatively more expensive option. Using energy storage systems, such as batteries or pumped hydropower, can help improve system efficiencies.

Regardless of the energy resources used for power generation, interconnecting with local electric grids requires extending transmission infrastructure. Losses occur in the distribution and transmission process. There are ways, such as reconductoring or using grid enhancement technologies, that can improve grid transmission capacities and efficiencies using existing high voltage infrastructure. Incorporating renewable and nuclear energy systems in distributed grid configurations helps bring new systems online while overcoming interconnection hurdles. Using smart grid technologies enables improved operational efficiencies. Smart microgrids can be used to supply electricity locally and serve to support their host grids. VPPs are potential solutions to enhance grid reliability and resilience [20]. They can be configured to make electrical generation more selectable and dispatchable. These methods of improving electrical generation and transmission capabilities provide pathways to the more responsive and resilient electrical grids of the future.

References

[1] Office of Nuclear Energy. Advanced small modular reactors. https://www.energy.gov/ne/nuclear-reactor-technologies/small-modular-nuclear-reactors, accessed 3 December 2024.

[2] Office of Nuclear Energy (2018, January 25). Department of Energy report explores U.S. advanced small modular reactors to boost grid resiliency. https://www.energy.gov/ne/articles/department-energy-report-explores-us-advanced-small-modular-reactors-boost-grid, accessed 3 December 2024.

[3] Rosen, L. (2013, April 7). The future of microgrids. http://hplusmagazine.com/2014/01/07/the-future-of-micrgrids, accessed 2 August 2019.

[4] Misra, G., Venkataramani, G., Gowrishankar, S., Ayyasam, E. and Ramalingam, V. (2017). Renewable energy based smart microgrids—a pathway to green port development. *Strategic Planning for Energy and the Environment*, 37(2), 17–32.

[5] Roosa, S. (2021). *Fundamentals of Microgrids—Development and Implementation*. Taylor and Francis: London, England. ISBN: 978-0-367-53539-1.

[6] Roosa, S. (2023, September). Nuclear energy—Lessons from Georgia's new power plant. *Alternative Energy and Distributed Generation Journal*, 6(3), 5–12.

[7] Union of Concerned Scientist (2008, July 13). Got water? Nuclear power plant cooling needs. https://www.ucsusa.org/resources/nuclear-power-plant-cooling-water-needs, accessed 22 January 2026.

[8] Gellings, C. (2015). *Smart Grid Planning and Implementation*. Fairmont Press, Inc.: Lilburn, Georgia.

[9] Alvin Power. What are smart microgrids? https://www.galvinpower.org/microgrids, accessed 31 October 2024.

[10] Conca, J. (2019, December 23). How the U.S. navy remains the masters of modular nuclear reactors. *Forbes*. https://www.forbes.com/sites/jamesconca/2019/12/23/americas-nuclear-navy-still-the-masters-of-nuclear-power, accessed 2 December 2024.

[11] Roosa, S. (2022, December). Heads up! The microgrids cometh! *International Journal of Strategic Energy and Environmental Planning*, 4(6), 5–8.

[12] Misra, G., Venkataramani, G., Gowrishankar, S., Ayyasam, E. and Ramalingam, V. (2017). Renewable energy based smart microgrids—a pathway to green port development. *Strategic Planning for Energy and the Environment*, 37(2), 17–32.

[13] North American Electric Reliability Council (2023, March). Understanding the grid. https://www.nerc.com/news/Documents/March%202023%20Understanding%20the%20Grid.pdf, 24 November 2024.

[14] Electrical 4 U (2024, April 18). Transmission lines: parameters, types and theory. https://www.electrical4u.com/transmission-line-in-power-system/#What-is-Transmission-Line, accessed 1 November 2024.

[15] American Electric Power (2011, June 4). Transmission facts, 4.

[16] Guarnieri, M. (2013). The alternating evolution of DC power transmission. *IEEE Industrial Electronics Magazine,* **7**(3), 60–63. doi:10.1109/MIE.2013.2272238.

[17] Kirk, K. (2022, October 24). Energy loss is single-biggest component of today's electricity system. Yale Climate Connections. https://yaleclimateconnections.org/2022/10/energy-loss-is-single-biggest-component-of-todays-electricity-system, accessed 1 November 2024.

[18] Next. Virtual power plant, how to network distributed energy resources. https://www.next-kraftwerke.com/vpp/virtual-power-plant, accessed 31 October 2024.

[19] Zurborg, A. (2010). Unlocking customer value: the virtual power plant. *Energy.gov*, accessed 15 February 2023.
[20] Brehm, K., Dyson, M., McEvoy, A. and Usry, C. (2023, January). Virtual power plants, real benefits. https://rmi.org/wp-content/uploads/dlm_uploads/2023/01/virtual_power_plants_real_benefits.pdf, 13-14, accessed 3 December 2024.
[21] Kashyap, S. (2017, October 12). What is the difference between micro-grids and virtual power plants? https://www.quora.com/What-is-the-difference-between-microgrids-and-virtual-power-plants, accessed 22 January 2026.
[22] Ziady, H. (2019, November 7). No wind? No sun? This power plant solves renewable energy's biggest problem. *CNN Business*. https://www.cnn.com/2019/11/07/business/statkraft-virtual-power-plant/index.html, accessed 31 October 2024.

13

Electrical Transmission System Modeling Within Utility Integrated Resource Planning to Accommodate Increasing Renewable Penetration

Andrew R. Kramer and Robert A. Kramer

Across the electric utility industry, traditional thermal electrical generation is being retired and replaced with renewable energy alternatives that are often supplemented with energy storage and transmission enhancement technologies. Utility long-range planning, also known as integrated resource planning (IRP), seeks to model system costs over a 20+ year time horizon. Marginal energy cost (MEC), a component of the locational marginal price (LMP), is forecast in this modeling for various scenarios and sensitivities. Within this analysis, the marginal cost of congestion (MCC), another key component of the LMP, is typically based on historic averages rather than forecasts. This can result in model costs deviating from true system costs when considering electrical generation retirements and replacement scenarios.

This research shows that the averages for the hourly standard deviations and maximum hourly peaks for the MCC have generally been increasing for the electric power hubs in the U.S. states of Illinois, Indiana, and Michigan, during recent winter seasons. An industrial power node near Chicago, Illinois, is used as an example to show significant swings in both the average of the hourly standard deviations and the average of maximum hourly peaks of the MCC during the spring season. Large electric power nodes of this type are particularly vulnerable to transmission system constraints. The results of this research indicate rising electric transmission costs due to the increasing penetration of intermittent electrical generation resources. Consequently,

DOI: 10.1201/9788743811312-13

it is advantageous for utilities to work with independent system operators (ISO) and other organizations such as the North American Electric Reliability Corporation (NERC) to forecast the MCC for IRPs when electrical generation retirement scenarios are being considered. This can lead to a more representative IRP process. It also improves forecasting accuracy in cases where utilities are unaware of potential neighboring generator retirements. Including MCC forecasts within IRPs may lead to reduced purchased electric power costs, expanded power sales opportunities, and improved system reliability.

13.1 Introduction

The U.S. electric power market is undergoing rapid transformation. In many cases, utilities are planning transitions from traditional thermal electric generation to renewable energy resources (e.g., solar and wind power) that are often supplemented by energy storage and transmission enhancement technologies. About 17% of total U.S. electricity production at utility-scale facilities is sourced from renewable energy. Of this share, 7.3% comes from wind power and 6.6% from hydropower. For the states selected in our assessment, just 8.2% of the electricity produced in Illinois, 7.1% in Indiana, and 7.8% in Michigan is sourced from renewable resources. An Illinois law requires that by the year 2025, at least 25% of the state's energy comes from renewable sources. In 2019, Chicago announced its public commitment to achieve 100% renewable energy by 2035 [1]. Transitioning to renewable energy presents opportunities to lower total system costs; however, if not done carefully, this can lead to increased electric power purchased at premium prices with less reliability.

When renewable energy resources are coupled with large electric storage systems, transmission issues may arise if the energy storage system is unable to charge to the levels needed during off-peak hours to support grid reliability during on-peak periods. Not forecasting the impacts of total electrical transmission costs due to local electric generation retirements in surrounding regions can lead to substantial unplanned costs.

13.1.1 Utility long-range planning

Utility long-range planning, also known as integrated resource planning (IRP), seeks to model the electric system's costs over a 20+ year horizon. As part of such modeling, it is common to include the costs of environmental and transmission system upgrades. However, the locational marginal price (LMP), which represents the electric power purchase and selling prices at a specific market node, often does not include electric transmission system adjustment scenarios associated with anticipated generation retirements and replacements. The LMP consists of the following components: marginal energy cost (MEC), marginal cost of congestion (MCC), and the marginal loss component (MLC). The LMP and its respective components can be modelled for both day-ahead and real-time scenarios. It is known that renewable energy systems and their associated energy storage systems may cause challenges when calculating LMPs [2].

Given the size and complexities of integrated bulk electric systems, it is not feasible to estimate the benefits of forecasting the MCC for every generation retirement scenario within an IRP. Regardless, it is recognized that electric transmission risks for renewable energy systems need to be managed due to the intermittent nature of these energy resources [3].

Published LMP data indicated that the average of the hourly standard deviations and the average of maximum hourly peaks for the MCC for electric power hubs in the U.S. states of Illinois, Indiana, and Michigan have been increasing during the recent winter seasons [4]. An industrial power node near Chicago, Illinois, displays large swings in both the average of the hourly standard deviations and the average of maximum hourly peaks for the MCC during the spring season. This is indicative of the rise in electric transmission costs due to increasing penetration of intermittent generation resources. The need to accommodate electric transmission considerations with renewable energy buildout is happening throughout the power industry. Examples of *red zones*, regions that require substantial electrical transmission upgrades to integrate renewable generation (e.g., those in Indiana and North Carolina), show how transmission limitations could delay the transition to renewables [5]. Such considerations support the need for long-range utility infrastructure planning.

13.1.2 Modeling MCC uncertainties

To mitigate model uncertainty, the MCC and potentially the MLC should be forecast within IRPs for critical retirement scenarios. To forecast the MCC and MLC, it is necessary to utilize electrical transmission system planning and potentially transient analysis software for a given retirement and replacement scenario. The pricing impacts would then be iteratively fed back into the IRP process as a modification to the input LMP forecast. Best practice would seek to incorporate potential retirements and replacements from high-impact electrical system transmission zones of adjacent regional interconnections. Electric power flows are determined by the laws of physics, not by contract path; the ability to include neighboring system detail improves model accuracy. This may result in increased model complexity for individual utility IRPs. It illustrates the importance of actively including independent system operator (ISO) information and input in the IRP process.

This modeling approach may contribute to improved long-term electric system reliability. As was recently highlighted by the Electric Reliability Council of Texas (ERCOT) and the Midcontinent Independent System Operator (MISO), an organization responsible for operating the power grid across 15 U.S. states and the Canadian province of Manitoba, potential shortages in capacity could impact the performance of the bulk electric system [6]. This is consistent with the "2022 Summer Assessment Report" prepared by the North American Electric Reliability Corporation (NERC) [7]. As thermal electric generation stations continue to be retired, capturing the total transmission costs during the IRP process informs planning decisions for cost mitigation options.

13.2 Market Data Analysis

Across the utility industry, research is being conducted at the local and system levels to understand the impacts of ongoing electric system changes, including thermal generation retirements. The historic marginal cost of congestion from MISO illustrates increasing congestion costs in the day-ahead market for the Illinois, Indiana, and Michigan hubs during the winter seasons [4]. These three hubs were chosen due to their adjacency and the shared electric power flows that move across their service territories. In addition, an industrial power node near Chicago was considered to show significant swings in MCC for the spring season. Transmission system congestion costs are extremely volatile and vary depending on outages and system loads. It is anticipated that such volatility will increase in the future as more spinning generation is retired and electrical power flows over greater distances from sources to loads. To maintain reliability and reduce transmission congestion for larger renewable energy buildout, upgrading existing electrical transmission systems is often required [8]. As such, understanding the uncertainties in MCC pricing at specific high-risk nodes becomes increasingly important as the electric system continues to evolve. This is particularly true if a highly variable node also serves as a major purchase and sales hub.

13.2.1 Applying statistical analysis

The standard deviations of the day-ahead MCC were calculated for every hour for the meteorological winter season (January 1 through March 31) for

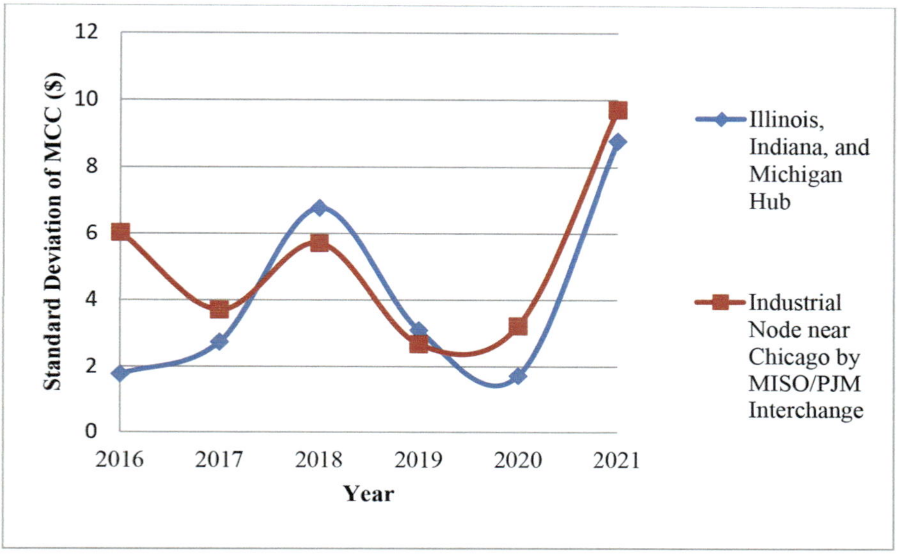

Figure 13.1 Average hourly standard deviation for day-ahead marginal cost of congestion for winter seasons.

the combined Illinois, Indiana, and Michigan hubs. These standard deviations were averaged as an indicator of the uncertainty in MCC during the winter season for a given year. For these hubs in the winter season, the maximum day-ahead MCC for every hour was averaged to indicate purchase power risk. This process was repeated for the industrial power node near Chicago, Illinois, for the meteorological spring season (April 1 through June 30). The summer season was also studied, but no definitive trend emerged using the data that were considered. Note that the MISO LMP calculation methodology can vary from year to year.

Figures 13.1 and 13.2 show an increasing trend in MCC risk for the Illinois, Indiana, and Michigan hubs for the winter season. An exception to this trend occurred in 2020, which is attributed to the COVID-19 outbreak. In addition, a deviation from this trend can be seen in the uncertainty of the MCC for these hubs in 2019. The decrease in 2019 could be attributed to numerous causes, including excess generation capacity, warmer temperatures during periods of reduced generation availability, and low transmission outages.

Figures 13.3 and 13.4 demonstrate that significant transmission cost risk is associated with large industrial nodes. These nodes are vulnerable to electric generation retirements, and care should be taken to understand the

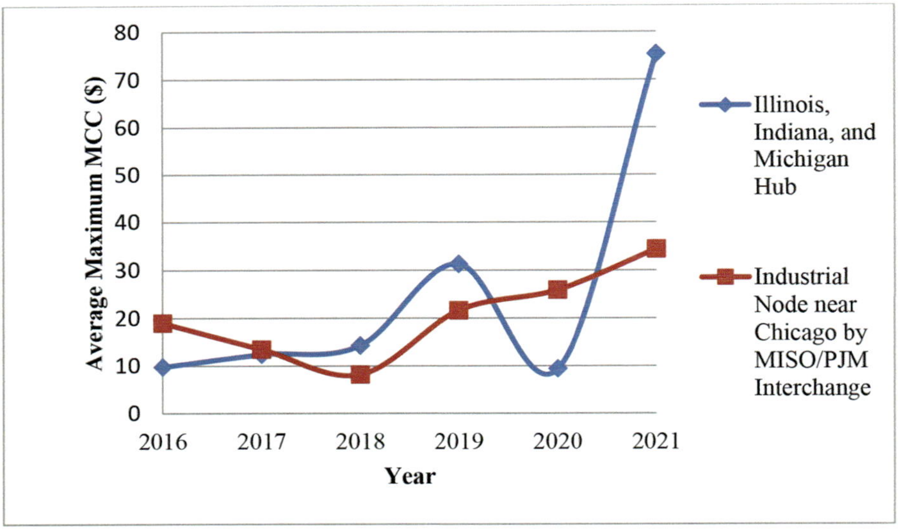

Figure 13.2 Average of maximum day-ahead marginal cost of congestion for winter seasons.

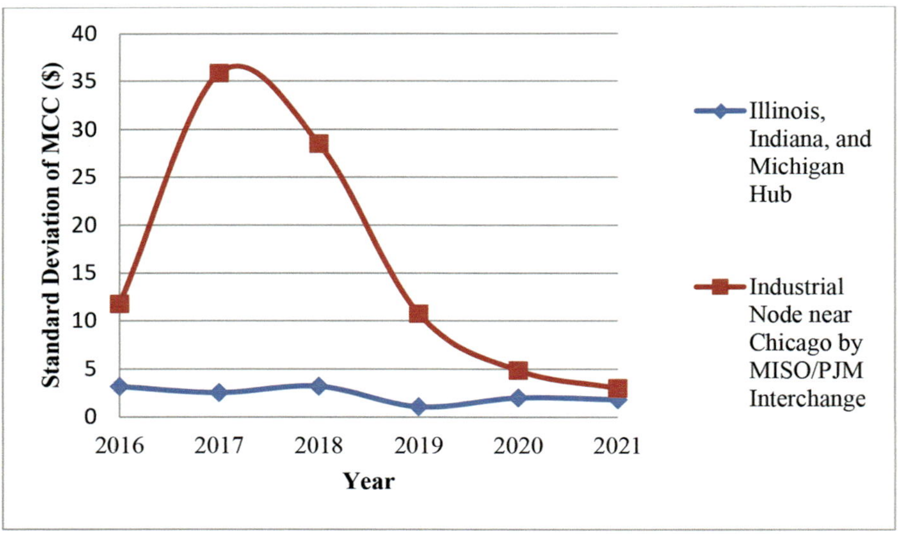

Figure 13.3 Average hourly standard deviation for day-ahead marginal cost of congestion for spring seasons.

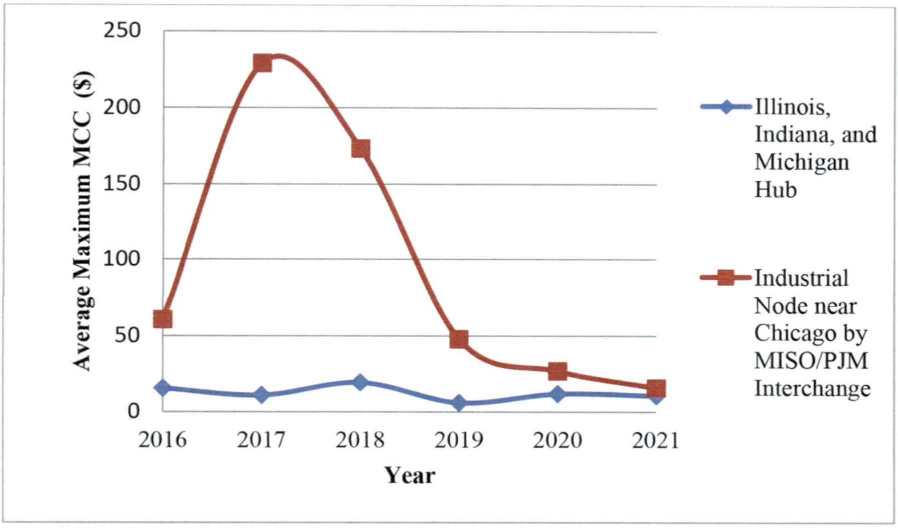

Figure 13.4 Average of maximum day-ahead marginal cost of congestion across all hours for spring seasons.

MCC risks for any retirement planning scenario. Special modeling within an IRP may be required for these nodes—especially if they are hubs for sales and purchases or have active participants in the ancillary or demand response markets.

13.3 Conclusions

As the electric power market evolves, new and improved modeling methods will be required to ensure forecasting accuracy and risk mitigation. For our study, an evaluation using integrated resource management was performed using data from three hubs in the midwestern U.S. We determined that there was an increasing trend in MCC risk for the Illinois, Indiana, and Michigan hubs for the winter seasons, except in 2020. We also demonstrated that significant transmission cost risk is associated with large industrial nodes. This research illustrates that forecasting transmission congestion costs can potentially lead to improvements in integrated resource planning.

Further work is needed to understand the impacts based on planned generation retirements and replacements. For Illinois, one study found that expanding renewables would save consumers $1.21 billion over 10 years, generate $8.2 billion in new economic activity, and create 53,000 Illinois construction jobs by 2033 [9]. To obtain these benefits, utilities should actively work with independent system operators to ensure that system-wide impacts from planned retirements and replacements do not negatively impact customer electric prices and system reliability. The effects and risks associated with the cost of electric transmission congestion and line losses can be captured; however, a broader look at the integrated resource planning process is needed. Future risks include the difficulties in financing infrastructure upgrades and the ability of electric companies to plan for microgrid integration that necessitates two-way transfer of electricity and information between the generators and the host utilities [10]. As electrical generation planning continues to increase in complexity, centralizing integrated resource planning within ISOs needs to be considered to ensure system reliability.

References

[1] Dzikiy, P. (2019, February 18). Chicago sets goal for 100% clean, renewable energy by 2035. https://electrek.co/2019/02/18/chicago-sets-100-percent-clean-renewable-energy-goal, accessed 20 January 2026.
[2] Bai, H., Yang, M., Song, D., He, Y., Su, T. and Jiang, X. (2019). A review of the application of locational marginal price theory in the new situation. Institute of Electrical and Electronics Engineers. https://ieeexplore.ieee.org/abstract/document/9061929, accessed 1 June 2022.
[3] Raikar, S. and Adamson, S. (2019). *Renewable Energy Finance.* Academic Press.
[4] Midcontinent Independent System Operator Market Reports. Historical annual day-ahead LMPs. www.misoenergy.org/markets-and-operations, accessed May 2022.
[5] Runyon, J. (2022, May 17). Grid upgrades in Duke Energy 'red zone' could cost $560M. *PowerGrid International.* www.power-grid.com/td/grid-upgrades-in-duke-energys-red-zone-will-cost-560m, accessed May 2022.
[6] Patel, S. (2022, May 6). ERCOT, MISO warn of potential power supply shortfalls. *Power: News and Technology for the Global Energy Industry.* www.powermag.com/ercot-miso-warn-of-potential-power-supply-shortfalls, accessed May 2022.
[7] North American Electric Reliability Corporation (2022). 2022 Summer reliability assessment.
[8] Gorman, W., Mills, A. and Wiser, R. (2019, December). Improving estimates of transmission capital costs for utility-scale wind and solar projects to inform renewable energy policy. *Energy Policy,* 135, 110,994.

[9] Misbrener, K. (2021, April 27). New study finds expanding Illinois renewable energy program would save consumers $1.21 billion over 10 years. https://www.solarpowerworldonline.com/2021/04/study-finds-expanding-illinois-renewable-energy-program-saves-consumers-money/#:~:text=The%20study%20finds%20the%20following%20impacts%20will%20occur,Funding%3A%20%2B%242.94%20billion%20Wholesale%20Energy%20Costs%3A%20-%241.76%20billion, accessed 31 July 2022.

[10] Roosa, S. (2021). *Fundamentals of Microgrids: Development and Implementation*, 201–214. Taylor and Francis Group.

14

Using Dynamic Line Ratings to Improve Electric Generation and Load Interconnection

Andrew R. Kramer, Robert A. Kramer, Omer Vejzovic, Nergui Rentsen, and Shannon Dancler

Much emphasis has been placed upon the use of grid-enhancing technologies (GETS) to assist with the ongoing evolution of electric grids in the United States. Dynamic line rating (DLR) is often associated with GETS because new technologies and equipment are needed to implement and validate dynamic transmission line ratings during system operations or within long-term planning models. Given the expected rapid load growth associated with data centers and electrification initiatives across the U.S., the ability to provide additional transmission capacity may help reduce or avoid reliability issues and restrictions on economic development. DLR can contribute to this by allowing system planners and operators to assume additional transmission capacity subject to site ambient conditions and electrical system configuration. DLR does not modify the physical structure of the transmission system but rather incorporates improved data into transmission rating values to allow for greater generation dispatch than would be captured with static, conservative rating values. A measurement of this increase in transmission capacity is reflected in the changes in transmission congestion and generation curtailment.

This research incorporates DLR transmission ratings into the nodal market simulation tool (PLEXOS®) to study the impact of these rating changes in terms of transmission congestion and generation curtailment. The power node studied was in the Pennsylvania-New Jersey-Maryland (PJM)

DOI: 10.1201/9788743811312-14

independent system operator (ISO) territory near the Indiana, Michigan, and Ohio borders. This region has potential for future data center load additions due to a robust transmission network and readily available supplies of natural gas. A theoretical data center load of 1.5 GW was considered in the model near the Saint Joseph Energy Center (SJEC) in New Carlisle, Indiana. In addition, a speculative, non-dispatchable 1.5 GW solar resource was also simulated near the energy center to provide power and offset carbon emissions. Upon analysis, two primary transmission binding constraints were identified. With the use of DLR for a transmission rating capacity increase of 10% (low case) and an increase of 40% (high case), the transmission congestion was reduced. The curtailment at the solar resource node was reduced by 53% and 99% respectively, for the low and high cases relative to the modelled baseline. The results of this research illustrate how DLR may be used to increase transmission line capacities while helping to defer reconductoring or other costly initiatives.

14.1 Introduction

As electric utilities continue to retire baseload thermal generation, complications arise because previously unplanned load growth may create new and aggravate existing reliability challenges. The evolving focus on artificial intelligence (AI) requires new types of energy-intensive data centers. Total data center growth is raising concerns regarding electric supply and transmission. The Pennsylvania-New Jersey-Maryland (PJM) independent system operator (ISO) may triple its rate of load growth through 2034, much of this due to new AI-focused data centers [1].

The distribution of the growth of data centers is not uniform across ISOs. Systems that contain large, dispatchable power sources and robust transmission networks feature the best support for these hyperscale electric loads. Locating dispatchable thermal generation near electric loads is ideal; however, existing natural gas transportation and supply infrastructure related to procurement of combined cycle gas (CCG) generation limits the opportunities for expansion. Also, carbon reduction policies and regulations may affect both capital expenditures and dispatch levels of electrical generation. Small module nuclear reactors may eventually provide an ideal localized thermal resource; however, the implementation timelines associated with these technologies, as well as optimization of construction cost, are highly speculative.

Traditionally, much of the U.S. data center growth has occurred in Northern Virginia, Chicago, Illinois, and Santa Clara, California [2]. Other states, including Utah and Nevada, are expected to show dramatic increases in electric loads due to data center developments [3]. Given an ample supply of natural gas, the integrated resource plans (IRPs) for utilities on the PJM-MISO seams near the Illinois, Indiana, Michigan, and Ohio borders show the potential to more than double or triple their current native load in less than seven years [4,5].

Elsewhere in the country, electrification initiatives have increased, but data center growth is less pronounced; renewable generation paired with battery energy storage systems (BESS) has the potential to mitigate the loss of traditional thermal baseload. Data centers are particularly challenging because they often must operate without interruption at a continuous load level. Renewable generation coupled with energy storage creates challenges when paired with data centers.

14.1.1 Grid-enhancement technologies

To accommodate future electrical load growth, additional energy resources will be required. In North America and particularly for the U.S., providing reliable power supplies and resilient electrical generation will be a formidable task [6]. While new generation and storage options have the potential to solve these challenges, an optimized solution must consider the electrical transmission system. Increasing transmission line capacity can allow for the interconnection of new generation resources and increased dispatch for existing generation. New transmission lines can be costly to construct and represent a considerable risk due to potential litigation, easement access, cost overruns, construction schedules, and other challenges. Transmission reconductoring, often proposed as a solution, has limitations that include material availability and construction costs [7]. The simplest and often least costly solution is to modify the system parameters to allow for greater flow ratings while respecting reliability requirements.

Historically, grid-enhancing technologies have included different options ranging from dynamic line rating (DLR) to equipment designed to redirect real-time flows, such as with phase-shifting transformers. As the electric grid continues to evolve and more non-dispatchable resources are added, especially in remote locations, GETs offer an alternative to constructing new transmission lines. DLR is an attractive option due to the minimal need for the installation of new equipment. DLR can utilize hourly or sub-hourly line ratings based upon ambient conditions, including temperature, wind speed, and electric configuration, plus adjust the ratings for day-ahead or real-time markets.

While DLRs, GETs, and new transmission interconnections provide important opportunities, they are not portrayed as complete solutions to load growth projections. The Midcontinent Independent System Operator (MISO) potentially faces up to 2.7 GW of electric generation shortage for fiscal year (FY) 2025/26 and up to a 14.4 GW shortage by FY 2029/30 [8]. These shortages are driven by load growth and new generation retirements. Additional transmission capacity cannot resolve this issue if neighboring ISOs have limited or no excess electricity available. Given the continued growth in data center electrical loads and other electrification initiatives, most proposed transmission upgrades or reconfiguration initiatives will provide only partial solutions.

14.2 Data Center Load and Generation Resources

Supplying data center electric loads is challenging when compared to other electrification initiatives. Electrical loads often follow on-peak and off-peak behaviors. Data centers require constant, uninterrupted, high-quality power. It is challenging to match data center load requirements with only intermittent and volatile renewable generation. Erroneous conclusions have been drawn when utilizing capacity factors to match renewable nameplate capacity to data center load requirements. A model that utilizes economic dispatch and commitments at hourly intervals or subintervals is required to evaluate the capacity needs for data centers. From a transmission planning and operations perspective, resolving transient stability, harmonics, and flicker is imperative when supplying electricity to data centers.

Renewable generation resources naturally benefit from DLRs because the electricity generated typically must be transmitted from rural areas to population or commercial centers. Long transmission distances between loads and generation can lead to transmission congestion and/or curtailment risks. Due to the limitations of matching renewable output to data center loads, renewable generation and associated DLRs are best utilized for carbon offsets as opposed to directly powering data centers, especially when there is a need to precisely match electrical loads to generation.

Given the scale of the expected electrical load growth associated with data centers, a study that compares powering these resources with renewable and storage vs traditional thermal generation assets is needed. To study the amount of generation resources required to power a 1 GW data center requiring constant load, PLEXOS® (a software platform from Energy Exemplar) was utilized. Its capacity expansion module enabled the model to evaluate electrical generation based on a production cost model utilizing hourly intervals.

Three theoretical scenarios were initially studied, the first powers the data center exclusively using renewables (i.e., wind and solar energy) and battery energy storage systems with an assumed 4-hour storage duration. The other two scenarios power the data center with combinations of natural gas generation, renewables, and a BESS. Within these scenarios, the data center exists as an island that is only served by the resources simulated by the software's capacity expansion module. Data centers have interconnected electric loads within the broader electric grid that typically import power from external sources. However, given the magnitude of projected data center growth across the U.S., these scenarios were selected to provide insights into data center load growth. The imported power for data centers is typically limited due to either electrical transmission and/or generation considerations. This analysis utilizes a methodology that has been successfully applied in previous studies [9].

14.2.1 Results of the capacity analysis

The analysis of electricity generation and BESS resources used relevant information preset within PLEXOS® or best engineering judgement localized for the Northern Midwest. Input data, such as commodity prices, was based on proprietary data sets or publicly available sources such as those from the U.S. Energy Information Administration. The capital expansion costs used were developed from the 2024 Lazard Lifecycle Cost of Energy (LCOE) + Report [10]. This report tends to understate the cost of renewable generation because it does not include paired transmission upgrades to move the electricity generated from remote locations to the site of the load [5]. No limitations were placed on the capacity of any specific energy resource that could be constructed.

Figure 14.1 shows the results of Scenario 1, which allows PLEXOS® to select from only renewable and BESS resources. As is seen in Figure 14.1, PLEXOS® allocates 3 GW of wind capacity, 5.4 GW of solar capacity, and 10 GW of BESS capacity with a 4-hour storage duration, based on a combined optimization of the three selectable resources. The total nameplate of generation and energy storage resources allocated exceeds the 1 GW load, with BESS providing by far the greatest contribution. To meet its load obligation, the model relies on the BESS to carry the data center through periods of reduced solar and wind generation, which occur frequently in the northern Midwest during the winter season. Due to the data center load obligation not allowing periods of off-peak charging, the model must overbuild both the renewable generation and BESS to ensure resource adequacy and prevent load shedding. To precisely match loads to generation, the model requires excess electricity to recharge the BESS, thus ensuring that the system is not over-generating. Due to this scenario's large amount of installed wind and solar capacity, there exists a possibility of conditions that enable excess generation. Therefore, a large capacity of BESS would be needed to balance the

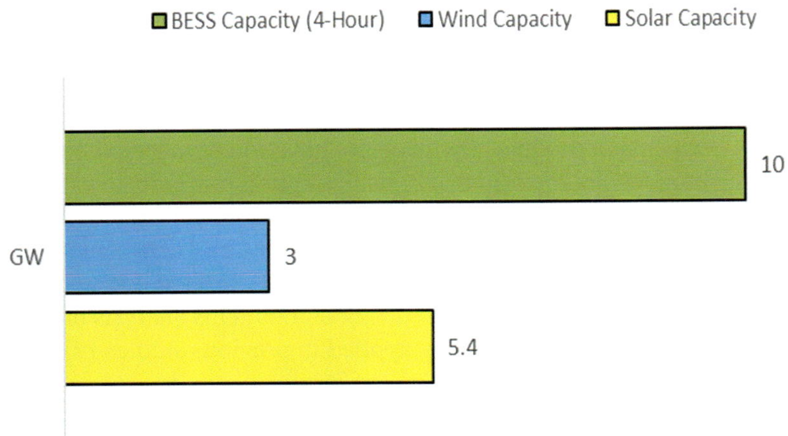

Figure 14.1 Capacity built for a 1 GW data center (renewable and BESS only).

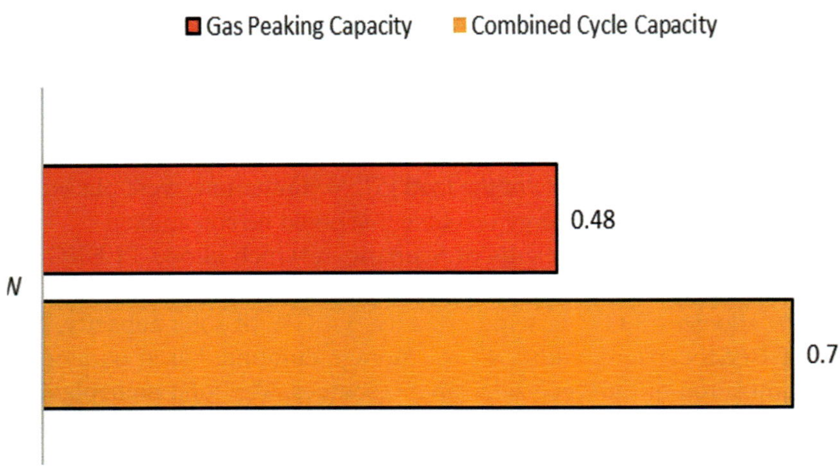

Figure 14.2 Capacity built for a 1 GW data center (all resources selectable).

system. It is often overlooked that too much generation is just as problematic as too little in terms of controlling system frequency.

The results of Scenario 2 are shown in Figure 14.2, which allows PLEXOS® to select from a combination of CCG generation, natural gas peaking units, renewables, and BESS. For this scenario, PLEXOS® built 0.48 GW of natural gas peaking capacity and 0.7 GW of combined natural gas generation based on the combined optimization of all resources being

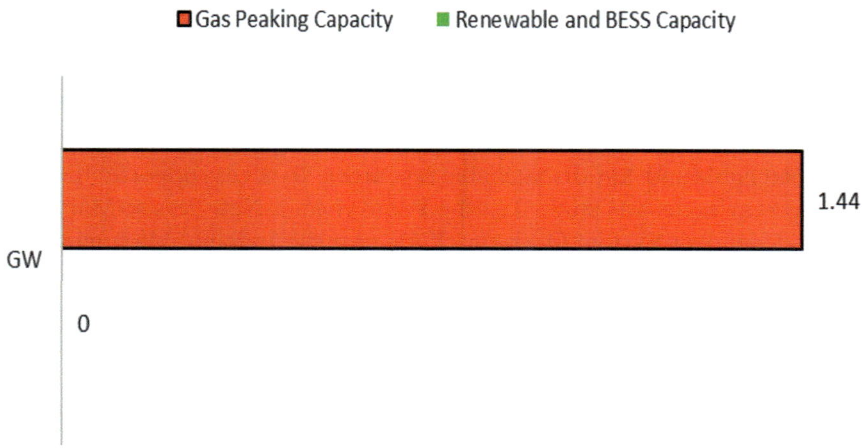

Figure 14.3 Capacity built for a 1 GW data center (all resources selectable except CCGTs).

selectable. The dispatchable nature of combined cycle gas generation enables the nameplate of the constructed resources to more closely match that of the load obligation. The advantages of pairing natural gas generation and data center load appear in many recent IRPs and similar plans [4,5]. This optimization does not include a carbon tax or cap.

Figure 14.3 shows the results of Scenario 3, which allows PLEXOS® to select from only natural gas peaking units, renewables, and BESS. To satisfy this requirement, PLEXOS® built 1.44 GW of natural gas peaking capacity based on a combined optimization of all resources being selectable except CCG generation. This demonstrates the benefits of dispatchable natural gas resources within the optimization model when trying to supply data center load; however, it does not include certain practical considerations. Natural gas-fired peaking units are not designed for continuous dispatch. Including drastically increased maintenance costs for this type of operation would likely change the optimization profile. Regardless, based on pure capital and dispatch costs, the advantages of natural gas generation are seen even when the efficiency of the generator is lower.

14.3 Utilizing DLR to Reduce Transmission Congestion and Curtailment

From an economic perspective, powering a data center exclusively with renewable generation and BESS degrades the value of a generation expansion plan. However, there remain opportunities to pair these resources with data center loads. Renewable generation and energy storage systems can be used for carbon offsets and to supply conforming loads, enabling thermal baseload resources to be available for other purposes. A robust utility resource planning portfolio should include a diversified set of resources to balance both risks and reliability.

Dynamic line rating can allow for improved transmission economics for both dispatchable and non-dispatchable resources. To access opportunities for DLR, a power node was studied in the Pennsylvania-New Jersey-Maryland (PJM) Independent System Operator (ISO) near the Indiana, Michigan, and Ohio border. This area is prime for current and future data center expansion plans due to a robust transmission network and readily available supplies of natural gas.

The nodal market simulation tool PLEXOS® and its 2024 Eastern Area Interconnection model (developed by Energy Exemplar) were used for this analysis. Scenario 4 simulates a theoretical data center load of 1.5 GW built into the model near the Saint Joseph Energy Center (SJEC) in New Carlisle, Indiana. This scenario served as a proxy to consider relevant load interconnections. It is preferable to locate data centers near electrical generation to reduce the need for transmission upgrades.

The SJEC already has a baseload combined cycle gas generator. An expansion site is available for the construction of a second CCG generator [11]. For this analysis, a solar energy resource with a 1.5 GW capacity was placed at the SJEC. This was done rather than adding an additional CCG generator to study the impact of non-dispatchable resources paired with dispatchable natural gas generation. It was beneficial to study a paired resource combination because using only non-dispatchable resources results in negative congestion, and potentially curtailment, depending upon the severity of the thermal stress; whereas, dispatchable resources simply avoid operating based on the severity of the economics.

14.3.1 Results of the transmission congestion and curtailment analysis

The results of the analysis showed that two critical transmission binding constraints were present near the study location when solar energy was included in the generation mix. Transmission binding constraints occur when a transmission line is at maximum capacity, which results in a non-zero shadow price. Generators will either increase or reduce the constraint, resulting in either negative or positive congestion for that specific line element. The SJEC node experienced negative congestion and curtailment due to the generation resources exceeding the capabilities of the local transmission system.

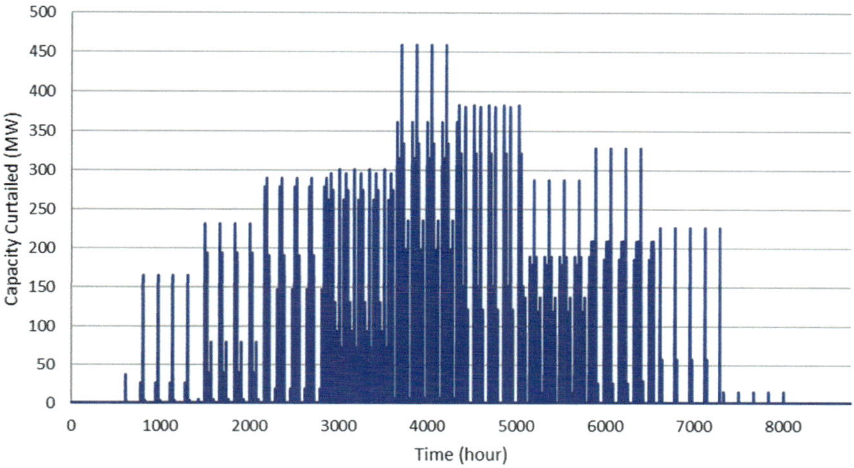

Figure 14.4 Curtailment of the 1.5 GW solar resource at SJEC, base rating.

Next, DLRs were used for the two primary binding constraints in place of the base ratings, enabling greater flows on the respective transmission lines. Using industry-based DLR models, an annual dynamic rating change of 10% was simulated for the low case and 40% for the high case. It is possible to utilize DLRs based upon seasonal or hourly intervals; however, that level of granularity was not required for our study.

The curtailment profile using the 1.5 GW solar resource built at the SJEC node is shown in Figure 14.4. The total congestion charge for the 1.5 GW solar resource was projected to be –\$4.4 million in 2027. The dispatch profile of the CCG generator was not shown due to the uncertainties surrounding the exact operational characteristics of the resource in both the day-ahead and real-time markets. Information such as generator ramp rates and cycling limitations was unknown; consequently, this analysis may overstate the ability of the CCG at SJEC to help accommodate the intermittent nature of the solar energy resource.

DLR reduced the negative congestion and curtailment levels of the solar facility at the SJEC. In the low case, the curtailment profile for the 1.5 GW solar resource built at the SJEC node is shown in Figure 14.5. The total congestion charge for the 1.5 GW solar resource was –\$3.6 million for 2027 (or total congestion savings of \$.8 million compared to the base case). In the high case, the curtailment profile for the 1.5 GW solar resource built at the SJEC node is shown in Figure 14.6. The total congestion charge for the 1.5 GW solar resource was –\$1.8 million for 2027 (or a total congestion savings of \$2.6 million compared to the base case).

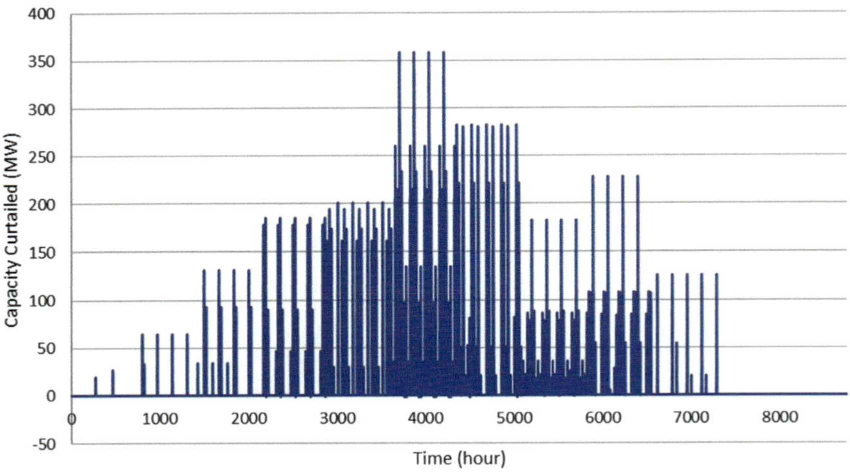

Figure 14.5 Curtailment of the 1.5 GW solar resource at SJEC, 10% dynamic rating increase.

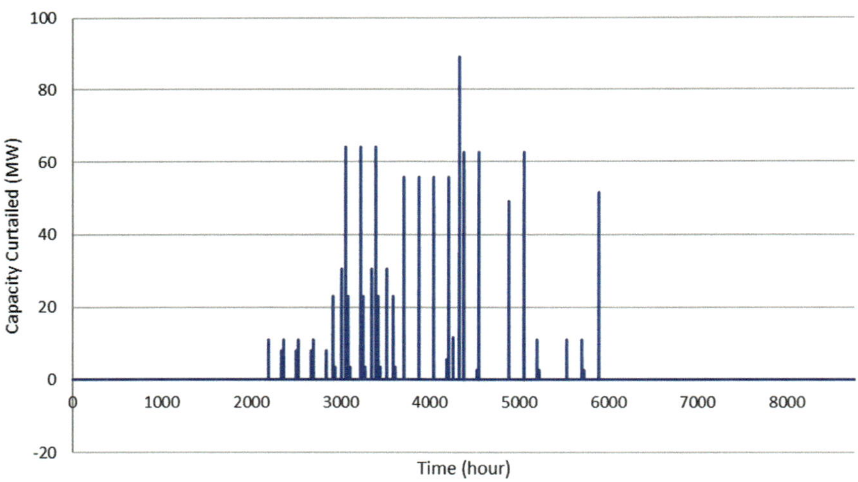

Figure 14.6 Curtailment of the 1.5 GW solar resource at SJEC, 40% dynamic rating increase.

Using a DLR for transmission rating capacity increases of 10% (low case) and 40% (high case), the transmission congestion was lowered, and the curtailment at the solar energy resource node was reduced by 53% and 99% respectively, relative to the model baseline. These results may change with the implementation of a security-constrained economic dispatch (SCED) and commitments. Typically, ISO day-ahead and real-time dispatch/commitment models must modify their optimal power flow solutions to accommodate a sequential ordering of relevant N-1 contingencies (N designates the number of pieces of equipment) and integrate to produce an SCED-based power flow solution. Reliability of an energy supply system usually requires that any single major unit failure leaves the system with enough resources to supply the current load [12]. The system that satisfies this requirement is described as meeting the N-1 contingency criterion [12]. Additional benefits may be realized if DLRs are paired with improved models that allow post-contingency adjustments rather than requiring the optimal power flow solution to adhere to all relevant N-1 contingencies.

14.4 Conclusions

DLRs present an important element to help build an enhanced electric grid. This research has shown the potential of DLRs to reduce the risks of congestion and curtailment if implemented for a limited number of key transmission binding constraints. While interesting, the benefits of DLR should not be overstated. The rapid growth of data centers and their massive electricity requirements are limiting factors; new resource capacity additions will be required to replace base load retirements and meet future electrical load obligations. Regardless, DLR is an important tool that can provide benefits, assuming reliability concerns related to system operations are considered.

Further work is needed to understand the impact of DLRs from both a seasonal and operational perspective. DLR cannot exist in a vacuum and must include real-time operation considerations. While DLR is beneficial in that higher transmission line ratings will allow for higher interconnection rates and dispatch levels, DLR could also reduce the reliability safety margin associated with reserving excess line capacity. As such, continuing discussions related to reliability metrics, North American Electric Reliability Corporation operation procedures, and Federal Energy Regulatory Commission policies are required.

14.4.1 Research constraints and limitations

The analysis for Scenarios 1, 2, and 3 has limitations. Along with not allowing any power imports, factors such as maintenance outages, forced outages, effective forced outage rates, transmission outages, planning reserve margin, and ISO seasonal capacity constructs are not considered in our analysis. A true operations-based model would have to include redundancy within the system to account for these factors beyond what this model produced. The price volatility of natural gas due to enhanced environmental regulations, restrictions based on carbon emissions, and other factors could significantly impact the results.

For Scenario 4, the flow limits for the primary transformer at the SJEC were removed, as it was assumed this would be upgraded to accommodate new resources. It should not be assumed from this analysis that a solar resource of this scale is practical at the studied location in terms of life-cycle economics or land availability. To determine this, additional work related to project economics, production output, and a rate impact analysis would be required.

Care should be taken as this analysis may represent an outlier case. Operations requirements may differ from the theoretical results, and it should not be assumed that the impact from DLR will be as significant for all situations.

Acknowledgments

The authors thank the teams at Ulteig, Energy Exemplar, and LineVision for partnering on this research. The information provided on PLEXOS® by Energy Exemplar and the dynamic line rating information provided by LineVision was critical in bringing this research to a successful conclusion. In addition, the authors would like to thank the Purdue University Energy Efficiency Center for its collaborative efforts. In the future, the broader team will be looking for partnerships with utilities to expand upon this analysis. Note that the views and opinions within this research are the authors' own and are not necessarily those of Ulteig.

References

[1] Howland, E. (2024, January, 9). PJM triples annual load growth forecast to 2.4% driven by data centers, electrification. https://www.utilitydive.com/news/pjm-interconnection-load-orecast-data-center-ev-dominion-firstenergy, accessed December 2024.

[2] Srivathsan, B., Sorel, M., Sachdeva, P., Bhan, A., Batra, H., Sharma R., Gupta, R., Choudhary (2024, October 29). AI power: expanding data center capacity to meet growing demand. https://www.mckinsey.com/industries/technology-media-and-telecommunications/our-nsights/ai-power-expanding-data-center-capacity-to-meet-growing-demand, accessed, accessed 22 January 2025.

[3] Wolfe, S. (2024, November, 11). Which cities are seeing the highest data center growth? https://www.power-grid.com/energy-business/policy-and-regulation/which-cities-are-seeing-the-highest-data-center-growth/, accessed 22 January 2026.

[4] Northern Indiana Public Service Company, LLC (November 2024). NIPSCO 2024 summary. https://www.nipsco.com/docs/librariesprovider11/rates-and-tariffs/irp/nipsco_2024-irp.pdf, accessed December 2024.

[5] Indiana Michigan Power (2024, September 2024). Indiana IRP stakeholder meeting #2. https://www.indianamichiganpower.com/community/improving-ourcommunity/projects/irp/stakeholder-engagement-process-Indiana, accessed December 2024.

[6] Roosa, S. (2025, March). Electricity generation in North America: the fuels that keep the lights on. *Alternative Energy and Distributed Generation Journal*, 7(1), 5–9.

[7] Roosa, S. (2025, June). Reinventing the electric grid. *Alternative Energy and Distributed Generation Journal*, 7(2), 25–48.

[8] Howland, E. (2024, June, 21). MISO faces possible capacity shortfall next summer in northern, central regions. https://www.utilitydive.com/news/miso-oms-midcontinent-capacity-survey-2024/719490, accessed 23 January 2025.

[9] Kramer, A. (2024). Andrew Kramer series: the necessity of gas turbines for data center growth. https://www.linkedin.com/feed/update/urn:li:-activity:7240952787277111296, accessed 22 January 2025.

[10] Lazard (June 2024). LCOE+ 2024 report. https://www.lazard.com/research-insights/levelized-cost-of-energyplus, accessed December 2024.

[11] Power Technology. Plant profiles: St. Joseph energy center, U.S. https://www.power-technology.com/marketdata/power-plant-profile-st-joseph-energy-center-us/?cf-view, accessed 22 January 2026.

[12] Wikipedia. Contingency (electrical grid). https://en.wikipedia.org/wiki/Contingency_(electrical_grid), accessed 22 January 2025.

15

Cutting Electricity Costs: A Guide for Organizations in Deregulated Markets

Shivani Saikar

In deregulated U.S. electricity markets, electricity costs are a substantial share of an organization's operating expenses. While the flexibility to choose providers and negotiate pricing structures presents cost-saving opportunities, it also brings challenges such as price volatility, contract complexities, and electrical demand charges. Without a defined strategy, businesses may face an energy transition fraught with volatile energy costs and financial uncertainties.

This chapter discusses key electricity pricing models, such as fixed-rate, variable-rate, and time-of-use plans, and outlines how organizations can manage costs through effective procurement strategies, energy efficiency improvements, and demand-side management. It also highlights the increasing importance of data-driven decision-making and renewable energy integration in reducing long-term electricity expenditures.

As energy markets evolve, adopting proactive strategies—like optimized procurement, real-time energy monitoring, and effective contract management—can help organizations reduce expenses, strengthen energy resilience, and improve their financial stability in competitive deregulated markets.

15.1 Introduction

In deregulated electricity markets, energy cost management is an operational concern and a key strategic financial decision. In regulated markets, electricity rates are fixed by a single utility. Deregulated markets allow organizations

DOI: 10.1201/9788743811312-15

to choose their energy supplier, negotiate contracts, and optimize procurement strategies. While this freedom creates opportunities for cost reduction and efficiency, it also introduces risks such as price volatility, complex contract terms, and demand-based charges. Unlike regulated markets, deregulated markets operate on competitive pricing models influenced by wholesale electricity markets, electric grid conditions, and policy changes. As a result, organizations must develop strategic approaches to energy procurement to mitigate financial risks and ensure cost predictability.

15.1.1 Key challenges in deregulated electricity markets

There are numerous challenges faced by deregulated markets, which include electricity price volatility, demand charges, contract complexity, and renewable energy considerations.

- Price volatility: Electricity rates fluctuate based on market demand, fuel prices, and extreme weather conditions, leading to unpredictable costs.
- Demand changes: Businesses with high peak electricity usage face higher costs, even if their overall consumption remains stable.
- Contract complexity: Retail electricity suppliers offer varied pricing models with different contract terms, fees, and risk factors, which require careful evaluation.
- Renewable energy considerations: Organizations looking to integrate clean energy into their procurement strategy must navigate pricing structures and grid reliability challenges.

To successfully manage electricity expenses in a deregulated market, organizations must understand pricing structures, leverage data-driven procurement strategies, and implement efficiency measures. This chapter provides a comprehensive guide to electricity cost optimization, covering:

- The types of electricity models available for deregulated markets such as fixed-rate, variable-rate, and time-of-use (TOU) pricing.
- Strategies to reduce electricity expenses through contract optimization, demand-side management, and renewable energy integration.
- Best practices for energy managers and procurement teams to mitigate risks and maximize cost savings.

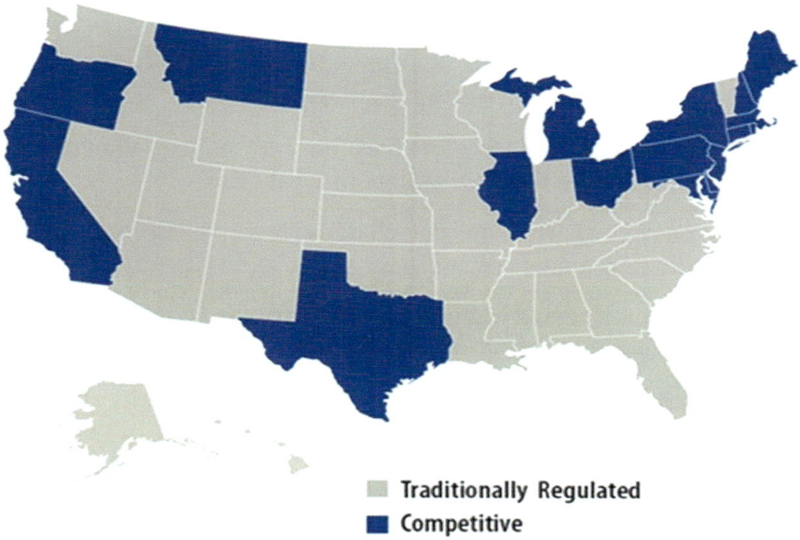

Figure 15.1 Retail electric power markets in the U.S [1].

By adopting proactive energy management strategies, organizations can reduce electricity costs, improve financial stability, and enhance long-term energy resilience in deregulated markets.

15.2 Electricity Pricing in Degregulated Markets

In deregulated electricity markets, organizations can select their electricity supplier instead of being bound to a monopoly utility provider. While this offers flexibility, it also requires careful evaluation of contract terms, market trends, and demand-based pricing models. States indicated in Figure 15.1 in darker shading have competitive (deregulated) retail electricity markets that allow customers to choose their electricity suppliers.

15.2.1 Optimizing electricity pricing

Electricity pricing is composed of multiple elements that influence the total cost paid by businesses. These components fluctuate based on market dynamics, supplier contracts, and local policy frameworks. The optimization strategies for these pricing components vary.

Wholesale energy cost: This is defined as the base cost of electricity purchased from power generators or energy markets.

- Impact on pricing: Wholesale prices fluctuate based on fuel prices, supply and demand, and generation capacity.
- Optimization: Procure energy through fixed-rate contracts or use market-based strategies to buy during low-price periods.

Transmission Charges: These are fees for transporting electricity over high-voltage power lines from generators to local grids.

- Impact on pricing: Transmission charges are regulated costs but subject to periodic adjustments by utilities.
- Optimization: Demand forecasting and distributed energy resources can reduce reliance on high-cost transmission.

Distribution charges: These are fees for delivering electricity from substations to businesses and households.

- Impact on pricing: Distribution fees are fixed per region but vary widely based on grid infrastructure investment.
- Optimization: Investing in microgrids or local renewable energy generation can help reduce dependency.

Demand charges*:* These are fees based on the highest electricity usage peak within a billing cycle.

- Impact on pricing: Demand charges are a major cost for businesses with high peak consumption.
- Optimization: Implement demand-side management strategies such as peak load shifting and battery storage.

Ancillary services fees*:* These are charges for grid stability, frequency control, and energy backup power services.

- Impact on pricing: Typically, ancillary services fees are a small portion of total costs but are necessary for reliable power.
- Optimization: Engage in demand response programs to help balance grid stability while receiving financial incentives.

Capacity charges*:* These are fees designed to cover investments in power plants and grid infrastructure to ensure future energy supply.

- Impact on pricing: Capacity charges are higher in areas where demand outpaces available generation.

- Optimization: Participating in peak shaving programs can reduce the need for excessive capacity charges.

Environmental and regulatory charges: These are costs related to renewable energy mandates, carbon pricing, and state-level energy policies.

- Impact on pricing: Charges vary by state and can increase with stricter carbon policies.
- Optimization: Enhance by using renewable energy credits (RECs) or power purchase agreements (PPAs) to meet compliance at a lower cost.

Retail supplier markup: This is the margin added by electricity providers for administrative costs and profit.

- Impact on pricing: Markups can be negotiated depending on contract terms.
- Optimization: Leverage competitive supplier bidding to lower retail markups.

Wholesale congestion charges: These charges are designed to reimburse for costs incurred due to bottlenecks in the transmission system that limit power flow.

- Impact on pricing: Fees charged are higher during peak demand periods.
- Optimization: Locate operations near lower-cost power sources or optimize grid interconnection strategies.

Balancing services charges: Fees for real-time adjustments that are needed to balance supply and demand fluctuations.

- Impact on pricing: Charges are based on grid stability needs.
- Optimization: Automated demand response can help reduce these costs.

Grid capacity expansion fees: These charges are passed on to businesses to fund the expansion or reinforcement of transmission networks.

- Impact on pricing: The prices charged vary based on the age of the transmission infrastructure and rates of amortization.
- Optimization: On-site power generation and storage can offset these costs.

Power factor penalties: These are fees for poor power factor correction, where reactive power disrupts grid efficiency.

- Impact on pricing: Higher fees are applied when a business, institutional, or industrial customer draws more apparent power than necessary.
- Optimization: Install capacitors or power factor correction devices to mitigate penalties.

Regulatory recovery fees*:* Surcharges are used to fund public benefit programs, low-income assistance, and renewable energy mandates.

- Impact on pricing: Pricing is fixed but varies by state and region.
- Optimization: Use RECs or direct PPAs to align with policy incentives.

Market settlement services: These are specific added costs related to financial settlements between generators and retail suppliers in wholesale markets.

- Impact on pricing: Typically, fees are low but can increase in volatile markets.
- Optimization: Fixed-rate contracts can protect against market fluctuations.

Ancillary reliability services: These are charges to maintain reserve capacity in the event of unexpected grid failures or outages.

- Impact on pricing: Small and variable impacts on pricing, but such fees are necessary for grid security.
- Optimization: Participating in demand response programs can reduce reliance on external reserve capacity.

Stranded asset costs: These fees are charged when utilities must recover investments in legacy power plants that become underutilized.

- Impact on pricing: The charges are more common in regions transitioning to renewables.
- Optimization: Businesses can hedge with self-generation to avoid dependency on stranded assets.

Several core components drive electricity pricing dynamics. Wholesale energy is often the largest contributor, with costs fluctuating based on fuel prices and grid demand. Demand charges, calculated from a facility's peak electricity

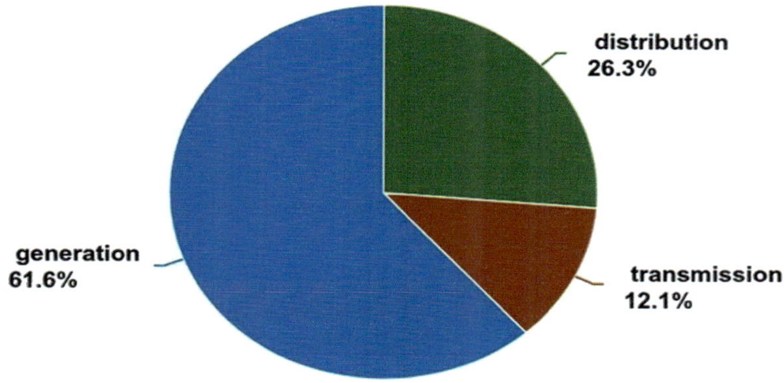

Figure 15.2 Major components of the U.S. average electricity price, 2022 [2].

usage, can increase overall costs if not carefully managed. Transmission and distribution costs cover the delivery of electricity through regional infrastructure, while congestion and capacity fees reflect investments needed to maintain and expand the grid. Additionally, regulatory and environmental fees vary by state and are shaped by renewable energy mandates, compliance obligations, and changes in policy.

Understanding and managing each of these components allows organizations to reduce costs through strategies that include demand-side management, competitive supplier bidding, and renewable energy contracting. Generation accounts for the largest share of electricity costs, followed by distribution and transmission (see Figure 15.2).

15.2.2 Choosing electricity pricing models

Organizations in deregulated markets can choose from various electricity pricing models. Selecting the appropriate model helps balance cost control with operational flexibility and risk tolerance. These pricing models (see Table 15.1) give organizations the opportunity to tailor electricity procurement strategies to their risk profile, financial goals, and load flexibility.

15.3 Procurement Considerations and Benefits

Once pricing structures are understood, organizations can deploy cost-reduction strategies across procurement, operations, and energy efficiency domains. There are ways to optimize procurement strategies. Strategic

Table 15.1 Electricity pricing models.

Pricing model	Description	Pros	Cons	Best for
Fixed-rate plan	A locked-in price per kWh over the contract term (e.g., 1-5 years).	Budget certainty, protection from price spikes.	No savings when market prices fall.	Organizations requiring predictability.
Variable-rate plan	Price per kWh fluctuates with wholesale market conditions.	Potential savings during low-price periods.	High risk of price spikes.	Businesses with flexible operations.
Time-of-use plan	Rates vary by peak and off-peak hours.	Encourages off-peak consumption; lower costs possible.	Expensive during peak hours.	Facilities with controllable load timing.
Indexed/ market-based plan	Prices tied to wholesale energy indices.	Cost advantages during market dips.	High volatility, complex planning.	Enterprises with energy procurement expertise.
Blended/block and index plans	Mix of fixed and variable pricing.	Some stability with potential for market gains.	Requires active oversight.	High-usage organizations managing energy risk.

procurement allows organizations to lock in favorable rates, mitigate risks, and leverage market trends. Key procurement strategies include:

- Competitive bidding: Soliciting quotes from multiple suppliers ensures better pricing and more favorable contract terms.
- Contract structuring: Customizing contracts using hybrid pricing models (e.g., fixed plus indexed) can provide risk mitigation and savings.
- Risk-based procurement: Some businesses hedge electricity purchases during low-price periods to protect against future volatility.
- Group purchasing (aggregation): Organizations can band together to increase buying power and reduce unit costs.
- Wholesale market participation: Sophisticated energy teams may purchase directly from wholesale markets, though this approach requires robust analytics and market insight.

By aligning procurement with usage patterns and market forecasts, businesses can gain greater control over electricity costs.

15.3.1 Demand-side management

Demand-side management (DSM) strategies reduce electricity consumption during peak demand periods, which can lower both energy and demand charges. Implementing DSM strategies can benefit both electrical energy consumers and providers. These strategies include:

- Load shifting: This approach involves rescheduling energy-intensive operations to times when electricity demand is lower, often during off-peak hours. By shifting activities such as manufacturing processes or large-scale data processing to these periods, organizations can take advantage of reduced electricity rates, thereby lowering overall energy expenses.
- Peak demand reduction: Utilizing on-site energy storage systems or deploying generation enables organizations to decrease their reliance on the electric grid during periods of high electricity demand. These practices reduce electrical demand charges, enhance energy independence and resilience, and ensure that critical operations can continue uninterrupted during peak times.
- Demand response participation: Engaging in demand response programs allows organizations to adjust their energy consumption in response to signals from utility providers, especially during grid emergencies or peak usage events. Participation in these programs can yield financial incentives, such as rebates or credits, and contribute to overall grid stability by alleviating stress on the electrical infrastructure during critical periods.

Examples of these benefits include:

- Cost savings: By optimizing and minimizing energy consumption during peak periods, organizations can achieve reductions in their electricity expenses.
- Enhance grid reliability: Collective participation in DSM programs will result in a more stable and efficient electricity grid, reducing the likelihood of outages and ensuring consistent energy availability.
- Environmental benefits: Reducing peak demand can decrease the need for additional power generation, leading to lower greenhouse gas emissions.

15.3.2 Energy efficiency improvements

Enhancing energy efficiency is a direct approach to reducing electricity consumption and costs. These measures help reduce operational costs while also promoting environmental sustainability by minimizing carbon emissions. Key strategies include:

- Upgrading equipment: Replacing outdated appliances and systems with energy-efficient models can lead to substantial savings.
- Process optimization: Streamlining operations to minimize energy waste enhances overall efficiency.
- Employee engagement: Educating staff on energy-saving practices fosters a culture of conservation within the organization.

15.3.3 Renewable energy integration

Incorporating renewable energy resources offers both economic and environmental benefits. These strategies not only offer potential cost savings but also enhance corporate social responsibility programs. Approaches include:

- On-site generation: Installing onsite renewables allows businesses to produce their own electricity, decreasing reliance on the electric grid and mitigating exposure to market price fluctuations.
- Power purchase agreements: Using PPAs enables consumers to purchase renewable energy at predetermined rates. PPAs provide price stability and support the growth of green energy projects.

15.3.4 Real-time energy monitoring and data-driven decision making

Utilizing real-time energy monitoring systems empowers consumers to make informed decisions based on accurate data. Effective use of real-time data enables organizations to boost efficiency, lower expenses, and lessen their environmental footprint. Benefits include:

- Identifying consumption patterns: Understanding when and how energy is used helps to pinpoint areas for improvement.
- Detecting anomalies: Real-time data can reveal irregularities, such as equipment malfunctions or energy leaks, allowing for prompt corrective action.
- Optimizing energy usage: Data analytics facilitate the development of strategies to shift energy-intensive operations to off-peak times, taking advantage of lower electric rates.

15.4 Best Practices for Energy Managers in Deregulated Markets

In deregulated energy markets, energy managers are instrumental in driving cost-effective procurement and efficient consumption. Their role is not only to manage supply contracts but also to strategically position their organizations to benefit from market opportunities. Expanding on previously outlined practices, the following are the key practices to enhance performance and reduce costs:

- Routine contract reviews: Regularly reviewing and reassessing energy contracts ensures that they are aligned with evolving market conditions and the organization's operational needs. This helps identify opportunities for renegotiation or switching to more favorable terms.
- Using energy analytics tools: Advanced analytics platforms provide real-time visibility into energy consumption and support accurate forecasting. These tools enable data-driven decision-making and help uncover inefficiencies that may otherwise go unnoticed.
- Deploying on-site renewables: Installing technologies such as solar panels, wind turbines, or battery energy storage systems can reduce dependence on grid electricity—particularly during peak pricing hours—while supporting sustainability goals.
- Engaging wholesale markets: For organizations with resources and expertise, direct participation in wholesale energy markets can offer access to more competitive pricing. This requires a proactive strategy and an understanding of market signals.
- Implementing demand response programs: Engaging in demand response initiatives allows organizations to adjust consumption during high-demand periods in exchange for financial incentives, ultimately reducing costs and supporting grid reliability.
- Ongoing education and market awareness: Staying current with regulatory updates, technological advancements, and energy market trends is essential. Continuous learning equips energy managers to make informed decisions and adapt strategies to a dynamic market environment.

15.5 Conclusions

Organizations operating in deregulated electricity markets must proactively manage their energy costs by understanding pricing structures, optimizing procurement strategies, and implementing demand-side management techniques. Key takeaways include:

- Selecting the appropriate electricity pricing model—whether fixed, variable, or time-of-use—is essential for cost control.
- Smart procurement and contract negotiation help mitigate market volatility risks.
- Demand-side management and renewable energy integration contribute to long-term cost reductions.
- Leveraging data-driven energy management tools enhances decision-making and operational efficiency.

By adopting a comprehensive electricity cost optimization strategy, organizations can improve financial stability, strengthen operational resilience, and advance long-term energy efficiency goals.

References

[1] U.S. Energy Information Administration. (2023). Retail electric power markets map. https://www.eia.gov/todayinenergy/detail.php?id=1170.

[2] U.S. Energy Information Administration (2023). Annual energy outlook 2023–Table 8: Electricity prices by sector and component. https://www.eia.gov/outlooks/aeo.

16

How Can Eco-friendly Power Be Supplied to Data Centers?

Stephen A. Roosa

Abstract

To support the growth of crypto-mining and artificial intelligence in the United States, companies have filed for permits to build new data processing centers. These data centers require substantial amounts of electricity that comes mainly from fossil fuel-fired generation. The larger ones use as much electricity and a medium-size U.S. city. The inefficient use of fossil fuels is inherently counter to the idea of reducing greenhouse gas emissions.

Decarbonization has been identified as one of the top three trends the energy sector [1]. This transformative concept challenges us to transition from our present reliance on fossil fuels to using more sustainable energy resources. As the energy transition gains momentum, data centers present a perplexing conundrum. How can the goals of the energy transition be achieved in the face of data center growth? Adding capacity to electric grids to meet the increased demand created by regional data centers is difficult. What are the alternative ways to provide electricity? Among the ways to provide this additional capacity is to use either renewables or nuclear energy. Regardless, natural gas is often chosen as a primary fuel to expand electrical generation in the U.S. This chpter explores the issues surrounding the use of natural gas to provide electricity for data centers and discusses their implications.

16.1 Introduction

Electric grid operators are seeing growth in electric demand due to the need for more air conditioning during summers, increased use of electricity by

DOI: 10.1201/9788743811312-16

electric vehicles (EVs), and the broader use of electric heat pumps. The electricity requirements of these applications are dwarfed by growth in crypto-mining and artificial intelligence (AI). Data centers need large amounts of power which present new electrification challenges. The U.S., China, and the European Union were the global leaders in data center development in 2022, with facilities collectively consuming approximately 500 terawatt-hours (TWh) of electricity [2]. The largest data center markets worldwide by power consumption capacity are in the U.S. states of Virginia (with over 500 data centers), Texas, Ohio, and Georgia. Other examples of central locations for data centers include Beijing, China, and London, England. Together, these hubs have a combined capacity need of 5,404 MW [2]. Companies that are pioneering new data center development include Amazon, Microsoft, Google, Digital Reality and Equinix. The rapid expansion of central data centers is being driven by cloud-based applications, expansion of 5G infrastructure, alternative intelligence computing, research, and virtual collaboration [3].

16.1.1 The problem of powering data centers

In a world primarily energized by fossil fuels, data center growth creates numerous challenges. Think of data centers as a highly-concentrated electrical loads. Data centers only rarely generate all their power requirements from onsite generation. They require both alternating current (AC) and direct current (DC) electricity. AC is used for operations, cooling, and grid interconnections. DC is stored in batteries to provide redundancy and backup power during outages. Newly constructed hyperscale data centers require power capacities of at least 100 MW which amounts to an annual electricity consumption equivalent to that of more than 400,000 EVs [2]. The electricity supplied must be reliable. The cost of downtime at data centers can average as much as $9,000/minute.

To support the growth of crypto mining and expansion of AI in the U.S., companies have filed for permits to build over 1,200 new data centers. The global power consumption of data centers is forecast to more than double between 2022 and 2026 [2]. For the U.S., data centers account for about 4% of the nation's total electricity usage; that number is expected to climb to as much as 12% by 2028 [3]. The electricity needed for new and existing data centers is expected to increase from 325 to 580 TWh by 2028 [2]. Existing energy supply infrastructure is presently inadequate to support this scale of growth.

Consider crypto mining. Without uninterrupted and inexpensive electricity, crypto mining is rarely profitable. For example, crypto miners use about 150,000 kWh to unlock a single Bitcoin—enough to power 170 average U.S. homes for a month [4]. If the needed electricity was priced at $.164/kWh (the average U.S. retail price in June 2025), the total energy cost would be $24,600. The coin gained varies in value due to market volatility. In February 2026 one Bitcoin was worth about $70,900.

Cryptocurrency mining operations are relatively mobile; their insatiable demand for electricity has driven crypto miners towards cheaper and dirtier sources of energy including restarting inefficient and aging coal-fired plants [5]. There has been an over-reliance on the existing U.S. electric grids, drawing on excess and reserve capacities for energy supplies. In parts of the country where electricity costs are deregulated, the added burden of increasing costs is being spread to residential and commercial customers.

16.2 Electricity from Clean Energy Resources

Where will this needed electrical capacity to support new data centers come from? Power generation is typically provided by fossil-fuel fired plants that use coal, natural gas, or oil. Others are supplemented by renewables such as hydropower, wind, or solar. While a fraction of the electricity is in some cases produced on site, most of the electricity needed for operations is normally supplied by the host utility grid which consists of the production, transmission, and distribution infrastructure. A key challenge is that data centers require 99.999% operations. This is difficult to achieve regardless of the energy resources used.

Some crypto mining operations are marketing their operations as eco-friendly yet green crypto mining remains an elusive proposition. This is due to financial considerations and their large demand for electricity. These operations also need large amounts of fresh water which creates hurdles for their development in dry regions of the U.S. So far, no true green crypto coin has yet emerged.

16.2.1 Renewable energy resources

The future of sustainable data centers requires collaboration, technological advancements, and decisive action to integrate cleaner energy resources [6]. Some renewable energy resources, such as solar PV and wind power, can provide supplementary electricity when available. Alone they are unable to provide the consistent base-load power needed by data centers. Others, such as geothermal and tidal power, hold potential to provide baseload generation but are rarely applied.

Many data centers already use renewable energy, especially solar, wind, and hydropower. However, the resources are typically not owned by the data centers but by operators who supply power to the electric grid. Adding renewable resources involves infrastructure integration. Solar and wind power are notoriously variable. Renewables require space for equipment and locations where the use of the resources can be optimized. Locational mismatches are the norm. The locations with the best solar insolation are often not the best places for data centers nor near transmission and distribution infrastructure.

To harmonize the electrical loads using renewables, some type of energy storage system is often needed. For utility-scale energy storage, pumped hydropower offers a high turn-around efficiency. Other solutions include battery energy storage systems particularly lithium-ion or sodium-sulfur batteries.

16.2.2 Nuclear energy

Nuclear reactors operate at full capacity more than other energy resources. The technology is designed to provide base load electrification. Nuclear plants operate 18 to 24 months at a stretch and refueling outages typically last only a few weeks. Regardless, a data center that relies solely on a nuclear plant may periodically require an alternative source of electricity.

There are no existing data centers that have their own nuclear reactors. In 2024, Microsoft and Constellation Energy reached a 20-year power purchase agreement to restart Three Mile Island Unit 1 to support Microsoft's data center operations [7]. Some companies are exploring placing data centers near existing nuclear plants for maximum efficiency and cost savings, including the Susquehanna Steam Electric Station in Pennsylvania and the Surry Nuclear Power Plant in northern Virginia [7]. Mirosoft and Meta are planning investments in nuclear energy to support some of their newest data centers. Google is planning a 50 MW small modular reactor in Oak Ridge, Tennessee, to power its infrastructure expansion.

There are challenges associated with the use of nuclear energy. The up-front costs are substantial. Design, construction, and commissioning require long lead times. Licensing, demonstrating, and deploying additional reactors require years [7]. Efforts to add more power using behind the meter stations have been placed on hold by the Federal Energy Regulatory Commission over concerns that such deals could allow data centers to benefit from transmission systems without paying for them, potentially raising costs for utility customers [7]. The U.S. domestic supply of uranium fuel to power the reactors of the future is not in place—particularly for high-assay low-enriched uranium which is required for many advanced reactor designs [7]. Spent nuclear fuel rods must be cooled after being used in a reactor and stored (typically onsite) until a permanent disposal location is established.

16.3 Crypto Mining Using Natural Gas from Abandoned Wells

Data centers are needed to support crypto mining operations. They require vast amounts of low cost, high-quality electricity. Cryptocurrency miners typically use proof-of-work models and employ vast arrays of computers to solve mathematical problems to verify transactions and earn income. Many crypto mining operations use AI for energy optimization, predictive maintenance, market analysis, and real-time decision making. Crypto mining companies attempt to use public funds in the form of lucrative tax credits and the exercise of eminent domain to benefit their interests and advance their projects. They often claim to bring much-needed development, high-paying employment, and added tax revenue to rural areas.

What does data center development have to do with abandoned oil and natural gas wells? Electricity can be produced using turbines fueled with natural gas from abandoned oil and gas wells. The technology is proven and commercially feasible. This application provides an excellent example for a case assessment. Though a fossil fuel is being used, it is often heralded by data center developers as having environmental benefits.

16.3.1 Background

Over the past hundred years, private landowners in partnership with U.S. oil and natural gas companies have drilled millions of wells [8]. When a well is no longer productive, most are abandoned and its operator moves on to richer deposits [8]. Drilling has left over three million abandoned oil and gas wells across the U.S. More than two million of these remain uncapped [9].

For most unproductive wells, the former owner(s) can be located. The term *abandoned wells* typically refers to those with a known owner or operator [8]. Being asset retirement obligations, they become a type of stranded industry liability [10]. *Orphaned wells* are inactive wells that are not reclaimed by the operators, have no identified liable party, and lack sufficient funding to cover the cost of environmental remediation and reclamation [11]. Of the total number of abandoned wells, about 125,000 are categorized as orphaned [12].

When not reclaimed, oil and natural gas wells create physical and environmental hazards. Not simply eyesores that present local environmental hazards, these abandoned wells gush millions of metric tons of methane—a greenhouse gas more potent than carbon dioxide over the time that it remains suspended in the atmosphere [9]. Plugging the wells prevents this and can reduce these fugitive emissions by as much as 99% [9]. These wells also pose risks of groundwater contamination by methane, brine, or other pollutants [11].

16.3.2 Reclamation

Reclamation is the process of restoring site stability and ecosystem functions. Disturbed lands are returned to their original use prior to disturbance, such as crop production, woodlands, or wildlife habitat. *Well reclamation* involves restoring the disturbed land areas to their natural state after oil and gas extraction has been completed and operations have ceased. After a drilling site has been safely abandoned, decommissioning involves removing as much infrastructure as possible from the site—this includes facilities, surface pipelines, wells, and related drilling equipment [13]. This process involves environmental assessments, site and equipment decommissioning and removal, soil replacement, revegetation, inspections, etc. [13]. After a well reclamation certificate is granted the oil and natural gas company remains responsible for surface issues related to reclamation, such as topography, vegetation, soil texture, and drainage for 25 years, and remains permanently responsible for contamination and any infrastructure left beneath the surface [11]. The reclamation responsibilities for orphaned wells are borne by government agencies, the public, and taxpayers [12].

Laws typically require that after production ceases, all oil and gas wells must be permanently plugged if abandoned [10]. In the U.S., state programs have allowed companies to set aside funds for the future estimated cost of capping wells [9]. However, funding is usually insufficient to retire and plug wells left behind by bankruptcy. If industry is unable to pay the costs, then its lenders, investors, creditors, and ultimately oil-producing states are eventually forced to cover such costs [10]. The exact per well costs of reclamation are often unknown yet commonly understated. Liabilities for reclamation are calculated based on an average cost of $20,000 to $40,000 per well; the actual expected cost for a modern shale well can be as much as $300,000 [10].

If not reclaimed, landowners must live with the consequences of millions of unplugged wells. This problem is more pervasive on federal lands, where drillers are required to set aside only $10,000 for a single well with a maximum of $150,000 for all wells drilled nationwide [9]. Louisiana alone is

Table 16.1 Number of at-risk oil and gas wells in selected U.S. states on federal lands [11].

	Colorado	Montana	New Mexico	Utah	Wyoming	Total
Orphaned wells	100	15	0	61	2,317	2,493
Unreclamed wells, inactive since 2015	537	497	529	3,609	3,609	8,781
Total	637	512	529	3,670	5,926	11,274

estimated to have about 4,600 abandoned wells with cleanup costs estimated to be $650 million dollars, or $141,300 per well [14]. Newer wells are often drilled deeper making them more technically difficult and costly to reclaim [11]. Often well site reclamation does not happen.

16.3.3 Environmental damage

The environmental damage from abandoned oil and gas wells goes beyond methane gas emissions. The wells can cause groundwater contamination and ecosystem degradation. Based on one study by the U.S. Government Accountability Office, most at-risk oil and gas wells on federal lands are in Wyoming with others in Colorado, Montana, New Mexico, Utah and Wyoming (see Table 16.1) [11]. Many have been inactive for over a decade. As time passes, site reclamation costs increase. These wells could cost up to $1.2 billion to fully reclaim—yet only an estimated $17 million is held in bonds for reclamation [11].

Abandoned oil and gas wells threaten local wildlife and their habitats. Over 5,000 at-risk well sites are in winter range or habitat restoration priority areas for big game animals [11]. Over 200 at-risk well sites are located on or near environmentally protected areas, including sensitive habitat for threatened or endangered species [11]. A total of 263 of these at-risk well sites are located within a mile of populated areas and 75 are located above the High Plains-Ogallala aquifer [11]. Held bond amounts are insufficient to fully cover modern reclamation costs. Abandoned or orphaned oil and natural gas wells are also found in Texas, Oklahoma, Pennsylvania, Kentucky, Ohio and elsewhere. There are environmentally conscious ways to reuse abandoned well sites. For example, the oil and gas industry has been active in reclaiming sites by using enhanced oil recovery.

16.3.4 Crypto mining operations

To indicate support for efforts that advance decarbonization, some crypto miners propose using the residual natural gas from abandoned well fields for the purpose of electrical generation. The concept of using abandoned natural gas operations and associated wells to power cryptocurrency mining rather than conventional energy resources supports the energy transition [15]. Natural gas-fired electrification can use highly efficient combined cycle generators. From an environmental perspective, this is preferable to using coal.

Locating crypto mining operations at or near the generating stations reduces distribution and transmission costs and loads on regional grids. However, many rural jurisdictions where abandoned natural gas wells are located, lack zoning for data centers and regulations that directly apply to their operations [16]. While not the essence of their business model, crypto miners can avoid this complication by applying to be classified as a private utility company [16]. If approved, this designation enables them to generate their own electricity, operating essentially as an independent microgrid. Since they only rarely sell the electricity produced, many opponents argue that this is a poor fit for existing zoning ordinances [16].

16.3.5 Using Natural Gas from Orphaned Well to Power Crypto Mining

Crypto mining at orphaned well sites is happening in Texas and South Dakota. The crypto miner's business model is to purchase old, inefficient natural gas wells, and extract the last bit of production from them to generate electricity for their operations [17]. The concept implies environmental benefits because the crypto mining operator is contractually required to cap the aging wells when they are completely depleted. This theoretically resolves the methane emission issue of site remediation. Unfortunately, many crypto mining operators fail to properly shut down and cap the natural gas wells, regardless of their commitments to do so. When their operations are no longer economical, they withdraw, and cease site activities, thus avoiding the costs of environmental remediation. Others file for bankruptcy protection prior to capping the wells as promised. This happens because the total cost of properly capping one natural gas well would likely far exceed the profit available from mining a single bitcoin.

There are numerous examples. Diversified Energy restarted the Longhorn Pad A site in Horton County, Pennsylvania in 2022 using its remaining natural gas reserves to power a crypto-mining facility [15]. Prior to starting operations in 2022, the company neglected to obtain a state-mandated air quality permit which it did not receive until December 2023 [17]. Despite committing to plug 14 wells after operations ceased, the firm failed to do so after moving elsewhere, invoking concerns regarding its history of neglecting cleanup duties for aging oil and gas plants [17]. The wells it used were left open which violated its commitments and regional laws governing operations [17]. Diversified Energy separately agreed to plug 3,000 additional wells by 2034, but state officials indicated the company has not provided updates on their efforts [17]. Using future environmental benefits to justify development and failing to deliver is a recurring problem in the data center industry. When the wells remain uncapped, the local stakeholders feel misled if not deceived. Rather that presenting an environmentally responsible public image, this constitutes a form of greenwashing.

Because of events like these some U.S. states are now against new energy-intensive crypto mining operations. Some argue that the high energy usage and electricity costs are ultimately subsidized by local consumers. This is because crypto mining owners are interconnected to a host electric grid which supplies power for peak demand. To attract development, industrial electricity tariffs are often lower than residential and commercial rates. Proposals for data center facilities have met with stiff public opposition during the approval process due to the potential of increased utility costs and lack of green energy products.

16.3.6 Climate-conscious approaches

The concept of using abandoned and orphaned natural gas wells to generate electricity for crypto mining operations in an environmentally appropriate manner is technically doable. Using energy efficient natural gas generators capable of capturing 98% of the leaking methane and using greenhouse gas offsets for any remaining fugitive methane emissions can be a workable and environmentally appropriate model [5]. Layering crypto mining into traditional oil and gas production operations could create additional sources of revenue for the industry [5]. However, while crypto miners may have digital programming prowess, they are not known to be environmental or energy experts. Effective regulatory policies must be in place else the power-hungry operators will simply abandon their sites when business conditions are no longer favorable.

By applying climate-conscious approaches, crypto mining at orphaned well sites can result in environmental advantages and support efforts that advance decarbonization [5]. Achieving this is difficult and costly if using natural gas from abandoned wells as the only resource to generate electricity. Without remediating the pollution from the oil and gas wells, the failure to deliver environmental benefits (i.e., reducing emissions and site remediation) represents an abject form of industrial greenwashing. When fossil fuels are used to generate electricity, crypto mining cannot rightly be called a green energy solution. A better use of abandoned and orphaned natural gas wells is to provide peaking power in combination with renewable generation.

16.4 Conclusion

The rapid growth of data centers, driven by crypto mining operations and advancements in AI, can overwhelm the demand for electricity [6]. The most advanced data centers and their AI microprocessors are useless without consistent and reliable electrical power. Given the drive toward decarbonization and the developing energy transition, the question is how can their insatiable demand for electricity be met? Most are supplied with power using fossil fuel-based generation. What are the alternatives?

Vantage Data Centers is developing a hyperscale data center in Texas that will have a capacity of 1.4 GW and uses advanced liquid cooling and a closed loop chiller to maximize operational efficiency [20]. While improving data center energy efficiency is important, meeting the growing electric demand for electricity by efficiency improvements alone is unlikely. From 2016 through to 2022, efficiency gains in processing capabilities were offset by increasing power demand from data centers. Afterwards this was no longer the case. Options to potentially improve energy efficiency include making use of the low-grade waste heat they tend to generate. Regardless, the electricity needed by data centers is outpacing efficiency improvements by a wide margin. Crypto mining and the increased use of AI are aggravating this problem.

Having exhausted efficiency improvements, the alternative is to provide more electricity. Natural gas generation reduces greenhouse gas emissions when compared to other fossil fuels. As shown by the example of using natural gas abandoned wells to generate electricity, it is difficult to supply data centers electrical loads without creating environmental issues.

There are alternatives to using fossil fuels to generate electricity for data center operations. Using renewables and nuclear power is possible yet both present challenging development scenarios in real-world settings. Platforms that float on stable bodies of water (i.e., lakes or fjords) and use renewables may offer a scalable, mobile, and rapidly deployable alternative to traditional land-based data centers plus overcome the challenges of power constraints, land scarcity, and permitting delays [21]. In these scenarios, water for cooling is readily available. Small modular nuclear reactors have potential to supply power in some situations [22]. The use of utility-scale nuclear power, such as Microsoft's scheme to reopen Three Mile Island, offers another solution.

16.4.1 Unresolved conundrums

Regardless of potential alternatives to supply electricity, the powering of data centers and their development present unresolved conundrums. Comprehensive solutions are needed. Achieving this is conceivable using a mix of enhancement initiatives and sustainable energy resources. Carefully studying data center site locations is an important first step. Locating data centers in cooler climates near large bodies of water (e.g., on the shores of the Great Lakes) will ultimately reduce their energy loads and operations costs. Using sites at or near renewable generation potential is an important consideration. Some of the better schemes involve the use of hybrid electrical generation platforms to resolve issues due to variable capacity.

The question remains as to how the electricity will be generated. Microgrids may offer solutions for some big data sites. To support data centers, they could be designed as private electric grids. The Hypergrid, near Amarillo, Texas, aims to integrate nuclear, natural gas, solar, and wind power sources for AI data centers generating as much as 11 GW [23]. There are numerous examples of data centers that provide some of their electricity using onsite renewable generation (e.g., roof-mounted solar PV systems). Research and demonstration of package nuclear energy technologies is another pathway. Battery energy storage systems have a role as do systems that enable longer term storage capacity such as pumped hydropower.

A more radical alternative is to use prefabricated, modular data center units in a decentralized configuration that are linked together in a data networking configuration. In this way, they could use excess electricity that is available locally on the host electric grid or generated nearby using renewables.

If the electricity can be generated, new infrastructure will be required for distribution and transmission. Despite the centralization of data center megaprojects, most of the infrastructure is not yet available to support their widespread development. Updating the electric grids and expanding regional generation are the only reasonable options. However, years are required for the completion of electrification projects. Once in place and operating, there are mismatches between the lifespan of data center and crypto mining operations. Lifespan estimates of these businesses based on optimal operating conditions average from 3 to 5 years, a period during which the mining hardware is expected to operate efficiently without significant loss in performance or failure [19]. Realizing longer-term sustainable energy solutions for their operations complicates investment profiles. Solar and wind farms have project lifecycles of 20 to 30 years.

16.4.2 Creating eco-friendly data centers

While some data centers use decarbonization to tout their sustainability and environmental initiatives, their operations are often linked to greenwashing schemes. Ultimately the responsibility lies with the data center owners and operators. They must find ways configure their facilities and operations to be more sustainable and environmentally appropriate.

An eco-friendly cryptocurrency mining operation that focuses on decarbonization might use renewable energy resources. It would employ best practices for electricity generation and use improved cooling systems that require less water consumption. Development would focus on enhanced scalability, environmentally sustainable methodologies that reduce energy use, and prioritize lowering greenhouse gas emissions [4]. Initiatives might entail the use of more energy-efficient mining hardware, the exploration of alternative consensus mechanisms like proof-of-stake, and software systems that optimize energy allocation and use. Data center owners must also accept financial responsibility for the needed electrical system infrastructure improvements.

References

[1] Marr, B. (2022, February 11). The 3 biggest future trends (and challenges) in the energy sector. https://www.forbes.com/sites/bernardmarr/2022/02/11/the-3-biggest-future-trends-and-challenges-in-the-energy-sector/?sh=71f4a87127b7, accessed 8 February 2026.

[2] Statista (2025, August 11). Data center power consumption, statistics and facts. https://www.statista.com/topics/13055/data-center-power/#topicOverview, accessed 20 August 2025.

[3] UtilityDive (2025, June 23). The rapid expansion of data centers across the U.S. and the critical role of renewable energy in powering the grid. https://www.utilitydive.com/spons/the-rapid-expansion-of-data-centers-across-the-us-and-the-critical-role-o/751006, accessed 21 August 2025.

[4] Moiner Bros (2023, May 16). What is green crypto mining? How does it help reducing environmental pollution? https://www.cryptominerbros.com/blog/what-is-green-crypto-mining, accessed 4 June 2025.

[5] Bassett, B. and Grument, S. (2022, January 28). Crypto mining could give orphaned natural gas wells a new lease on life. https://www.global-counsel.com/insights/research/crypto-mining-could-give-orphaned-natural-gas-wells-new-lease-life, accessed 4 June 2025.

[6] Whire, R. (2024, October 18). Renewable energy in data centers: drivers, challenges, and solutions. https://keyfive.com/thecurrent/renewable-energy-in-data-centers-drivers-challenges-and-solutionsnbsp, accessed 4 August 2025.

[7] U.S. Department of Energy (2025, April 8). Advantages and challenges of nuclear-powered data centers. https://www.energy.gov/ne/articles/advantages-and-challenges-nuclear-powered-data-centers, accessed 21 August 2025.

[8] Turrentine, J. (2021, July 26). Millions of leaky and abandoned oil and gas wells are threatening lives and the climate. https://www.nrdc.org/stories/millions-leaky-and-abandoned-oil-and-gas-wells-are-threatening-lives-and-climate, accessed 2 June 2025.

[9] Energy Innovation: Policy and Technology (2020, September 21). Plugging abandoned oil wells is one 'Green New Deal' aspect loved by both republicans and democrats. https://www.forbes.com/sites/energyinnovation/2020/09/21/plugging-abandoned-wells-the-green-new-deal-jobs-plan-republicans-and-democrats-love/?sh=24abee-a2e100, accessed 2 July 2025.

[10] Schuwerk, R. and Rogers, G. (2020, June 18). It's closing time: the huge bill to abandon oilfields comes early. https://carbontracker.org/reports/its-closing-time, accessed 22 October 2025.

[11] Western EcoSystems Technology, Inc. (2021, March 1). Inactive oil and gas wells on federal lands and minerals: potential costs and conflicts. https://www.nwf.org/-/media/Documents/PDFs/Press-Releases/2021/03-17-21_Inactive-Oil-and-Gas-Wells-on-Federal-Lands-and-Minerals-Report, accessed 2 June 2025.

[12] Boutot, J., Peltz, A., McVey, R. and Kang, M. (2022, October 18). Documented orphaned oil and gas wells across the United States. *Environmental Science and Technology*, 56(20), 14228–14236.

[13] Alberta Energy Regulator. Reclamation process and criteria for oil and gas sites. https://www.aer.ca/regulations-and-compliance-enforcement/site-closure-requirements/reclamation/oil-and-gas-sites/reclamation-process-and-criteria-oil-and-gas-sites, accessed 22 October 2025.

[14] Oldani, J. (2025, February 11). Turning abandoned oil wells into clean energy. https://www.thedriller.com/articles/93303-turning-abandoned-oil-wells-into-clean-energy, accessed 3 June 2025.

[15] Turi, J. Officials sound alarm after crypto firm quietly disappears controversial energy site: 'Disaster waiting to happen'.
[16] Gloeicki, M. (2025, June 6). New site proposed for data center. *Courier Journal.*
[17] Adedoyin, P. (2025, May 8). Crypto miner leaves unplugged wells in Pennsylvania, environmental concerns grow. https://www.tronweekly.com/crypto-miners-unplugged-wells-in-pennsylvania, accessed 4 June 2025.
[18] LG&E KU (2023, November 6). We're creating a more sustainable energy future with the right mix—responsible, affordable and reliable. https://lge-ku.com/future#:~:text= The%20KPSC%20noted %20the%20need,%2Dthe%2Dclock%20energy%20needs, accessed 5 June 2025.
[19] KryptyHex (2024, June 17). The lifespan of an ASIC miner: How long do ASIC miners last? https://d-central.tech/the-lifespan-of-an-asic-miner-how-long-do-asic-miners-last/#:~:text=The%20evolution%20of%20ASIC, accessed 22 October 2025.
[20] Sinha, S. (2025, August 21). Oracle to back massive 1.4 GW data center in Texas with gas power: report. *Interesting Engineering.*
[21] Walker, J. (2025, June 8). Japanese shipping firm to develop 'world's first' integrated floating data center platform. https://www.datacenterknowledge.com/next-gen-data-centers/japanese-shipping-firm-to-develop-world-s-first-integrated-floating-data-center-platform, accessed 18 September 2025.
[22] Roosa, S. (2024, September). Nuclear energy—lessons from Georgia's new power plant. *Alternative Energy and Distributed Generation Journal*, 6(3), 5–12.
[23] Curviello, M. (2025, June 27). Texas Tech, Fermi America announce world's largest AI energy campus near Pantex. *Amarillo Globe-News.*

17

Policy Design for Prosumer Viability: Evaluating Solar PV Grid Connection Mechanisms in Jordan's Energy Transition

Samer Adnan Zawaydeh

This chapter evaluates the economic and environmental impacts of three electric grid connection mechanisms—net billing, zero-to-grid, and buy-all/sell-all—on residential solar photovoltaic (PV) prosumers in Jordan. Drawing on empirical data from Jordan's Ministry of Energy and Mineral Resources, transmission system operators, distribution system operators, and international energy reports, a techno-economic analysis was conducted to assess the viability and performance of distributed solar PV systems under evolving regulatory frameworks.

Findings indicate that residential solar PV systems contributed approximately 9% of Jordan's total electricity generation in 2024, avoiding an estimated 1.026 million metric tons of carbon dioxide emissions annually. An economic evaluation reveals significant variation across connection mechanisms: net billing yields the highest internal rate of return (IRR) at 27.7% for high-consumption households, while buy-all/sell-all offers up to 15.6% IRR, particularly under declining PV system costs. The zero-to-grid mechanism emerges as advantageous for consumers with substantial daytime usage due to Jordan's tiered tariff structure.

A critical insight from the analysis concerns the adverse impacts of regulatory delays on market uptake. A delay in implementing Bylaw 58, 2024, resulted in a 40% decline in new installations during mid-2024, underscoring the importance of timely and transparent policy implementation. Inconsistencies in reporting practices across distribution companies are also

DOI: 10.1201/9788743811312-17

identified, which hinder accurate monitoring and planning. The study further explores the emerging role of battery energy storage systems (BESS) in enhancing prosumer self-sufficiency. With the levelized cost of energy for solar plus storage reaching $0.0592/kWh, such systems are now competitive with grid-supplied electricity ($0.0726/kWh), signaling a shift toward greater decentralization.

This research provides a comprehensive assessment of prosumer economics and policy design in an emerging market context, offering actionable insights for accelerating the transition to more sustainable energy resources. Policy recommendations are presented that aim to harmonize regulatory frameworks, promote standardized data collection, incentivize energy storage integration, and enable equitable access to solar PV adoption.

17.1 Introduction

The global energy landscape has recently undergone a profound transformation. This is driven by technological innovation, falling costs of renewable energy technologies, and urgent calls to decarbonize electricity systems. According to the International Energy Agency's Global Energy Review 2025, global electricity demand surged in 2024 at nearly twice the average growth rate of the previous decade, largely fueled by increased consumption for cooling, industrial operations, electric vehicles (EVs), and data centers [1]. Renewable energy sources accounted for the largest share of this demand growth, supported by rapid advancements in solar photovoltaic (PV) technology and manufacturing particularly in China. As reported by SolarPower Europe (2025), the world took 68 years to reach the first terawatt (TW) of cumulative installed solar PV capacity, but only two additional years to add the second TW [2]. By 2024, global installed solar PV capacity had reached 596 gigawatts (GW), with projections indicating that annual installations will soon exceed 1 TW per year before the end of the decade.

Solar PV technology not only contributes to meeting rising energy demands but also has a central role in mitigating climate change. The International Energy Agency (IEA) estimates that low-emission energy sources accounted for over 80% of incremental electricity demand growth in 2024, helping to decouple economic growth from carbon emissions [1]. However, extreme weather events and fossil fuel dependencies continue to exert upward pressure on global carbon dioxide (CO_2) emissions, which reached record levels in 2024. Against this backdrop, distributed solar PV systems have emerged as a viable solution for reducing reliance on centralized fossil-fuel-based generation, improving energy access, and empowering consumers to become prosumers.

In Jordan, a country heavily reliant on imported energy and facing growing electricity demand, the adoption of residential solar PV systems has gained momentum. The country issued its renewable energy and energy efficiency Law 13, 2012, with a net metering scheme for all sectors. This law was updated to Law 12, 2024, which substituted the net metering mechanism with three different connection mechanisms. As of 2024, solar PV contributed approximately 9% of total electricity generation in the country, avoiding an estimated 1.026 million metric tons of CO_2 emissions annually. This shift toward decentralized energy production aligns with Jordan's national energy strategy (2020–2030) and international climate commitments under the Paris Agreement. However, the pace and pattern of adoption are influenced by the regulatory frameworks governing electric grid connection mechanisms.

From an economic perspective, these mechanisms shape investment incentives, consumer behavior, and grid stability. They reflect different approaches to valuing distributed energy resources (DERs) and managing the bidirectional flow of electricity between households and utilities. Each mechanism introduces distinct pricing signals, cost recovery structures, and equity considerations, all of which affect the internal rate of return (IRR), simple payback period, and overall attractiveness of solar investments for households. For instance, net billing offers high IRRs for large consumers due to Jordan's tiered tariff structure, while the buy-all/sell-all mechanism provides a more predictable revenue stream, albeit at lower rates. The zero-to-grid model benefits consumers with significant daytime usage but requires accurate load matching and time-of-use (TOU) alignment.

The transition to prosumer-driven energy systems carries important social and environmental implications. For households, solar PV adoption reduces electricity bills, enhances energy independence, and supports the electrification of essential services such as heating, cooling, and EV battery charging. Nationally, it strengthens energy security by reducing Jordan's dependence on imported fuels and volatile international markets. It also lowers transmission and distribution losses, defers infrastructure investments, and creates local employment opportunities. Globally, widespread deployment of solar PV contributes to climate mitigation efforts, reduces air pollution, and supports progress toward the UN Sustainable Development Goals (SDGs), particularly SDG7 (Affordable and clean energy), SDG8 (Decent work and economic growth), and SDG 13 (Climate action).

Despite economic and social benefits, the integration of prosumers into the electricity grid presents policy challenges related to regulation, pricing, and technical coordination. A notable example occurred in mid-2024 when delays in implementing updated regulatory bylaws led to a 40% drop in new solar PV installations in one of Jordan's distribution zones. Such disruptions highlight the importance of timely policy implementation, transparent communication, and standardized reporting practices across distribution system operators (DSOs). Moreover, the increasing affordability of battery storage systems is reshaping the prosumer value proposition, enabling greater self-consumption and reducing reliance on grid buy-back rates.

SolarPower Europe (2025) estimates that global annual solar PV capacity additions will soon exceed 1 TW annually, signaling rapid technological advancement, improved efficiency, and falling deployment costs [2]. This projected growth will require a parallel expansion of the global clean energy workforce to meet installation demands. With global electricity consumption reaching 27,000 terawatt-hours (TWh) in 2023, the world's CO_2 emissions from electricity generation alone are about 13.6 gigatons (Gt), yielding an average grid emission factor of 0.5 kg CO_2/kWh [3]. With the average annual energy yield of solar PV systems worldwide reported to be 1,300 kWh/kWp in 2023, each installed terawatt (TW) of solar PV capacity generates an average 1,300 TWh of clean electricity annually—avoiding roughly 650 million metric tons of CO_2 emissions annually [4].

17.2 Impacts of Prosumer Solar PV

Solar photovoltaic technology delivers substantial economic, social, and environmental benefits across multiple levels—ranging from individual prosumers to global sustainability outcomes (see Table 17.1). These benefits reinforce the importance of integrating distributed solar energy into modern electricity systems, particularly in emerging economies such as Jordan.

17.2.1 Economic impacts

For households and businesses, solar PV reduces electricity costs by generating on-site power at a lower levelized cost of energy (LCOE) than grid-supplied electricity. This enables consumers to reallocate savings toward essential needs such as healthcare, education, and nutrition. The integration of solar PV also facilitates the electrification of energy services, supporting efficient heating, cooling, and EV battery charging without relying on fossil fuels.

On a national scale, harnessing solar energy leverages domestic natural resources, enhancing energy security by reducing dependence on imported

Table 17.1 Economic, social, and environmental benefits for distributed energy resource owners.

	Benefits from Distributed Energy Resources – Solar PV		
	Owner Benefits	**National Benefits**	**Global Benefits**
Economy	Reduced cost of electricity	Use natural resources	Reduce shipping of energy
	Transition into electrification of energy	Energy security	Reduce pipelines to transmit energy
	Increased spending on other essential needs	No investment in generation	Reduce losses in resources
	Cover heating and cooling electricity needs	Increase revenues from taxes	Increase national welfare
	Cover EV electricity needs	Reduce cost of electricity	Reduce global poverty
		Reduce budget deficit	Increase access to electricity
		Reduce impact of inflation in international oil prices	Increase financing for green projects
		Achieve national green finance strategies	Help achieve climate finance for $300 billion/year for developing countries
		Reduce losses in distribution, transmission and generation	
Social	Available local manpower	Create local jobs	Create local jobs
	Capacity-building programs	Export professional experience	Achieve UN SDGs
		Clean energy workforce development	Reduce health and safety risk at job sites
		Develop clean energy local supply chains	Support sustainability projects
Environment	Reduce carbon footprint	Reduce national CO_2 emissions	Reduce global CO_2 emissions
	Less urban heat island effect	Achieve net-zero targets	"CO_2-free" energy growth
	Increased share of energy	Reduce electricity CO_2 emissions	Reduce global climate change
	Less air pollution	Reduce urban pollution	Reduce demand on cooling
	Less noise	Reduce cost of sick days	Achieve Paris Agreement
		Increase employee productivity	Improve global health
	Better health	Achieve NDC targets	Preserve bio-diversity

fuels and volatile international oil markets. This transition lowers the need for government investment in conventional power generation infrastructure and minimizes distribution, transmission, and generation losses. This contributes to a more efficient and resilient energy system. Additionally, reduced electricity costs for consumers can stimulate economic activity and increase tax revenues, supporting public finances and helping to reduce budget deficits.

Globally, the widespread adoption of solar PV diminishes the need for long-distance energy shipping and extensive pipeline networks, thereby cutting resource losses and associated environmental impacts. These shifts contribute to increased national welfare, support poverty reduction efforts, and improve access to reliable and affordable electricity—particularly in underserved and remote regions. This supports policies that promote inclusive and sustainable global development.

17.2.2 Social impacts

For owners, solar PV systems can be installed and maintained using local manpower, fostering community-based employment opportunities and supporting local economic development. Deploying solar PV also encourages capacity-building programs that enhance technical skills and promote the development of clean energy supply chains, strengthening regional resilience and self-sufficiency.

On a national scale, the solar PV industry drives job creation across manufacturing, installation, maintenance, and research and development sectors. This growth supports the export of professional experience and best practices in renewable energy technologies, positioning countries as leaders in the global clean energy transition. National investments in solar PV also contribute to the development of a skilled clean energy workforce, aligning labor markets with future energy demands.

At the global level, the expansion of solar PV supports the creation of sustainable employment opportunities in both developed and developing economies, thus contributing to poverty reduction and inclusive economic growth. By promoting cleaner and safer working environments compared to fossil fuel industries, solar PV helps reduce health and safety risks at job sites. Furthermore, its alignment with decentralized and modular deployment models enhances access to modern energy services, directly supporting the achievement of several United Nations Sustainable Development Goals (SDGs), particularly SDG 7 (Affordable and clean energy), SDG 8 (Decent work and economic growth), and SDG 13 (Climate action).

17.2.3 Environmental impacts

At the owner level, adopting solar PV reduces the carbon footprint by generating clean, renewable electricity, which decreases reliance on fossil fuels. It increases the owner's share of self-produced energy, leading to greater energy independence and lower utility bills. Solar PV systems produce no on-site pollution or noise, contributing to a quieter and healthier living environment, which can improve overall well-being.

On a national scale, widespread adoption of solar PV helps reduce CO_2 emissions and lowers greenhouse gas emissions, specifically from the electricity sector. This contributes to cleaner air in urban areas, reducing smog and other pollutants that affect public health. It also supports nations in meeting their nationally determined contributions (NDCs) under international climate agreements.

At the global level, expanding solar PV has a crucial role in reducing global CO_2 emissions, which mitigates the pace and severity of climate change. By displacing fossil fuel-based power generation, solar energy also reduces the overall demand for cooling associated with heat-intensive energy production processes. This global transition supports efforts to meet the goals of the Paris Agreement, particularly limiting global temperature rise. Furthermore, by curbing pollution and land degradation associated with fossil fuel extraction and combustion, solar PV contributes to preserving biodiversity and helps maintain healthy ecosystems.

17.3 Methodology

This study evaluates the economic and environmental impacts of solar PV systems installed by primarily residential prosumers in Jordan under the three grid connection mechanisms: net billing, zero-to-grid, and buy-all/sell-all. The methodology integrates empirical data, regulatory frameworks, and techno-economic modeling. Limitations are due to data gaps in the annual report and websites.

17.3.1 Data collection and sources

Energy demand and solar PV deployment

- Electricity consumption trends (2010–2024) from Jordan's Ministry of Energy and Mineral Resources (MEMR) facts and figures report [5].
- Electricity consumption from the annual report of the transmission system operator, National Electric Power Company (NEPCO) [6].

Table 17.2 Residential electricity tariff in Jordan as of April 2020.

Monthly electricity consumption (kWh)	Supported group tariff ($/kWh)	Unsupported group tariff ($/kWh)
Below 300	$0.071	$0.169
301–600	$0.142	
601–1,000	$0.284	
Over 1,000	$0.284	$0.212

- Electricity consumption per sector from distribution operators' annual reports: Jordan Electric Power Company (JEPCO), Irbid Distribution Company (IDECO), and Electricity Distribution Company (EDCO) [7–9].
- Global solar PV capacity growth and cost trends from IEA and SolarPower Europe reports (2023–2025) [2,4].

Regulatory policies

- Tariff structures for all sectors from the Energy and Mineral Regulatory Commission (EMRC) annual report and website [10].
- The electricity tariff for the residential supported sector is shown in Table 17.2.
- Connection mechanisms, DC:AC ratio, daily tariff periods, and grid fees for each sector after September 2024, from MEMR website Bylaw 58, 2024 [11].

17.3.2 Technical parameters and assumptions

Solar PV systems

- Degradation rate: Linear decline from 99% (year 1) to 88.85% (year 30), based on the current 2025 available solar PV data sheet specifications.
- Degradation rate: Linear decline from 99% (year 1) with a reduction of 0.4% up to year 30, based on the old solar PV data sheet specifications.
- Energy yield: 1,560 kWh/kWp/year in Jordan, and a global average of 1,300 kWh/kWp/year based on the IEA-PVPS 2023.

- Inverter changed every 8 years at a cost of \$150[1]/KWac.
- Salvage value of solar PV system at year 30 is 15% of the initial cost.
- The LCOE represents the average cost per unit of electricity generated over the lifetime of a power system, accounting for all costs (capital, operations, maintenance, fuel, etc.) and discounted to present value.

Grid losses

- Distribution (13%), transmission (2%), and generation (50%) losses, derived from operator reports (NEPCO, JEPCO 2023) and author estimates [6,7].
- Assumed uniform distribution losses (13%) across all regions.

17.3.3 Economic analysis framework

Key metrics

- Net present value (NPV): The discount rate of 3% was based on the low-interest-rate loan of the Clean Energy Central Bank of Jordan.
- Internal rate of return (IRR): The IRR was calculated using spreadsheet formulas.
- Simple payback period (SPP): A solar PV system lifespan of 30 years is assumed.

Sensitivity Analyses

- Varied system sizes (1.29–14.8 kWp), solar PV costs (\$450–\$600/kWp), and electricity tariffs (\$0.056–\$0.282/kWh).
- CO_2 emissions avoided: Calculated using Jordan's grid emission factor (0.4585 kg CO_2/kWh, SBUR 2020) [13].
- Jordan's overall CO_2 emissions per Fourth National Communication on Climate Change are 32.647 million tons CO_2/year [16].
- Annual savings converted to barrels of oil equivalent (1 BOE = 11,630 kWh) and monetary value (Brent crude at \$83/barrel).

---

[1] All currency values are in U.S. dollars unless otherwise indicated.

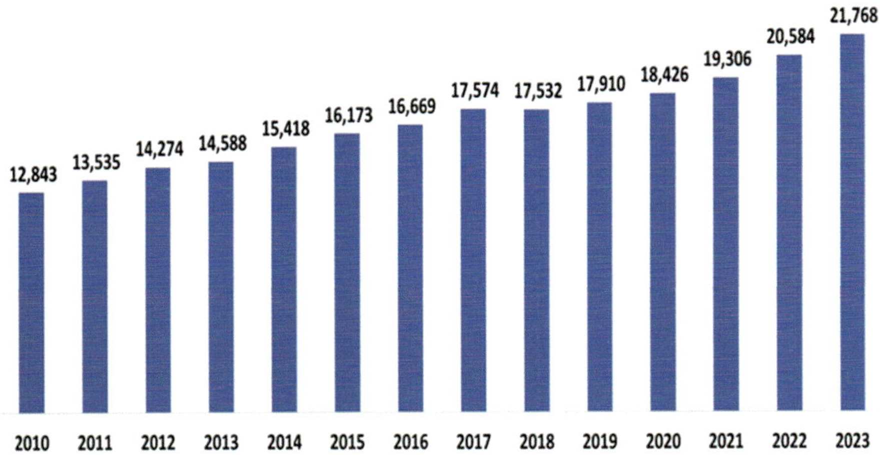

Figure 17.1 Total electricity consumption in Jordan from 2010 to 2024 (GWh) [5].

17.4 Electricity Status in Jordan

From 2010 to 2023, the total electricity consumption in Jordan is shown in Figure 17.1 [5,14]. The compound annual growth rate (*CAGR*) is calculated using Equation 17.1.

$$CAGR = (\text{Starting value}/\text{Energy value})^{1/n} - 1 \qquad (17.1)$$

For Equation 17.1, the starting value is 12,843 GWh, the ending value is 21,768 GWh, the number of years is 13, and the annual growth rate is 3.97%. The residential sector annual electricity consumption during the same period is shown in Figure 17.2.

Based on the annual report for the DSO in the northern part of Jordan, IDECO, the electrical consumption in the residential sector from 2012 to 2024 is shown in Figure 17.3 [8]. It had an average of 3,233 kWh/year (270 kWh/month), a mean of 3,231 kWh/year, and a standard deviation of 119 kWh.

As documented in the 2024 annual reports of Jordan's three DSOs, the cumulative installed capacity of solar photovoltaic (PV) systems integrated into their distribution grids by the end of 2024 is presented in Figure 17.4 [7–9].

The economic and environmental benefits for the 1,196 MWac on the three distribution grids are calculated over the next 20 years using the following assumptions:

- Solar PV degradation is 0.99 at year 1, with a 0.4% decrease annually.

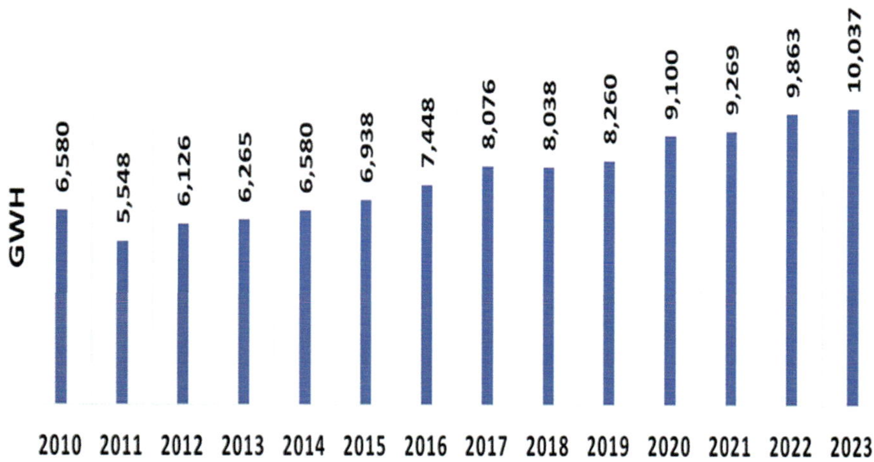

Figure 17.2 Residential sector annual electricity consumption in Jordan from 2010 to 2023.

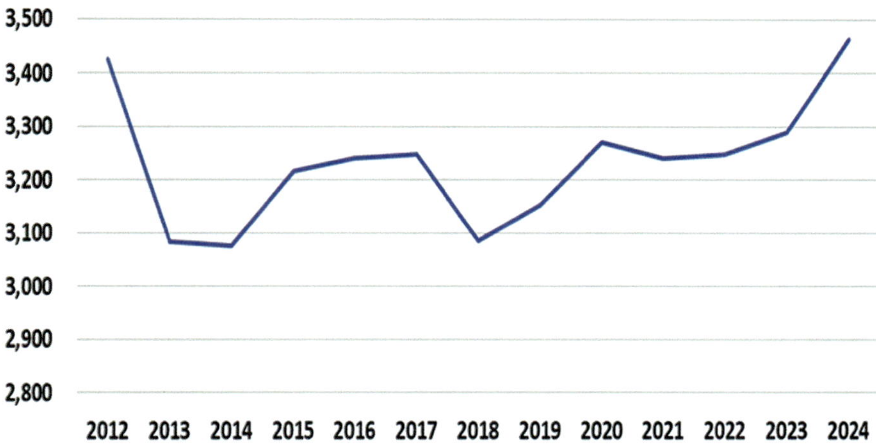

Figure 17.3 Residential sector average annual electricity consumption in IDECO region from 2012 to 2024 (kWh/year) [8].

- The DC:AC ratio is 1.2:1.0.
- The average annual specific production for the solar PV is 1,560 kWh/kWp.
- The losses on the distribution grid are 13%.

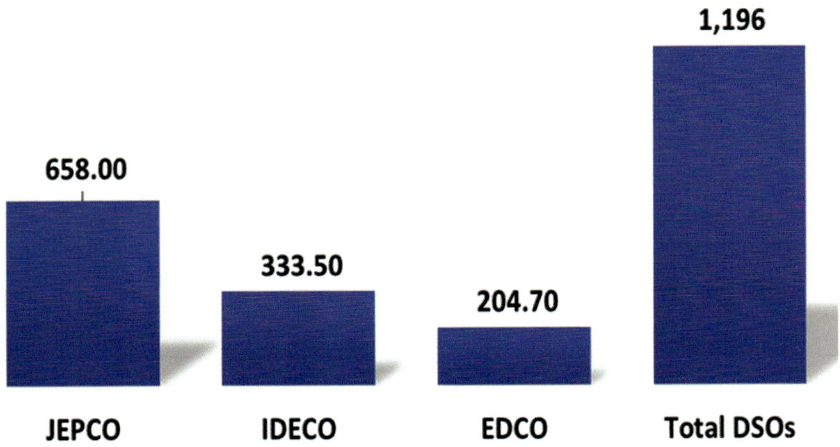

Figure 17.4 Total installed solar PV capacity on distribution grids in Jordan by the end of 2024 (MWac).

- The losses on the transmission grid are 2%.
- The losses in generation are 50%.
- 1 ton oil equivalent is equal to 11,630 kWh.
- 1 ton oil equivalent is equal to 7.33 barrels of oil equivalent.
- The cost of a barrel of oil in the base year is an average $83.
- The rate of inflation is 1% annually.
- The interest rate is 3%.

The annual consumed electricity by prosumers is estimated at 2.2 TWh, and overall savings after calculating the distribution, transmission, and generation losses at 2.2 TWh. The reduced imports are 3.28 million barrels of oil equivalent annually, estimated savings of $264.1 million, and a 20-year net present value of $4.2 billion. The estimated CO_2 reduction is 2.38 million tons with a 20-year sum of 45.8 million tons of CO_2 as shown in Table 17.3.

17.4.1 Impact of change in connection mechanism regulations

The renewable energy and energy efficiency Law 13, 2012, was in force until the update of Law 12, 2024, was issued in June 2024 [15]. Unfortunately, the regulations to implement this law were several months late and Bylaw 58, 2024, was not issued until September 2024 [11]. A sharp drop in solar PV

Table 17.3 Annual economic and environmental impact of 1,196 MW installed solar PV capacity in Jordan.

Year	Degradation	Annual consumed electricity by prosumers (TWh)	With added 13% losses in distrib. grid avoided (TWh)	With added 2% losses in trans. grid avoided (TWh)	With added 50% losses in Gen. avoided (TWh)	Reduced imports (million ToE)	Reduced imports (million BoE)	Reduced imports [1% inflation] ($ USD)	Reduced imports ($ million USD)	Present value reduced imports [i=3%] ($ million USD)	Annual CO_2 savings (million Ton CO_2/ year)
1	0.990	2.22	2.55	2.60	5.20	0.447	3.28	$ 83.0	$ 272.0	264.1	2.38
2	0.986	2.21	2.54	2.59	5.18	0.445	3.26	$ 83.8	$ 273.6	257.9	2.37
3	0.982	2.2	2.53	2.58	5.16	0.443	3.25	$ 84.7	$ 275.2	251.9	2.36
4	0.978	2.19	2.52	2.57	5.14	0.442	3.24	$ 85.5	$ 276.8	246.0	2.36
5	0.974	2.18	2.51	2.56	5.12	0.440	3.22	$ 86.4	$ 278.5	240.2	2.35
6	0.970	2.17	2.50	2.55	5.09	0.438	3.21	$ 87.2	$ 280.1	234.6	2.34
7	0.966	2.16	2.49	2.54	5.07	0.436	3.20	$ 88.1	$ 281.7	229.1	2.33
8	0.962	2.15	2.48	2.53	5.05	0.434	3.18	$ 89.0	$ 283.4	223.7	2.32
9	0.958	2.14	2.47	2.52	5.03	0.433	3.17	$ 89.9	$ 285.0	218.4	2.31
10	0.954	2.14	2.46	2.51	5.01	0.431	3.16	$ 90.8	$ 286.7	213.3	2.30
11	0.950	2.13	2.44	2.49	4.99	0.429	3.14	$ 91.7	$ 288.3	208.3	2.29
12	0.946	2.12	2.43	2.48	4.97	0.427	3.13	$ 92.6	$ 290.0	203.4	2.28
13	0.942	2.11	2.42	2.47	4.95	0.425	3.12	$ 93.5	$ 291.6	198.6	2.27
14	0.938	2.10	2.41	2.46	4.93	0.424	3.10	$ 94.5	$ 293.3	193.9	2.26
15	0.934	2.09	2.40	2.45	4.91	0.422	3.09	$ 95.4	$ 295.0	189.3	2.25
16	0.930	2.08	2.39	2.44	4.88	0.420	3.08	$ 96.4	$ 296.6	184.9	2.24
17	0.926	2.07	2.38	2.43	4.86	0.418	3.07	$ 97.3	$ 298.3	180.5	2.23
18	0.922	2.06	2.37	2.42	4.84	0.416	3.05	$ 98.3	$ 300.0	176.2	2.22
19	0.918	2.06	2.36	2.41	4.82	0.415	3.04	$ 99.3	$ 301.7	172.0	2.21
20	0.914	2.05	2.35	2.40	4.80	0.413	3.03	$ 100.3	$ 303.4	168.0	2.20
										4,254.1	45.80

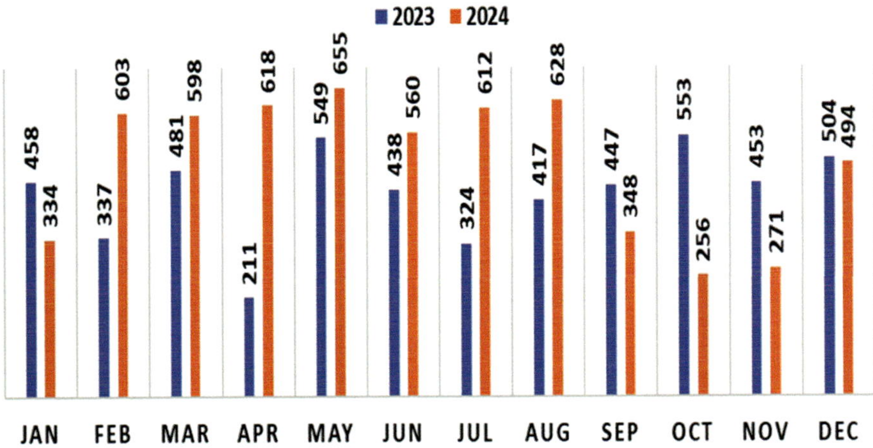

Figure 17.5 Monthly connected renewable energy systems on JEPCO distribution grid from 2023 to 2024.

Table 17.4 Solar PV installations in the residential sector during 2023–2024 in Jordan.

Distribution system operator (DSO)	Total number of residential projects in 2024	Total residential in 2024 (MWac)	Total number of residential projects in 2023	Total residential in 2023 (MWac)	Installed number of residential projects in 2024	Installed residential in 2024 (MWac)	Average capacity on 2024 (MWac)
JEPCO	29,161	176.89	27,544	143.10	1,617	33.79	20.89
IDECO	29,320	113.58	25,717	99.47	3,603	14.11	3.92
EDCO	-N/A-	-N/A-	11,594	45.26			
Total DSOs			64,855	287.83			

installations occurred during this period (June-August 2024) because of the lack of a regulatory framework to be applied. The total number of connected renewable energy systems during the first eight months was 576 systems/month; it dropped by 40% to 342 systems/month during the last four months of the year (see Figure 17.5) [7]. Nevertheless, the overall annual number of connected renewable energy systems grew by 16% from 431 systems/month in 2023 to 498 systems/month in 2024. The solar PV systems that were installed in the residential sector across three distribution grids between 2023 and 2024 are detailed in Table 17.4.

In 2024, the number of solar PV systems installed on the JEPCO grid was 1,617, with a total capacity of 33.79 MWac and an average capacity of 20.89 kWac [7]. Similarly, the IDECO grid had 3,603 installed solar PV

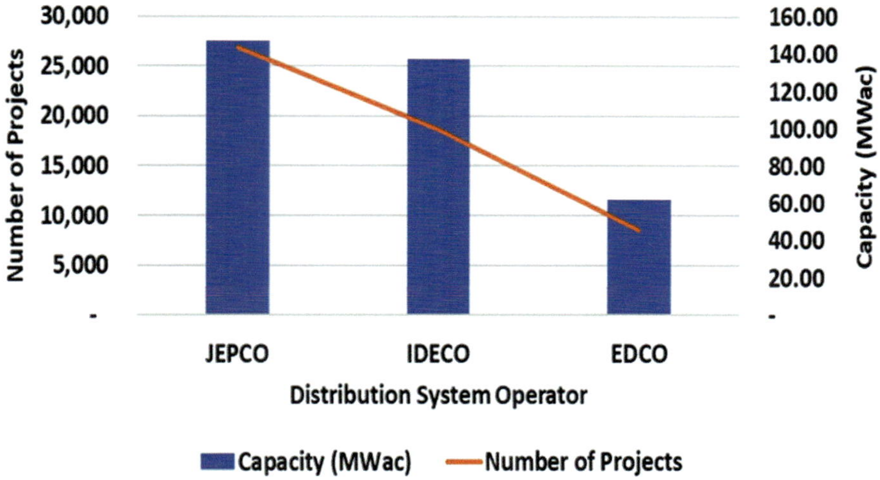

Figure 17.6 Residential sector solar PV installations on distribution grids in Jordan in 2023.

systems in 2024, with a total capacity of 14.11 MWac and an average capacity of 3.92 kWac [8]. Regrettably, similar data from EDCO regarding the overall distribution of solar PV systems was unavailable [9]. The solar PV systems installed in the residential sector on the distribution grids in Jordan are shown in Figure 17.6.

Consumers in the northern region of Jordan exhibit lower electricity consumption levels, which can be adequately provided by small-scale solar PV systems; this drives an increasing trend to adopt such systems to reduce electricity costs. This trend is further reinforced by the prevalence of single-family homes with available roof space suitable for solar PV installations. Conversely, in the central region of Jordan, where multistory buildings are prevalent, the installation of solar PV systems is more challenging because municipal regulations mandate approval from all building residents to utilize communal roof areas. The average size of solar PV systems installed on the JEPCO distribution grid in the residential sector in 2024 was 20.89 KWac. This indicates a preference among single households residing in independent dwellings with high annual electricity consumption levels to opt for solar PV systems to meet their annual electricity requirements.

DSOs are transitioning to smart meters per Jordan Energy Strategy 2020–2030. The EMRC announced early 2025 that Jordan has reached market penetration of 80% smart meters and is expected to complete the transition by the end of 2025 [12].

Following the implementation of the dynamic tariff in July 2024, coupled with the global proliferation and lower cost of lithium-ion (Li-Ion) batteries, solar PV companies began offering prosumers the option of adding BESS. This allows prosumers to store excess electricity generated by their solar PV systems during the day, rather than selling it back to the grid at a low rate of $0.071/kWh. Residential prosumers can then utilize this stored PV electricity during the night. This avoids the need to purchase electricity at higher rates, which can exceed $0.282/kWh (for residential consumers with a supported tariff) for monthly consumption exceeding 600 kWh during nighttime hours.

This cost disparity, outlined in Bylaw 58, 2024, reflects a price four times higher than the rate at which the grid purchases electricity from residential solar PV systems [11]. As a result, larger residential consumers who rely on electricity for all their energy needs, particularly heating and cooling, face substantial monthly electricity expenses. The integration of BESS technology presents a viable solution for such consumers.

The National Electrical Power Company (NEPCO), which is the transmission system operator in Jordan, reported that the overall 2024 electricity consumption was 22,723 GWh [6]. However, this figure does not account for the electricity generated by the 1,196 MWac solar PV systems integrated into the three distribution grids in Jordan. Considering an average DC:AC ratio assumed in the methodology, the total approximate installed capacity by the end of 2024 was calculated to be 1,435.2 MWdc. With an estimated annual specific production of 1,560 kWh/kWp/year, it is projected that the residential solar PV prosumers consumed approximately 2,239 GWh, accounting for 9% of the total electricity generation.

Based on a grid emission factor of 0.4585 KG/kWh, the avoided CO_2 emissions are an estimated 1.026 million tons per year, which is equivalent to a 3.14% reduction of emissions as reported in the Fourth National Communication [16].

Adopting an estimated distribution loss of approximately 13% for all three distribution grid operators based on the reported distribution losses in the JEPCO 2023 annual report, the utilization of prosumer-generated electricity resulted in avoided losses of approximately 303 GWh [7]. This is equivalent to 26,018 tons or 190,710 barrels of oil. Using an average annual price of Brent crude oil at $83 per barrel in 2023, this translates to an approximate annual savings of $15.8 million [17].

17.5 Connection Mechanisms

According to Bylaw 58, 2024, issued by the Ministry of Energy and Mineral Resources, three connection mechanisms are available to the residential sector in Jordan as of September 2024 [11]. These mechanisms are net billing, zero-to-grid, and buy-all/sell-all.

The 2022 data show that 58% of the residential sector customers consume less than 300 kWh/month, 35% consume between 301 and 600 kWh/month, 6% consume between 601 and 1,000 kWh/month, and 1% consume over 1,000 kWh/month [10]. The residential sector consumers were divided into two groups after the EMRC modified the tariff in April 2022. Local Jordanian residential consumers can opt to register one electricity meter to benefit from the supported tariff if their average monthly electricity consumption is less than 937 kWh/month. Large residential consumers and non-Jordanians use the unsupported residential tariff. For this study, the economic parameters for all three connection mechanisms were calculated based on the following parameters:

- PV module size: 645 W.
- Number of modules: varied between 2 to 23.
- Size of system (DC): Equal to the module size multiplied by the number of modules (KW_p).
- DC:AC ratio: Maximum limit allowed for the residential sector outlined in methodology.
- Size of system (AC): System sizes vary. They are equal to system size in DC divided by DC:AC ratio (KWac).
- Cost of 1 KWp: $600/KWp (average price in May 2025 in Jordan).

The degradation of top-quality commercial solar PV modules as of May 2025 was considered in the calculations. The degradation was 0.99 at year one, and continued in a linear decline to reach 88.85% at year 30 per the data sheet for a top-quality commercial solar PV module. Tables showing the detailed calculations for each mechanism are available in Appendix C. A spreadsheet was developed to calculate the economic parameters net present value (NPV), internal rate of return (IRR), benefit cost ratio (BCR) and the simple payback period (SPP) for scenarios with sensitivity analysis for some cases.

17.5.1 Buy-all/sell-all connection mechanism

The prosumer who opts to use this connection mechanism will install a solar PV system with a dedicated smart meter for the PV system. All the electricity generated from the PV system will be fed to the grid and sold at a fixed rate for the residential sector at $0.071/kWh and for all other sectors at $0.0571/kWh according to Bylaw 58, 2024 [11]. The prosumer will continue

Table 17.5 Economic parameters for prosumers using the buy-all/sell-all mechanism starting September 2024.

Number of modules	Size (AC)	Size (DC)	Monthly electricity from solar PV (kWh)	NPV ($ USD)	Simple payback period (year)	IRR	BCR
2	0.86	1.290	192	$ 1,754.61	4.74	15%	3.27
3	1.29	1.935	287	$ 2,631.92	4.74	15%	3.27
4	1.72	2.580	383	$ 3,509.22	4.74	15%	3.27
5	2.15	3.225	479	$ 4,386.53	4.74	15%	3.27
6	2.58	3.870	585	$ 5,263.84	4.74	15%	3.27
7	3.01	4.515	670	$ 6,141.14	4.74	15%	3.27
8	3.44	5.160	766	$ 7,018.45	4.74	15%	3.27
23	9.89	14.835	2,203	$ 20,178.04	4.74	15%	3.27

Table 17.6 Economic parameters for prosumers using the buy-all/sell-all mechanism at $0.0564 /kWh as of September 2024

Number of modules	PV system size (AC) (Kwac)	PV system size (DC) (KWp)	1st year monthly electricity from solar PV (kWh)	NPV ($ USD)	Simple payback period (year)	IRR	BCR
2	1.29	1.55	192	$ 1,144.75	5.97	10.3%	2.48
3	1.94	2.32	287	$ 1,717.12	5.97	10.3%	2.48
4	2.58	3.10	383	$ 2,289.49	5.97	10.3%	2.48
5	3.23	3.87	479	$ 2,861.86	5.97	10.3%	2.48
6	3.87	4.64	575	$ 3,434.24	5.97	10.3%	2.48
7	4.52	5.42	670	$ 4,006.61	5.97	10.3%	2.48
8	5.16	6.19	766	$ 4,578.98	5.97	10.3%	2.48
23	14.84	17.80	2,203	$ 13,164.58	5.97	10.3%	2.48

to purchase their monthly electricity needs based on the applied tariff for each sector.

Residential prosumers that consume less than 300 kWh/month will buy electricity and sell electricity at the same rate of $0.071/kWh. Because there is no monthly grid fee applied to this connection mechanism, there is no negative impact on these prosumers. The economic parameters are calculated and presented as shown in Table 17.5, which shows a sensitivity analysis based on the size of their solar PV system. The IRR was 15% in this case. The table used for this calculation is shown in Appendix C.

This interconnection mechanism functions as an investment opportunity for all participating consumers, enabling electricity sales at a fixed rate of $0.0564/kWh. The DC:AC ratio complies with the specifications outlined in Bylaw 58, 2024.

A sensitivity analysis was conducted by systematically varying the system size and evaluating the resulting economic parameters, as presented in Table 17.6. The analysis yielded an internal rate of return (IRR) of 10.3%.

Table 17.7 Economic parameters with sensitivity analysis for buy-all/sell-all mechanism at $0.0564/kWh as of September 2024.

Number of modules	Size (AC)	Size (DC)	Monthly electricity from solar PV (kWh)	NPV ($ USD)	Simple pay-back period (year)	IRR	BCR
24	12.90	15.480	$ 600.0	$ 13,736.95	5.97	10.3%	2.48
24	12.90	15.480	$ 575.0	$ 14,106.05	5.72	11.0%	2.58
24	12.90	15.480	$ 550.0	$ 14,475.15	5.47	11.7%	2.70
24	12.90	15.480	$ 525.0	$ 14,844.25	5.22	12.5%	2.83
24	12.90	15.480	$ 500.0	$ 15,213.36	4.97	13.5%	2.97
24	12.90	15.480	$ 475.0	$ 15,582.46	4.73	14.5%	3.12
24	12.90	15.480	$ 450.0	$ 15,951.56	4.48	15.6%	3.29

The complete dataset and calculation methodology supporting these results are provided in Appendix C.

With the increase in the global solar PV manufacturing and drop in prices, another sensitivity analysis was performed for all consumers using this mechanism and selling the electricity at $0.056 /kWh by varying the prices of the PV system between $450-$600 as shown in Table 17.7. The IRR has increased from 10.3% to 15.6%; the table used for this calculation is shown in Appendix C.

17.5.2 Zero-to-grid connection mechanism

The residential consumer with a supported tariff was considered in these calculations. The sensitivity study was conducted by varying the electricity consumption and installing the solar PV system to cover the electricity consumption during the day.

Figure 17.7 shows the variation of the internal rate of return (IRR) with the average monthly electricity consumption during the day.

The increase in the IRR is related to the stepped-up tariff for the (supported) residential sector, which increases the electricity prices with the increase in consumption. With the residential sector consuming more than 600 kWh/month, it pays fourfold (from $0.071 to $0.282/kWh) and the return on investment becomes more attractive. The residential prosumer opting to use this mechanism pays a monthly grid fee which equals $1.41/kWac/month. The table used for this calculation is shown in Appendix C.

17.5.3 Net billing connection mechanism

The residential consumer with large electricity consumption (estimated 1% consume above 1,000 kWh according to 2022 EMRC statistic) can install only

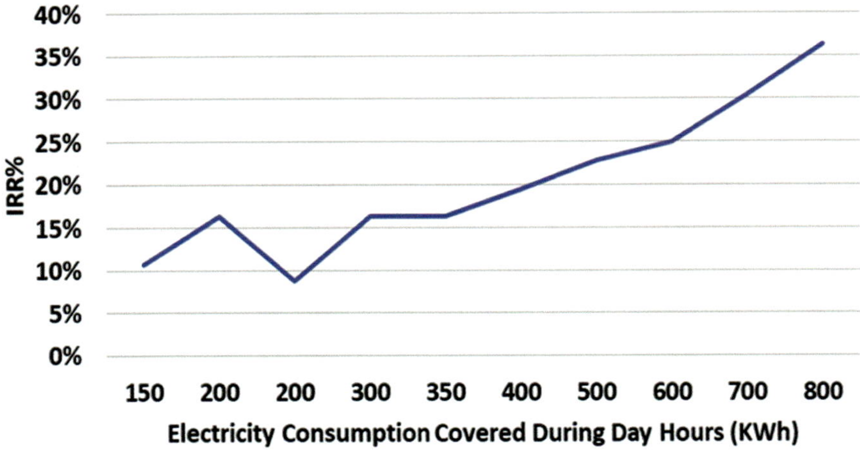

Figure 17.7 Zero-to-grid connection mechanism economic advantage.

3.68 kWac/phase or 11.0 kW with the maximum DC:AC ratio of 1.5; the maximum installed solar PV system is 16.5 kWp [10]. Based on current available technologies in the market, the distribution system operators are allowing 10.0 kWac inverters with a 15 kWp solar PV module and the annual specific production in Bylaw 58, 2024 set at 1,800 kWh/kWp. The maximum expected generation by residential prosumers from solar PV is 27,000 kWh [11].

PVwatts online software provided by the National Renewable Energy Laboratory (NREL) was used to predict the solar PV generation for 1 kWp at an area in Phoenix, Arizona with the system size of 5.4 kWp [18,19]. Figure 17.8 shows the monthly electricity generated from solar PV in green and the monthly residential consumption in red.

During the winter and summer months, the demand for heating and cooling will be high, whereas in autumn and spring seasons it is low. The assumed behavior for this monthly consumption and solar PV generation is as follows:

- Period 1: From 5 am to 5 pm (40% is consumed, 90% of solar PV electricity generated).
- Period 2: From 5 pm to 11 pm (40% is consumed, 10% of solar PV electricity generated).
- Period 3: From 11 pm to 5 am (20% is consumed, 0% of solar PV electricity generated).

Using the net billing connection mechanism, the daily generated electricity from solar PV will be directly consumed by the prosumer, and the excess is

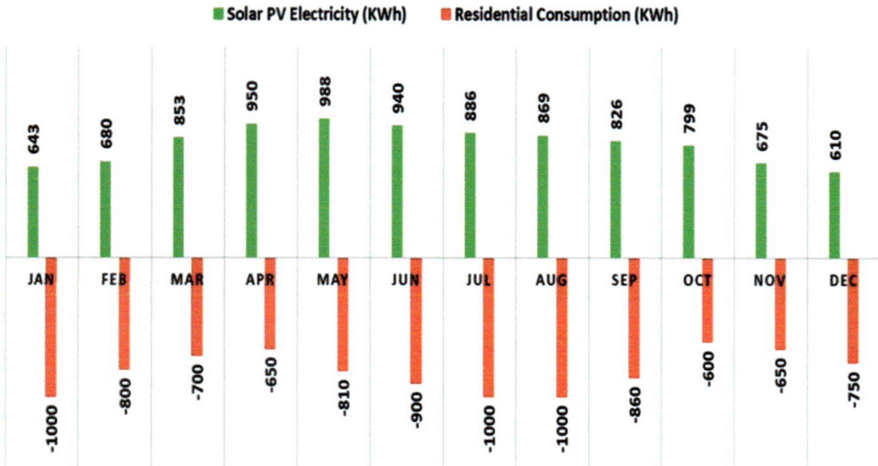

Figure 17.8 Net billing connection mechanism—monthly solar PV generation vs. consumption.

fed back to the grid; the accounting with the electricity distribution company is done on a monthly basis. Based on these assumptions, the annual savings were $1,168, the electricity generated over 30 years was 235,702 kWh, and the LCOE was $0.0184/kWh. The economic parameters for this option were: the NPV equaled $14,130, simple payback period of 2.96 years, IRR was 27.7%, and the BCR was 5.36.

17.5.4 Solar PV with storage

Solar PV technology is advancing rapidly, and the cost for L-I battery storage has dropped over 90% compared to six years ago, according to current market prices. As an example, the LCOE was calculated to compare the LCOE for solar PV with storage compared to the grid electricity tariff for the residential supported category.

- Solar PV module size: 645 W; the system cost is $600/kWp.
- Degradation of the battery was 2% per year.
- The battery was replaced twice, 1 replacement at 75% initial cost, second replacement at 50% initial cost. $200/year annual inspection fee was considered for most years. No salvage value was considered for the battery.
- Battery: 28.6 kWh; BESS cost is $150/kWh.

- Number of modules: 8. Solar PV system size (AC) is 3.44 KW, PS system size (DC) is 5.16 kWp

The LCOE for solar PV with storage over the 30-year lifetime was $0.0592, whereas the grid electricity LCOE without considering any changes in current prices was $0.0726. Therefore, solar PV with storage is considered a competitive option for grid electricity and will attract many customers to use this option.

17.6 Recommendations

17.6.1 Policy recommendations

Streamline regulatory frameworks

- Ensure the timely publication of regulatory updates to avoid disruptions in project pipelines.
- Harmonize connection mechanisms across all distribution companies for consistency and clarity. Residential sector consumers with less than 300 kWh/month consumption must opt to use the buy-all/sell-all connection mechanism and not any other.

Increase uptake of distributed energy resources

- DERs owned by prosumers provide electricity without any capital cost to governments. These systems reduce the losses on the grid and avoid fluctuation in international oil prices.
- Invest in grid infrastructure and energy storage to achieve higher energy security.
- Integrate higher shares of variable renewable energy to reduce the cost of electricity.
- Increase the solar PV installations to reach 30% in each sector.

Revise building and municipal regulations

- Amend building codes to allow installation of solar PV systems on the rooftops of multifamily buildings without requiring unanimous tenant approval.
- Encourage local municipalities to offer fast-track permits for residential solar installations.

- Develop and implement a mandatory solar roof for all new construction.

Mandate solar readiness in new construction

- Enforce mandatory rooftop solar readiness or installation in new residential and commercial buildings.
- Include provisions for EV charging stations and solar water heaters in national building codes.

Standardize reporting across distribution companies

- Require all distribution system operators to publish uniform reports on DERs approvals, installed capacity, and connection mechanisms used.

Develop dynamic pricing

- Implement time-of-use (TOU) tariffs that reflect generation patterns and incentivize prosumer behavior aligned with grid stability.
- Prepare for adopting a seasonal dynamic tariff to solve the impact of seasonal variation in DERs on the grid.
- Oil-exporting countries will need to develop DER regulatory frameworks and increase their energy tariffs with the transition away from fossil fuels.

17.6.2 Technical infrastructure

Upgrade grid infrastructure

- Invest in smart grid technologies to accommodate high shares of distributed energy resources.
- Enhance bidirectional power flow capabilities and real-time monitoring systems.
- Retrofit and support grids to integrate various storage technologies.

Support battery storage integration

- Update regulations to include battery storage within DER frameworks.

- Provide technical standards and safety guidelines for residential battery installations.

Expand mini- and off-grid solutions

- Promote decentralized energy solutions in rural and remote areas where grid extension is economically unfeasible.
- Support community-based mini-grids powered by solar PV and storage.
- The installation of solar PV systems for the low-income residential sector will reduce the cost of electricity, which reached a LCOE of $.015/kWh in 2025.

Financing grid infrastructure development projects

- Upgrade the skills of national financial corporations to meet the requirements of climate fund financing; secure loans to finance infrastructure development goals.
- The costs of integrating more solar PV on national grids need to be identified and needed projects must be financed.

Align tariff structures with prosumer benefits

- Review and revise electricity tariffs to better reflect the value of solar generation and reduce disparities between buy-back rates and consumption costs.

17.6.3 Social engagement and equity

Launch public awareness campaigns

- Educate consumers about the different connection mechanisms, their benefits, and how to choose the most suitable one.
- Educate solar PV companies and the public to understand the advantages of different connection mechanisms.
- Use social media, workshops, and local outreach to increase awareness.
- Develop off-grid solar PV with storage best practices, because adopting this system will continue to increase in the future.

Support regional knowledge sharing

- Collaborate with neighboring countries to exchange best practices in solar integration, regulation, and financing.
- Participate in regional interconnection projects.

Conduct periodic impact assessments

- Evaluate the effectiveness of solar PV policies in social terms every two years.

17.7 Conclusion

This study evaluated the impact of different grid connection mechanisms on prosumer adoption and performance of residential solar PV systems in Jordan. The findings yielded a set of doable recommendations intended to improve the market uptake of distributed renewable energy systems. The analysis revealed that solar PV systems contributed approximately 9% of total electricity generation in 2024, avoiding an estimated 1.026 million tons of CO_2 emissions annually. These systems also reduced distribution losses by 303 GWh/year, equivalent to over $15.8 million in annual savings.

A regulatory delay in implementing Bylaw 58, 2024, led to a 40% drop in new installations during mid-2024, highlighting the importance of timely policy updates and transparent communication. Disparities in reporting practices among distribution companies underscore the need for standardized data collection and transparency to support future planning and investment decisions.

Among the three permitted connection mechanisms—net billing, zero-to-grid, and buy-all/sell-all—net billing was found to be the most economically attractive for high-consumption households due to Jordan's tiered tariff structure, offering an IRR of 27.7% and a simple payback period of less than 3 years. The buy-all/sell-all mechanism provided an IRR of up to 15.6%, particularly beneficial under declining system costs. The zero-to-grid mechanism showed strong potential for consumers with notable daytime usage.

The integration of BESS further enhances the economic viability of solar PV systems. With the LCOE of solar plus storage at $0.0592/kWh, it has become competitive with grid electricity ($0.0726/kWh). This signals a growing trend toward off-grid adoption as technology improves and costs decline.

The findings emphasize the need for updated building codes, awareness campaigns, and infrastructure modernization to support higher shares of distributed energy resources. Policy recommendations include facilitating solar PV adoption in multifamily buildings, promoting smart meter deployment, and incentivizing energy storage integration.

Jordan's transition toward decentralized solar energy offers substantial economic, environmental, and social benefits. This will position Jordan as a regional model for emerging economies aiming to meet climate goals while enhancing energy security and affordability.

References

[1] International Energy Agency. Global energy review 2025. https://iea.blob.core. windows.net/assets/ 5b169aa1-bc88-4c96-b828- aaa50406ba80/GlobalEnergyReview2025.pdf.

[2] Solar Power Europe (2025). Global market outlook for solar power (2024–2029). https://www.solarpowereurope.org.

[3] Statista. Global electricity consumption 1980-2023. https://www.statista.com/statistics/280704/world-power-consumption.

[4] International Energy Agency Photovoltaic Power Systems Programme (2023). Trends in photovoltaic applications. https://iea-pvps.org/wp-content/uploads/2023/10/PVPS_Trends_Report_2023_WEB.pdf.

[5] Ministry of Energy and Mineral Resources. Facts and figures reports, 2010–2023.

[6] National Electric Power Company (2023). Annual report. Amman, Jordan.

[7] Jordan Electric Power Company (2023,2024). Annual reports. Amman, Jordan.

[8] Irbid District Electricity Company (2023,2024). Annual reports. Amman, Jordan.

[9] Electricity Distribution Company (2023,2024). Annual reports. Amman, Jordan.

[10] Energy and Minerals Regulatory Commission. Tariff guide 2025. https://www.emrc.gov.jo.

[11] Ministry of Energy and Mineral Resources. Regulation of Connecting Renewable Energy Sources Facilities on the System Electrical, Exempting Renewable Energy Sources Systems and Rationalizing Energy Consumption, Bylaw 58, 2024. https://memr.gov.jo.

[12] The Jordan Times (2025). EMRC to replace all electric meters with smart ones by 2025. https://www.jordantimes.com/news/local/emrc-replace-all-electric-meters-smart-ones-2025.

[13] Jordan's Second Biennial Update Report (2020, December) Under the United Nations Framework Convention on climate change. https://unfccc.int/sites/default/files/resource/Jordan%E2%80%99s%20Second%20Biennial%20Update%20Report%20for%20web%2010-5.pdf.

[14] Ministry of Energy and Mineral Resources. Annual report 2023. Amman, Jordan.

[15] Ministry of Energy and Mineral Resources (2024). Law (12) year 2024 Amended Law on Renewable Energy and Energy Conservation Law.

[16] Jordan's Fourth National Communication on Climate Change, 2023. https://www.undp.org/jordan/publications/jordans-fourth-national-communication-climate-change.

[17] U.S. Energy Information Administration. Brent crude oil prices. https://www.eia.gov/todayinenergy/detail.php?id=61142 #.

[18] National Renewable Energy Laboratory. PVWatts calculator. https://pvwatts.nrel.gov/pvwatts.php.

[19] Frisvold, G., Patton, W. and Reynolds, S. (2009, January 9). Arizona solar energy and economics outlook. *Arizona Solar Energy and Economics Summit*. https://climas.arizona.edu/sites/climas.arizona.edu/files/migrated_media/pdfarizonasolarenergyandeconomicoutlook_5.pdf.

Appendix C

Table C17.1 Buy-all/sell-all mechanism (residential sector selling at $0.071/kWh).

PV Module	**645 Watt**	**DC:AC ratio**	**1.5**
# of Modules	**8**	**Size (AC)**	**3.440 KW**
Size (DC)	**5.160 KWp**	**Cost**	**$ (600.00) USD/KWp**

Buy-All / Sell All Mechanism (Residential sector selling at $0.071 USD/ KWh)

Year	Payment ($ USD)	Degradation	Annual Electricity from PV (KWh)	Average Monthly Electricity from PV (KWh)	Revenue from Buy All - Sell All [$0.071 USD/ KWh] (USD)	Annual Payments (USD)	Present Value [i=4%] Annual Payments (USD)	Present Value [i=4%] Costs (USD)
0	$ (3,096.00)					(3,096.00)	$ (3,096.00)	$ (3,096.00)
1		0.9900	9,195	766	$ 652.85	652.85	$ 627.74	$ -
2		0.9865	9,163	764	$ 650.55	650.55	$ 601.47	$ -
3		0.9830	9,130	761	$ 648.24	648.24	$ 576.28	$ -
4		0.9795	9,098	758	$ 645.93	645.93	$ 552.14	$ -
5		0.9760	9,065	755	$ 643.62	643.62	$ 529.01	$ -
6		0.9725	9,033	753	$ 641.31	641.31	$ 506.84	$ -
7		0.9690	9,000	750	$ 639.01	639.01	$ 485.59	$ -
8	$(516.00)	0.9655	8,968	747	$ 636.70	120.70	$ 88.19	$ (377.04)
9		0.9620	8,935	745	$ 634.39	634.39	$ 445.71	$ -
10		0.9585	8,903	742	$ 632.08	632.08	$ 427.01	$ -
11		0.9550	8,870	739	$ 629.77	629.77	$ 409.09	$ -
12		0.9515	8,838	736	$ 627.46	627.46	$ 391.91	$ -
13		0.9480	8,805	734	$ 625.16	625.16	$ 375.45	$ -
14		0.9445	8,773	731	$ 622.85	622.85	$ 359.68	$ -
15		0.9410	8,740	728	$ 620.54	620.54	$ 344.56	$ -

16	$ (516.00)	0.9375	8,708	726	$ 618.23	102.23	$ 54.58	$ (275.50)
17		0.9340	8,675	723	$ 615.92	615.92	$ 316.20	$ -
18		0.9305	8,642	720	$ 613.62	613.62	$ 302.90	$ -
19		0.9270	8,610	717	$ 611.31	611.31	$ 290.15	$ -
20		0.9235	8,577	715	$ 609.00	609.00	$ 277.94	$ -
21		0.9200	8,545	712	$ 606.69	606.69	$ 266.24	$ -
22		0.9165	8,512	709	$ 604.38	604.38	$ 255.02	$ -
23		0.9130	8,480	707	$ 602.08	602.08	$ 244.28	$ -
24	$ (516.00)	0.9095	8,447	704	$ 599.77	83.77	$ 32.68	$ (201.30)
25		0.9060	8,415	701	$ 597.46	597.46	$ 224.12	$ -
26		0.9025	8,382	699	$ 595.15	595.15	$ 214.66	$ -
27		0.8990	8,350	696	$ 592.84	592.84	$ 205.61	$ -
28		0.8955	8,317	693	$ 590.54	590.54	$ 196.93	$ -
29		0.8920	8,285	690	$ 588.23	588.23	$ 188.62	$ -
30	$ 464.40	0.8885	8,252	688	$ 585.92	1,050.32	$ 323.83	$ -
			261,713		NPV		$ 7,018.45	$ (3,949.8)
					Payback Period (yr)		4.74	
					IRR		15.0%	
					BCR		3.27	
					LCOE			$ 0.0151

Table C17.2 Buy-all/sell-all mechanism (all sectors selling at $0.0564/ kWh).

	PV Module	**645 Watt**		**DC:AC ratio**		**1.5**	
	# of Modules	**23**		**Size (AC)**		**9.890 KW**	
	Size (DC)	**14.835 KWp**		**Cost**	**$**	**(600.00) USD/KWp**	
Buy-All / Sell-All Mechanism (all sectors selling at $0.0564 USD/ KWh)							
Year	**Payment ($ USD)**	**Degradation**	**Annual Electricity from PV (KWh)**	**Average Monthly Electricity from PV (KWh)**	**Revenue from Buy-All/Sell-All [$0.0564 USD/ KWh] (USD)**	**Annual Payments (USD)**	**Present Value [i=4%] Annual Payments (USD)**
0	$ (8,901.00)					(8,901.00)	$ (8,901.00)
1		0.9900	26,436	2,203	$ 1,490.99	1,490.99	$ 1,433.64
2		0.9865	26,343	2,195	$ 1,485.72	1,485.72	$ 1,373.63
3		0.9830	26,249	2,187	$ 1,480.45	1,480.45	$ 1,316.11
4		0.9795	26,156	2,180	$ 1,475.18	1,475.18	$ 1,260.99
5		0.9760	26,062	2,172	$ 1,469.90	1,469.90	$ 1,208.15
6		0.9725	25,969	2,164	$ 1,464.63	1,464.63	$ 1,157.52
7		0.9690	25,875	2,156	$ 1,459.36	1,459.36	$ 1,108.99
8	$ (1,483.50)	0.9655	25,782	2,148	$ 1,454.09	(29.41)	$ (21.49)
9		0.9620	25,688	2,141	$ 1,448.82	1,448.82	$ 1,017.92
10		0.9585	25,595	2,133	$ 1,443.55	1,443.55	$ 975.21
11		0.9550	25,501	2,125	$ 1,438.28	1,438.28	$ 934.28
12		0.9515	25,408	2,117	$ 1,433.01	1,433.01	$ 895.05
13		0.9480	25,314	2,110	$ 1,427.73	1,427.73	$ 857.46
14		0.9445	25,221	2,102	$ 1,422.46	1,422.46	$ 821.44
15		0.9410	25,128	2,094	$ 1,417.19	1,417.19	$ 786.92

16	$ (1,483.50)	0.9375	25,034	2,086	$ 1,411.92	(71.58)	$ (38.22)
17		0.9340	24,941	2,078	$ 1,406.65	1,406.65	$ 722.14
18		0.9305	24,847	2,071	$ 1,401.38	1,401.38	$ 691.76
19		0.9270	24,754	2,063	$ 1,396.11	1,396.11	$ 662.65
20		0.9235	24,660	2,055	$ 1,390.84	1,390.84	$ 634.76
21		0.9200	24,567	2,047	$ 1,385.57	1,385.57	$ 608.03
22		0.9165	24,473	2,039	$ 1,380.29	1,380.29	$ 582.42
23		0.9130	24,380	2,032	$ 1,375.02	1,375.02	$ 557.88
24	$ (1,483.50)	0.9095	24,286	2,024	$ 1,369.75	(113.75)	$ (44.38)
25		0.9060	24,193	2,016	$ 1,364.48	1,364.48	$ 511.84
26		0.9025	24,099	2,008	$ 1,359.21	1,359.21	$ 490.25
27		0.8990	24,006	2,000	$ 1,353.94	1,353.94	$ 469.57
28		0.8955	23,913	1,993	$ 1,348.67	1,348.67	$ 449.75
29		0.8920	23,819	1,985	$ 1,343.40	1,343.40	$ 430.76
30	$ 1,335.15	0.8885	23,726	1,977	$ 1,338.12	2,673.27	$ 824.22

NPV	**$ 13,778.27**
Payback Period (yr)	**5.97**
IRR	**10.6%**
BCR	**2.55**

Table C17.3 Buy-all/sell-all mechanism (all sectors selling at $0.0564/ kWh) – price sensitivity.

PV Module	**645 Watt**	**DC:AC ratio**	**1.2**
# of Modules	**24**	**Size (AC)**	**12.900 KW**
Size (DC)	**15.480 KWp**	**Cost $**	**(450.00) USD/KWp**

Buy-All / Sell-All Mechanism (all sectors selling at $0.0564 USD/ KWh) -Price Variation

Year	Payment ($ USD)	Degradation	Annual Electricity from PV (KWh)	Average Monthly Electricity from PV (KWh)	Revenue from Buy-All/Sell-All [$0.0564 USD/ KWh] (USD)	Annual Payments (USD)	Present Value [i=4%] Annual Payments (USD)
0	$ (6,966.00)					(6,966.00)	$ (6,966.00)
1		0.9900	27,585	2,299	$ 1,555.81	1,555.81	$ 1,495.98
2		0.9865	27,488	2,291	$ 1,550.31	1,550.31	$ 1,433.35
3		0.9830	27,390	2,283	$ 1,544.81	1,544.81	$ 1,373.33
4		0.9795	27,293	2,274	$ 1,539.31	1,539.31	$ 1,315.81
5		0.9760	27,195	2,266	$ 1,533.81	1,533.81	$ 1,260.68
6		0.9725	27,098	2,258	$ 1,528.31	1,528.31	$ 1,207.85
7		0.9690	27,000	2,250	$ 1,522.81	1,522.81	$ 1,157.21
8	$ (1,935.00)	0.9655	26,903	2,242	$ 1,517.31	(417.69)	$ (305.20)
9		0.9620	26,805	2,234	$ 1,511.81	1,511.81	$ 1,062.18
10		0.9585	26,708	2,226	$ 1,506.31	1,506.31	$ 1,017.61
11		0.9550	26,610	2,218	$ 1,500.81	1,500.81	$ 974.90
12		0.9515	26,513	2,209	$ 1,495.31	1,495.31	$ 933.97
13		0.9480	26,415	2,201	$ 1,489.81	1,489.81	$ 894.74
14		0.9445	26,318	2,193	$ 1,484.31	1,484.31	$ 857.15
15		0.9410	26,220	2,185	$ 1,478.81	1,478.81	$ 821.13

16	$ (1,935.00)	0.9375	26,123	2,177	$ 1,473.31	(461.69)	$ (246.50)
17		0.9340	26,025	2,169	$ 1,467.81	1,467.81	$ 753.53
18		0.9305	25,927	2,161	$ 1,462.31	1,462.31	$ 721.84
19		0.9270	25,830	2,152	$ 1,456.81	1,456.81	$ 691.46
20		0.9235	25,732	2,144	$ 1,451.31	1,451.31	$ 662.36
21		0.9200	25,635	2,136	$ 1,445.81	1,445.81	$ 634.47
22		0.9165	25,537	2,128	$ 1,440.31	1,440.31	$ 607.75
23		0.9130	25,440	2,120	$ 1,434.81	1,434.81	$ 582.14
24	$ (1,935.00)	0.9095	25,342	2,112	$ 1,429.31	(505.69)	$ (197.28)
25		0.9060	25,245	2,104	$ 1,423.81	1,423.81	$ 534.09
26		0.9025	25,147	2,096	$ 1,418.31	1,418.31	$ 511.57
27		0.8990	25,050	2,087	$ 1,412.81	1,412.81	$ 489.98
28		0.8955	24,952	2,079	$ 1,407.30	1,407.30	$ 469.30
29		0.8920	24,855	2,071	$ 1,401.80	1,401.80	$ 449.49
30	$ 1,044.90	0.8885	24,757	2,063	$ 1,396.30	2,441.20	$ 752.67
					NPV		**$ 15,951.56**
					Payback Period (yr)		**4.48**
					IRR		**15.6%**
					BCR		**3.29**

Table C17.4 Zero-to-grid mechanism (all sectors selling at $0.0564/ kWh).

PV Module	**645 Watt**		**DC:AC ratio**		**1.5**		
# of Modules	**8**		**Size (AC)**		**3.440**	**KW**	
Size (DC)	**5.160**	**KWp**	**Cost**		**$ (600.00)**	**USD/KWp**	
Buy-All / Sell-All Mechanism							

Year	**Payment ($ USD)**	**Degradation**	**Annual Electricity from PV (KWh)**	**Average Monthly Electricity Consumption during day (KWh)**	**Average Monthly Electricity from PV (KWh)**	**Savings from Zero -to-Grid [(1-300 $0.071 301-600 $0.142 USD/ KWh] (USD)**	**Annual Payments (USD)**	**Present Value [i=4%] Annual Payments (USD)**
0	$ (3,096.00)						(3,096.00)	$ (3,096.00)
1		0.9900	9,195	800	766	$ 1,329.42	1,329.42	$ 1,278.29
2		0.9865	9,163	800	764	$ 1,320.26	1,320.26	$ 1,220.65
3		0.9830	9,130	800	761	$ 1,311.09	1,311.09	$ 1,165.55
4		0.9795	9,098	800	758	$ 1,301.92	1,301.92	$ 1,112.89
5		0.9760	9,065	800	755	$ 1,292.75	1,292.75	$ 1,062.55
6		0.9725	9,033	800	753	$ 1,283.59	1,283.59	$ 1,014.44
7		0.9690	9,000	800	750	$ 1,274.42	1,274.42	$ 968.45
8	$ (516.00)	0.9655	8,968	800	747	$ 1,265.25	749.25	$ 547.47
9		0.9620	8,935	800	745	$ 1,256.09	1,256.09	$ 882.51
10		0.9585	8,903	800	742	$ 1,246.92	1,246.92	$ 842.37
11		0.9550	8,870	800	739	$ 1,237.75	1,237.75	$ 804.02
12		0.9515	8,838	800	736	$ 1,228.58	1,228.58	$ 767.37
13		0.9480	8,805	800	734	$ 1,219.42	1,219.42	$ 732.35
14		0.9445	8,773	800	731	$ 1,210.25	1,210.25	$ 698.89
15		0.9410	8,740	800	728	$ 1,201.08	1,201.08	$ 666.92

16	$ (516.00)	0.9375	8,708	800	726	$ 1,191.92	675.92	$ 360.88
17		0.9340	8,675	800	723	$ 1,182.75	1,182.75	$ 607.19
18		0.9305	8,642	800	720	$ 1,173.58	1,173.58	$ 579.31
19		0.9270	8,610	800	717	$ 1,164.41	1,164.41	$ 552.68
20		0.9235	8,577	800	715	$ 1,155.25	1,155.25	$ 527.24
21		0.9200	8,545	800	712	$ 1,146.08	1,146.08	$ 502.94
22		0.9165	8,512	800	709	$ 1,136.91	1,136.91	$ 479.73
23		0.9130	8,480	800	707	$ 1,127.74	1,127.74	$ 457.56
24	$ (516.00)	0.9095	8,447	800	704	$ 1,118.58	602.58	$ 235.08
25		0.9060	8,415	800	701	$ 1,109.41	1,109.41	$ 416.16
26		0.9025	8,382	800	699	$ 1,100.24	1,100.24	$ 396.85
27		0.8990	8,350	800	696	$ 1,091.08	1,091.08	$ 378.40
28		0.8955	8,317	800	693	$ 1,081.91	1,081.91	$ 360.79
29		0.8920	8,285	800	690	$ 1,072.74	1,072.74	$ 343.98
30	$ 464.40	0.8885	8,252	800	688	$ 1,063.57	1,527.97	$ 471.10
						NPV		**$ 17,338.60**
						Payback Period (yr)		**2.33**
						IRR		**36.3%**
						BCR		**6.60**

Table C17.5 Net billing connection mechanism—residential unsupported.

Annual Saving: **$ 1,167.6**
Size (DC) **5.4** **KWp**
Cost **$(600.00)** **USD/KWp**

Net Billing Connection Mechanism - Residential Un-Supported.

Year	Payment (USD)	Degradation	Annual Electricity from PV (KWh)	Annual Electricity Savings from PV ($ USD)	Annual Grid Fee [$1.41 USD/KWac/ month] ($ USD)	Annual Payments ($ USD)	Present Value [i=4%] Annual Payments ($ USD)	Annual Costs (USD)	Present Value [i=4%] Annual Costs ($ USD)
0	$ (3,240.00)					(3,240.00)	$ (3,240.00)	(3,240.000)	(3,240.000)
1		0.9900	9,623	$ 1,156.0	$ (60.91)	1,095.04	$ 1,052.92	(60.912)	(58.569)
2		0.9865	9,589	$ 1,151.9	$ (60.91)	1,090.95	$ 1,008.65	(60.912)	(56.317)
3		0.9830	9,555	$ 1,147.8	$ (60.91)	1,086.87	$ 966.22	(60.912)	(54.151)
4		0.9795	9,521	$ 1,143.7	$ (60.91)	1,082.78	$ 925.57	(60.912)	(52.068)
5		0.9760	9,487	$ 1,139.6	$ (60.91)	1,078.69	$ 886.61	(60.912)	(50.065)
6		0.9725	9,453	$ 1,135.5	$ (60.91)	1,074.61	$ 849.28	(60.912)	(48.140)
7		0.9690	9,419	$ 1,131.4	$ (60.91)	1,070.52	$ 813.51	(60.912)	(46.288)
8	$ (540.00)	0.9655	9,385	$ 1,127.3	$ (60.91)	526.43	$ 384.66	(600.912)	(439.081)
9		0.9620	9,351	$ 1,123.3	$ (60.91)	1,062.35	$ 746.39	(60.912)	(42.796)
10		0.9585	9,317	$ 1,119.2	$ (60.91)	1,058.26	$ 714.92	(60.912)	(41.150)
11		0.9550	9,283	$ 1,115.1	$ (60.91)	1,054.17	$ 684.77	(60.912)	(39.567)
12		0.9515	9,249	$ 1,111.0	$ (60.91)	1,050.09	$ 655.88	(60.912)	(38.045)
13		0.9480	9,215	$ 1,106.9	$ (60.91)	1,046.00	$ 628.20	(60.912)	(36.582)
14		0.9445	9,181	$ 1,102.8	$ (60.91)	1,041.91	$ 601.68	(60.912)	(35.175)
15		0.9410	9,147	$ 1,098.7	$ (60.91)	1,037.83	$ 576.27	(60.912)	(33.822)

16	$ (540.00)	0.9375	9,113	$ 1,094.7	$ (60.91)	493.74	$ 263.61	(600.912)	(320.832)
17		0.9340	9,078	$ 1,090.6	$ (60.91)	1,029.65	$ 528.60	(60.912)	(31.271)
18		0.9305	9,044	$ 1,086.5	$ (60.91)	1,025.57	$ 506.25	(60.912)	(30.068)
19		0.9270	9,010	$ 1,082.4	$ (60.91)	1,021.48	$ 484.84	(60.912)	(28.911)
20		0.9235	8,976	$ 1,078.3	$ (60.91)	1,017.39	$ 464.33	(60.912)	(27.799)
21		0.9200	8,942	$ 1,074.2	$ (60.91)	1,013.31	$ 444.67	(60.912)	(26.730)
22		0.9165	8,908	$ 1,070.1	$ (60.91)	1,009.22	$ 425.85	(60.912)	(25.702)
23		0.9130	8,874	$ 1,066.0	$ (60.91)	1,005.13	$ 407.81	(60.912)	(24.714)
24	$ (540.00)	0.9095	8,840	$ 1,062.0	$ (60.91)	461.05	$ 179.86	(600.912)	(234.429)
25		0.9060	8,806	$ 1,057.9	$ (60.91)	996.96	$ 373.98	(60.912)	(22.849)
26		0.9025	8,772	$ 1,053.8	$ (60.91)	992.87	$ 358.12	(60.912)	(21.970)
27		0.8990	8,738	$ 1,049.7	$ (60.91)	988.79	$ 342.93	(60.912)	(21.125)
28		0.8955	8,704	$ 1,045.6	$ (60.91)	984.70	$ 328.38	(60.912)	(20.313)
29		0.8920	8,670	$ 1,041.5	$ (60.91)	980.61	$ 314.44	(60.912)	(19.532)
30	$ 486.00	0.8885	8,636	$ 1,037.4	$ (60.91)	1,462.53	$ 450.92	425.088	131.063
			273,885		**NPV**		**$ 14,130.11**		**$ (5,037.00)**
					Payback Period (yr)		**2.96**		
					IRR		**27.7%**		
					BCR		**5.36**		
					LCOE				**$ (0.0184)**

Table C17.6 Solar PV with storage vs. grid electricity.

PV Module	**645**	**Watt**	**DC:AC ratio**		**1.5**			**Battery (KWh)**
# of Modules	**8**		**Size (AC)**		**3.440**	**KW**		**28.6**
Size (DC)	**5.160**	**KWp**	**Cost**		**$ (600.00)**	**USD/KWp**		
Solar PV with Storage vs Grid Electricity								

Year	Payment ($ USD)	PV Module Degradation	Battery Degradation	Annual Electricity from PV (KWh)	Average Monthly Electricity from PV (KWh)	Cost of Equivalent Electricity from Grid ($ USD)	Present Value Solar PV + Storage ($ USD)	Present Value Equivalent Electricity from Grid (USD)
0	$ (3,096.00)						(3,096.00)	
0	$ (4,290.00)						(4,290.00)	
1	$ (200.00)	0.9900	0.9900	9,103	759	$ 1,298.50	(192.31)	$ 1,248.55
2	$ (200.00)	0.9865	0.9700	8,888	741	$ 1,237.72	(184.91)	$ 1,144.35
3	$ (200.00)	0.9830	0.9500	8,674	723	$ 1,177.32	(177.80)	$ 1,046.63
4	$ (200.00)	0.9795	0.9300	8,461	705	$ 1,117.28	(170.96)	$ 955.06
5	$ (200.00)	0.9760	0.9100	8,249	687	$ 1,057.61	(164.39)	$ 869.28
6	$ (200.00)	0.9725	0.8900	8,039	670	$ 998.31	(158.06)	$ 788.98
7	$ (200.00)	0.9690	0.8700	7,830	653	$ 939.37	(151.98)	$ 713.84
8	$ (516.00)	0.9655	0.8500	7,622	635	$ 880.80	(377.04)	$ 643.59
9		0.9620	0.8300	7,416	618	$ 822.59	-	$ 577.94
10	$ (3,217.50)	0.9585	0.8100	7,211	601	$ 764.76	(2,173.63)	$ 516.64
11	$ (200.00)	0.9550	0.9900	8,781	732	$ 1,207.71	(129.92)	$ 784.51
12	$ (200.00)	0.9515	0.9700	8,572	714	$ 1,148.77	(124.92)	$ 717.52
13	$ (200.00)	0.9480	0.9500	8,365	697	$ 1,090.20	(120.11)	$ 654.75
14	$ (200.00)	0.9445	0.9300	8,158	680	$ 1,032.00	(115.50)	$ 595.95
15	$ (200.00)	0.9410	0.9100	7,953	663	$ 974.16	(111.05)	$ 540.92

16	$ (516.00)	0.9375	0.8900	7,750	646	$ 916.69	(275.50)	$ 489.43
17	$ (200.00)	0.9340	0.8700	7,547	629	$ 859.59	(102.67)	$ 441.29
18	$ (200.00)	0.9305	0.8500	7,346	612	$ 802.85	(98.73)	$ 396.31
19	$ (200.00)	0.9270	0.8300	7,146	596	$ 746.48	(94.93)	$ 354.31
20	$ (2,145.00)	0.9235	0.8100	6,948	579	$ 690.48	(978.95)	$ 315.13
21	$ (200.00)	0.9200	0.9900	8,460	705	$ 1,116.93	(87.77)	$ 490.15
22	$ (200.00)	0.9165	0.9700	8,257	688	$ 1,059.82	(84.39)	$ 447.20
23	$ (200.00)	0.9130	0.9500	8,056	671	$ 1,003.09	(81.15)	$ 406.98
24	$ (516.00)	0.9095	0.9300	7,856	655	$ 946.72	(201.30)	$ 369.34
25	$ (200.00)	0.9060	0.9100	7,658	638	$ 890.71	(75.02)	$ 334.12
26	$ (200.00)	0.9025	0.8900	7,460	622	$ 835.08	(72.14)	$ 301.20
27	$ (200.00)	0.8990	0.8700	7,264	605	$ 779.81	(69.36)	$ 270.45
28	$ (200.00)	0.8955	0.8500	7,070	589	$ 743.27	(66.70)	$ 247.86
29	$ (200.00)	0.8920	0.8300	6,876	573	$ 716.00	(64.13)	$ 229.59
30	$ 464.40	0.8885	0.8100	6,684	557	$ 688.92	143.18	$ 212.41
				235,702		**NPV**	**$ (13,948.1)**	**$ 17,104.28**
						Payback Period (yr)		**-**
						IRR		
						BCR		
						LCOE	**$ (0.0592)**	**$ 0.0726**

18

An Asset Manager's Conundrum: Sunsetting Solar Farms

Stephen A. Roosa

This chapter considers the process of sunsetting distributive generation infrastructure from the perspective of an asset manager. While there are ways to extend their lifecycles, when renewable energy developments such as solar farms reach the end of their efficient operations, they must at some point be decommissioned. When projects are being developed, the costs of decommissioning and equipment disposal at the end of the development's useful lifecycle are often ignored. What is involved in this decommissioning process? What sort of costs must be incurred? The example of decommissioning solar farms is presented to highlight the challenges involved. Planning for sunsetting assets considers decommissioning costs, types of regulation, and the available financial mechanisms. The decommissioning process for distributed generation technologies, such as those used for solar arrays, is evolving. In some jurisdictions in the United States, this process is regulated. Asset managers are encouraged to survey the changing landscape and include equipment decommissioning in their lifecycle planning. There are opportunities for related industries to establish standards, bolster recycling capacity, and advocate for common regulations across the United States.

18.1 Managing the Asset

Asset management is the systematic process of developing, operating, maintaining, upgrading, and disposing of assets in the most cost-effective manner (considering all costs, risks, and performance attributes). It involves project development, management, operations and maintenance (O&M),

DOI: 10.1201/9788743811312-18

measurement and verification (M&V), and ultimately component disposal. Typically, the initial cost of the project will include site acquisition, permitting, site preparation, design, equipment purchase, delivery, installation labor, landscaping, utility interconnections, punch lists, and commissioning. The development and buildout costs are usually the project's initial cash investment and financing costs. A lifecycle cost estimate is often prepared, which considers the life of the equipment, annual O&M costs, warranty costs, insurance, financing expenses, depreciation, cash flow, and M&V if applicable. Managing risks and limiting costs are essential parts of the asset manager's business model.

18.1.1 Project End-of-Life

To enhance the asset value of a project, future risks and the ultimate sunsetting of equipment and systems must be addressed and mitigated. *Sunsetting* refers to the automatic expiration or termination of a development at the end of a fixed period unless renewed. What happens when the project is permanently closed? For asset management, this is not the close-out that occurs when the project is initially developed and commissioned for operation; it is the financial close-out at the end of the investment lifecycle when the project is ready to be decommissioned. In the excitement to develop a new project, the cost of decommissioning is often overlooked. This omission increases financial risks and results in the project's cash flow being overstated.

Engineers often disparage certain business aspects of asset management, especially how developments end and what is required to close them out. A more appropriate view is to consider a project's sunsetting process both strategically and realistically, or from a perspective that is often called *strategic realism*. When development projects are permanently closed, the final end-of-project-life phase is decommissioning. An intentional decommissioning plan is preferable to in situ abandonment of the infrastructure, leaving the local community and landscape with an eyesore and potential safety hazards. While it is difficult to estimate decommissioning costs since they are incurred many years into the future, this final project phase can be surprisingly expensive.

Full decommissioning involves the permanent closure of the project. Asset managers typically consider the following options: 1) sell the project site as is; 2) remove all equipment and retain the site for a similar development in the future; 3) subcontract the removal of equipment and sell the site to a new owner; or 4) in situ abandonment. Each of these approaches will likely yield a different end-of-life value. Land use options may be limited by zoning, economics, environmental regulations, or other site constraints. However, if not being redeveloped, restoring the site to its original state is often preferred.

In the utility and their supply chain businesses, site closeout costs for projects that are known to be expensive to decommission are often ignored (e.g., coal-fired plants with ash storage issues, nuclear plants with radioactive materials stored on site, methane leaks from natural gas fields, etc.). The end-of-life business plan involves carving off the entity as a separate enterprise, forcing it to fail, and abandoning the site to the governments (i.e., taxpayers) to manage. For many renewable energy projects, the final decommissioning and site closeout process is often not regulated; yet their owners are held to higher standards. As a result, local governments in some states require that project owners or developers have a sunsetting plan in place to remove solar project components at the end of their lifecycles [1].

Decommissioning distributed generation infrastructure creates more questions than answers. How can the project sites be decommissioned at the end of their natural lifecycle? What are the steps in this process? What costs are involved? Do the project components have residual value? If so, how can their value be estimated? What components of the project can be recycled, and what are the associated costs? Will there be markets in the future for the residual components? What are the costs associated with changing the use of the site? These are the types of questions asset managers face. To complicate the definition of residual value, all future values are simply engineering estimates; they are somewhat theoretical and influenced by many factors, including how the developer defines the decommissioning process.

18.2 Solar Farms

Solar arrays that use photovoltaics (PV) rather than solar thermal technologies to generate electricity offer a heuristic example of the uncertainties associated with decommissioning distributed energy systems. Their warranted lifecycle is often based on the type of solar modules that are installed to generate the electricity required. Most commercially available modules use crystalline silicone PV technologies and have warranties that might range from 25 years or more [2].

Solar farms are large solar installations that often consist of thousands of ground-mounted solar panels. Monocrystalline modules are often preferred for use in utility-scale solar farms as they have higher efficiencies and tend to have greater longevity than thin-film or polycrystalline technologies. Higher panel efficiencies mean that less array area is required to generate the needed power, less racking structure is needed, and ultimately, there is less facility to be maintained.

18.2.1 The lifecycle of solar farms

Most solar farms in the U.S. are a decade or more away from their decommissioning dates; in many cases, there is still time to plan for this process [3]. One benefit from the asset manager's perspective is that utility-scale solar arrays are generally field- or canopy-mounted rather than mounted on the roofs of buildings. Therefore, project life is more predictable and not a function of the lifecycle of a building's roofing system. Most large-scale commercial solar farms have a predicted lifecycle of 25 years or more.

Determining project lifecycles and end dates can be challenging. Utility-scale solar arrays are often phased developments constructed in rural areas. Decommissioning dates for the equipment will vary based on when each specific phase of the project was completed. Determining the life of the project is confounded by economics. The output of solar panels varies over the life of the project, as does the amount of electricity produced. Though the output of silicone-based commercial solar panels degrades incrementally at a relatively slow rate, after 25 years, the total electrical output of the solar farm will likely decline by as much as 20%; some components may become unsafe to use. Declining total output, cost variability, O&M costs, and safety concerns threaten the economic viability of the development. This means a shorter-than-anticipated project life is possible.

There is typically a lifecycle mismatch in the types of components used in solar farms. While some solar PV modules may have a potential depreciated life of as much as 30 years, the life of the power electronics is less, especially for batteries and inverters. This mismatch requires that lifecycle analysis must factor in periodic costs for maintenance and component replacement.

Being grid-interconnected, these projects supply electricity directly to their host grids the instant it is produced. However, if the project includes an on-site energy storage facility that uses batteries, the lifecycle of the batteries will likely vary from that of the solar arrays, meaning that their decommissioning dates also vary. For example, deep-cycle, lead-acid batteries are on the low end of the lifecycle spectrum (as little as a one-year limited warranty) while lithium-ion (L-I) and most flow batteries have much longer manufacturer warranties. More appropriate battery systems for on-site distributed applications include sodium-sulfur (NaS), zinc-bromide (ZnBr), and vanadium-redox (VRB) batteries [4]. A sodium-sulfur battery is a type of molten metal battery that has a high energy density, high efficiency of charge/discharge (89%–92%), long cycle life, and is fabricated from inexpensive materials that can be recycled [4]. In sum, these attributes make NaS batteries a preferred alternative, yet many new solar farms in the U.S. rely on L-I battery banks.

18.2.2 Extending the project lifecycle

Alternatives to decommissioning the site are redevelopment or extending its useful life by refurbishing or replacing its equipment. In cases when it is feasible, redeveloping the site can extend the project's lifecycle. Even so, there are costs that will be incurred to enable these options. Racking systems are seldom the primary concern. Fixed racking systems, if made from stainless steel or otherwise maintained, can outlast the other system components. Those with tracking capabilities have moving parts, but if properly maintained, they will also outlast the solar panels and other components.

The components that have shorter lifecycles are solar PV modules and power electronic devices such as charge regulators or inverters. The solar panels have a life cycle of 20 to 30 years (excluding thin film technologies), which varies depending on the composition of the panels and ambient environmental conditions. While they will continue to produce electricity, their efficiency declines. The degradation rate of a solar panel is the reduction in solar panel electrical output over time, which varies depending on the type of technology used (see Table 18.1) [5]. The life of solar PV modules can be lengthened with periodic cleaning, routine inspections, protection from extreme temperatures, applying protective coatings, and sealing gaps that are found in the panels. The life of the module array can be further extended if damaged or non-performing panels are detected and replaced. Such problems can be discovered by using infrared detection equipment as part of a periodic O&M program.

Among the least durable components of a solar array are its inverters. The lifetime of the inverters is a maximum 10 to 12 years; their lifecycle is reduced by incorrect installation and exposure to heat [6]. The inverter converts the direct current (DC) electricity from solar panels into the alternating current (AC) used by the electric grid. Inverters use several component technologies—passive filters, active switches, thermal management, and packaging [7]. These components experience wear and tear due to everyday use, faults, physical damage, or unusual occurrences such as electromagnetic disturbances [7]. These degrade the performance of inverters and limit their lifecycle, causing the need for part replacements and creating higher operating

Table 18.1 Representative annual solar PV degradation rates [5].

Type of solar PV module	Degradation rate % per year
Amorphous silicon (a-Si)	0.87
Monocrystalline silicon (sc-Si)	0.36
Multicrystalline silicon (mc-Si)	0.64
Cadmium telluride (CdTe)	0.40
Copper indium gallium selenide (CIGS)	0.96
Concentrator	1.00

costs [7]. The best way to maximize the solar array's lifecycle might be to consider an all-DC system that serves only DC loads and devices (e.g., in a microgrid configuration).

Extending the life of the solar farm has obvious benefits. They include taking advantage of the availability of the existing site and utility infrastructure while deferring the cost of decommissioning the site. This approach suggests redeveloping the existing site by replacing the solar modules and inverters, and reusing the existing foundations, racking systems, and electrical transmission infrastructure.

18.3 Decommissioning Solar Farms

Solar farms provide many benefits to local communities. They provide decentralized electricity, local employment, educational opportunities, property tax base, and may help meet local policy goals. Regardless, rural communities that allow solar farms fear that such projects could be abandoned in place at the end of their useful lifecycle. In situ abandonment creates large-scale solar array eyesores that require substantial funds to mitigate and eliminate. If subject to regulation, a local governmental entity often determines the fate of the solar farm and may take enforcement actions. This might involve imposing civil penalties or fines, mandating the removal of the infrastructure, and imposing liens on the property to recover the associated costs.

Solar farm decommissioning is a multiphase process. It is essentially redoing the project development and component installation process in reverse [1]. The larger the solar farm, the more complicated, time-consuming, and costly the process. The first stage involves planning, which includes evaluating and documenting the process for project closure. If a decommissioning plan was created, it would need to be reviewed and updated periodically over the full lifecycle of the development. Components with end-of-life value have the potential to be harvested and sold to interested third parties. Components that are to be recycled must be separated and packaged.

To complicate the sunset date of a solar panel's lifecycle, some photovoltaic modules can be refurbished to extend their usable life and resold; others cannot and must be recycled or sent to landfills. However, the default action of sending PV panels to the local landfill needs to be replaced with specialized decommissioning procedures and ways to ensure that funds are available to support sustainable end-of-project processing.

Often, the project documentation will indicate that the solar panels will be returned to the manufacturer for recycling [8]. If this option is available, the manufacturer accepts delivery, disassembles the panels, and recycles the semiconductor materials (e.g., glass and metals). In such cases, who incurs the packaging and transportation costs? It is possible that the manufacturer (possibly located in China) may refuse to accept the obsolete modules or may no longer be in operation [8].

18.3.1 Developing a decommissioning plan

The decommissioning plan needs to be customized for each individual project. The process varies somewhat depending on the types of infrastructure. Though most solar PV systems are field or canopy-mounted, some arrays float on ponds or lakes. Racking systems could be flat, fixed tilt, or have tracking systems. The panels might be single-faced, bi-facial, or hybrid. Some may have mounted micro-inverters. Some installations may have battery energy storage equipment that must also be removed and excessed.

The greater the complexity of the system, the greater the time and costs of decommissioning. Decommissioning a solar farm is a deconstruction process. It typically includes removing rack wiring and solar panels, dismantling and removing racking systems, removing electrical wiring and equipment, removing concrete pads and foundations, removing power poles, dismantling security fences, site grading, seeding, landscaping, hauling material to disposal sites and recycling centers, substation removal, etc. [3].

These future costs of site decommissioning can be substantial and challenging to estimate. The solar industry believes that there is a significant salvage value for solar arrays, but cannot ensure that there will be a market to salvage the equipment when removed [9]. The feasibility of realizing salvage value may depend on who removes the equipment—the developer, operator, the tenant, or the landowner—and how each estimates costs and perceives environmental value [9].

The National Renewable Energy Laboratory (NREL) and its partner labs encourage owners to develop the decommissioning plan for the site prior to construction. Solar decommissioning policies in the U.S. occasionally require that the project developer draft and submit a decommissioning plan to federal, state, and/or local authorities, depending on the location of the site [10]. Solar decommissioning plans typically include the following information:

- The projected lifespan of the solar project.
- A site reclamation plan, a restoration plan, or both.
- Plans for transportation, salvage, and disposal of the solar project equipment and infrastructure (e.g., modules, inverters, supports, and cables).
- Cost estimates for solar system decommissioning (e.g., costs for removal of solar project equipment and infrastructure, land restoration and reclamation, and insurance requirements).
- Estimates of the salvage value for solar project equipment and infrastructure (e.g., modules, inverters, supports, and cables).
- A description of any expected impacts on natural resources.
- Estimated timelines for completion of decommissioning activities [11].

18.3.2 Regulations in the U.S.

Regulations regarding the process of decommissioning solar projects are varied. At the federal level, the Bureau of Land Management is presently the only U.S. agency with a written policy that applies to projects on federally-managed lands. U.S. national laboratories have published best practices for O&M of photovoltaic and energy storage systems [5]. Notwithstanding, most U.S. states lack regulatory policies regarding the decommissioning of solar sites. If the site is used as a *microgrid*, most state public utility commissions lack not only regulations, but many have not yet defined the term.

As of this writing, twelve U.S. states have mandatory regulations regarding solar site decommissioning (identified in Figure 18.1). Local governmental entities may also have specific regulatory requirements that impact the decommissioning of these projects.

18.4 Decommissioning Costs

All costs associated with decommissioning are best estimates. It is not unusual for project developers to claim that the future value of the materials used will be greater than the cost of removal and restoration of the site. However, the evidence supporting this is scant, and the costs of future labor and disposal

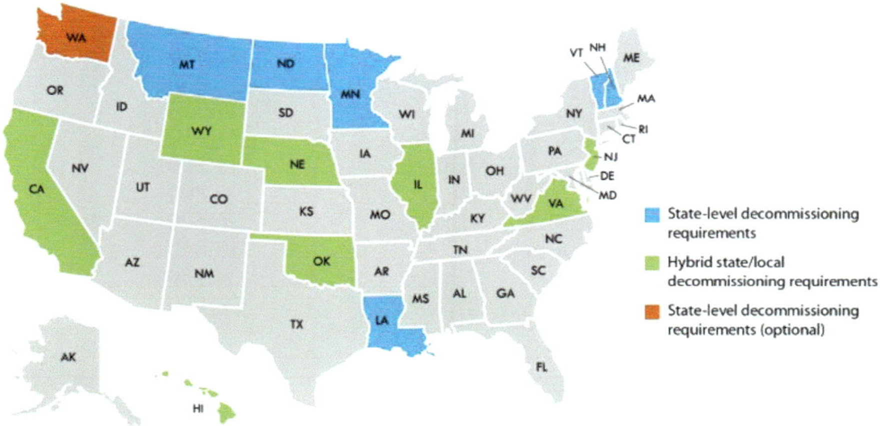

Figure 18.1 U.S. state solar decommissioning polices [10].

are unknown. It is known that total decommissioning costs for solar arrays can be as much as 40% of their salvage value [12].

Newer solar panels are manufactured using less aluminum, copper, and silver, which makes them less valuable to recycle. Local labor costs, a key component of site decommissioning, vary widely and are difficult to predict. In the U.S., the infrastructure for recycling solar PV panels is notably absent. First Solar, the sole U.S. panel manufacturer with an up-and-running recycling initiative, has a global capacity of two million panels per year, but it only recycles its own products. With the current capacity, it costs an estimated \$20 to \$30 to recycle one solar PV panel [13]. The alternative is to send that same panel to a landfill, which would cost \$2 or less [13]. To further confound this issue of disposal, some panels (i.e., those known to contain lead and cadmium) are classified as hazardous wastes [14].

The states of California, Hawaii, New Jersey, North Carolina, and Washington have already enacted regulations and policies impacting solar panel waste [14]. Developers will likely consider new regulations as a factor that increases decommissioning costs. Original equipment manufacturers of solar modules and regulators must consider a cost/benefit analysis comparing the cost of disposal in landfills versus the environmental benefits and recycling costs. Collectively, there is more that industry, asset owners, and government can do to provide components recycling options. There is high potential for material recovery rates for the glass, aluminum, silicon, silver, and copper used in solar panels—all exceeding 90%. Steel and aluminum racking systems can be reused or recycled as scrap metal. Inverters and other

Table 18.2 List of decommissioning tasks and estimated costs [15].

Decommissioning task	Costs ($s)
Remove rack wiring	$2,459
Remove panels	2,450
Dismantle racks	12,350
Remove electrical equipment	1,850
Break up and remove concrete pads or ballasts	1,500
Remove racks	7,800
Remove cable	6,500
Remove ground screws and power poles	13,850
Remove fencing	4,950
Site grading	4,000
Seed disturbed areas	250
Truck to recycling center	2,250
Total	$60,209
Total after 20 years (3.0% inflation rate)	$108,744

electronic waste can also be recycled. These components have some residual value, which may offset some decommissioning costs.

It is challenging to predict costs that will be incurred far into the future. The way that engineers typically approach this problem is to estimate present costs and apply a fixed annual escalation percentage to each year through to the project's end year. The specific estimate of average annual inflation is normally stated and used to gauge future costs. The future costs vary depending on the decommissioning strategy and the actual impact of inflation over the project's financial life.

Past estimates for full decommissioning of solar farms have ranged from $9,000 to $56,300 per MW [3]. The New York State Energy Research and Development Authority (NYSERDA) has provided a summary listing the types of tasks involved in the decommissioning process for a sample 2 MW solar project and a preliminary estimate of task costs (see Table 18.2).

To further obfuscate asset management, the decommissioning costs will vary based on site accessibility, local conditions, types of equipment installed, etc. Estimates are often cited from $0.03/W to as much as $0.20/W; these estimates often fail to indicate if they are referring to module output in DC (normally assumed) or AC wattages [16]. Though variable, we are seeing total present decommissioning costs of roughly $0.035/W [17]. It is somewhat easier to obtain decommissioning costs from specific projects after the process is concluded, but case studies with publicly available data are rare. Recent examples include:

- New York state suggests costs if about $0.03/Wac [17].
- The 100 MWAC/126 MW (DC) Three Rivers Solar Power Plant in Hancock, Maine, estimated decommissioning costs at $0.018¢/Wac [17].

Another common way of estimating decommissioning costs is to count the number of solar panels to be deactivated and disposed of and estimate a cost per panel that includes all associated project costs. Key cost variables are the average hourly cost of labor and the number of panels that can be safely removed per labor hour. For a ground-mounted project in Athol, Massachusetts, it was estimated that 20 panels could be removed in one labor hour. The 15,000-panel project (about 5MW/DC) was estimated to cost $175,630 or $11.71/panel [17].

Given the wide variability of costs, it is imperative that asset managers develop a specific cost estimate for each site. The NYSERDA estimates average decommissioning costs of about $30,000/MW in present-day costs [2]. Using this figure, a 250 MW project would cost about $7.5 million to decommission, excluding the effects of inflation. Future values are perplexing because inflation forecasting tends to be volatile, especially for projects with lifecycles greater than 10 years.

18.5 Financial Mechanisms

What surety options are available to guarantee that financial resources will be available when the time comes to decommission the solar farm? Letters of credit are often used, but there are other options. They can assure the landowner of a specific maximum payment, given that certain conditions are met if the project developer fails to remove the system [15].

Decommissioning provisions can also be contractually stated in land-lease agreements. If a decommissioning plan is mandated, public or private landowners can include decommissioning clauses in the land-lease agreement [15]. The specifics of these clauses depend on the preferences of the landowner, developer, and asset manager. Solar farm site developers can establish a cash account or trust fund for decommissioning purposes. Applying this approach, the developer makes a series of payments during the project's lifecycle until the fund reaches the estimated cost of decommissioning [15]. In third-party financial arrangements, the end-user acts as the project host and agrees to pay specified fees for all-inclusive solar electric services using a power purchase agreement [18]. Under this scenario, the third party might bear the responsibility for the decommissioning costs depending on the terms of the contract [18].

Typical forms of surety for solar projects include cash escrow, insurance bonds, collateral bonding, self-bonds, or letters of credit held by a federally insured financial institution. Having to initially set aside large sums of cash for future decommissioning often makes projects financially infeasible. Other forms of surety can be difficult to enforce. Determining the financial stability of surety companies has been challenging in many instances [19]. Consider, for example, sureties that have been set aside to reclaim land from strip mining sites; they have been highly ineffective and failed to provide adequate funds for the needed environmental remediation. Surety funds sit unused in escrow because they are inadequate to cover the site remediation costs [19]. It is not unusual for coal mining companies to simply file bankruptcy and dissolve with their sureties unenforced, leaving nearby communities to suffer from the environmental damage [20].

One way asset managers can avoid or mitigate such future expenses is to apply some type of financial mechanism that distributes associated costs over the life of the project, instead of during the initial development [3]. This might take the form of a small initial cash set-aside augmented annually by a percentage of the proceeds from the project's income streams (e.g., sales of KW and kWh) with periodic adjustments for inflation. In this case, the funds would need to be fenced and held by a trusted third party.

Financial assurance is not universally mandated by regulation. In the U.S., there are mandatory statewide financial assurance requirements for projects in the states of California, Illinois, Louisiana, North Dakota, Minnesota, Montana, New Hampshire, Vermont, Virginia, and Washington [10]. These policies typically require compliance with solar decommissioning standards that include the submission of a plan and proof of financial assurance as a condition of approval for the initial construction or operation of the solar project [10].

18.6 Conclusion

For asset managers to be successful and reduce project risks, decommissioning plans must be considered early in the project development stages. These plans need to be customized based on the project scope. Once distributed generation projects are developed and operational, decommissioning costs can be deferred as there are ways to extend project longevity.

The process of project decommissioning is like project development but operates in reverse. Solar farms provide a heuristic example of how this sunsetting process works. Site activities include the deconstruction, removal, and transport of solar panels, racking systems, and other site equipment, and preparing system components for recycling and disposal. This process is challenging as various project components have different lifecycles.

Successful, state-of-the-art standards for decommissioning distributive generation technologies, such as those used in the development of solar farms, are evolving. Environmentally sensitive areas may require special protection during the process of dismantling the solar farm [21]. The costs for decommissioning are widely variable and difficult to estimate, as they will occur far into the future. Project size is directly related to the decommissioning budget [21]. Skilled labor is a major cost component of the decommissioning process. The larger sites with more modules and extensive infrastructure require more labor hours and equipment to dismantle [21]. Location is also crucial, as transportation distances to recycling facilities can appreciably affect total costs. Transporting dismantled equipment and materials to recycling facilities or disposal sites is a widely variable cost based on distance and the costs of fuel [21]. When projects are initiated, these costs are often overlooked, which results in overstated cash flows.

It is becoming increasingly important to identify and use project components that can be more easily dismantled, sorted, and recycled without incurring substantial shipping costs. This means that component recycling needs to be less centralized. Less restrictive site redevelopment options would be supportive. Such efforts will help reduce decommissioning costs at the end of a project's lifecycle and potentially make such costs less difficult to estimate.

The regulatory environment in the U.S. is developing. Some states already have decommissioning and financial assurance regulations. Policies requiring decommissioning are also being established by some local governmental entities. To protect the interests of host communities, workable financial mechanisms are needed to ensure that funds will be available to cover the eventual costs for project decommissioning.

References

[1] Ludt, B. (2019, March 11). How to decommission a solar array, and why it's important to plan ahead. https://www.solarpowerworldonline.com/2019/03/how-to-decommission-a-solar-array-and-why-its-important-to-plan-ahead, accessed 25 January 2023.

[2] Kolbeck-Urlacher, H. (2022, June). Decommissioning solar energy systems resource guide. https://www.cfra.org/sites/default/files/publications/Decommissioning%20solar%20energy%20systems%20WEB.pdf, accessed 23 January 2026.

[3] Engel, J. (2022, June 22). How solar plans can protect rural communities. https://www.renewableenergyworld.com/solar/how-solar-decommissioning-plans-can-protect-rural-communities/?fbclid=IwAR2NdAiX-hEPzCSRq4RUn59uVDFI24HFZhcsRLyxY-jR0aO_YeMRtaG8ZT3Q, accessed 22 August 2022.

[4] Roosa, S. (2021). *Fundamentals of Microgrids: Development and Implementation*, 66–121. CRC Press: Boca Raton, Florida.

[5] National Renewable Energy Laboratory (2018). Best practices for operation and maintenance of photovoltaic and energy storage systems, 3rd edition, accessed 29 January 2023.

[6] Frank, J. (2022, August 22). When should I replace my solar inverter (the average life)? https://thepowerfacts.com/when-should-i-replace-my-solar-inverter, accessed 25 January 2023.

[7] Miller, J. and Krishnaswami, H. (2018, May 9). Extending solar energy system lifetime with power electronics. U.S. office of Energy Efficiency and Renewable Energy. https://www.energy.gov/eere/solar/articles/extending-solar-energy-system-lifetime-power-electronics, accessed 25 January 2023.

[8] Virginia Smart Solar. Disposing of solar panels. https://vasmartsolar.com/decommissioning-disposal-of-solar-panels, accessed 25 August 2022.

[9] Virginia Smart Solar. Decommissioning costs. https://vasmartsolar.com/decommissioning-costs/#:~:text=Decommissioning%20can%20cost%20millions%20in%20today%E2%80%99s%20dollars.%20The,a%20market%20to%20salvage%20the%20equipment%20when%20removed, accessed 22 August 2022.

[10] Curtis, T., Smith, L., Buchanan, H. and Heath, G. (2021, December). A survey of federal and state-level solar system decommissioning policies in the United States. National Renewable Energy Laboratory. https://www.nrel.gov/docs/fy22osti/79650.pdf, ccessed 23 January 2026.

[11] New York State Energy Research and Development Authority (2020). Decommissioning solar panel systems: information for local governments and landowners on the decommissioning of large-scale solar panel systems. https://www.nyserda.ny.gov/-/media/NYSun/files/Decommissioning-Solar-Systems.pdf.

[12] Maamari, L. Decommissioning of solar sites: a key consideration of the project. https://solunesco.com/2018/09/10/decommissioning-of-solar-sites-a-key-consideration-of-the-project, accessed 25 August 2022.

[13] Atasu, A., Duran, S. and Van Wassenhove, L. (2021, June 18). The dark side of solar power. *Harvard Business Review*. https://hbr.org/2021/06/the-dark-side-of-solar-power, accessed 25 January 2023.
[14] U.S. Environmental Protection Agency (2022, August 28). End-of-life solar panels: regulations and management. https://www.epa.gov/hw/end-life-solar-panels-regulations-and-management, accessed 27 January 2023.
[15] New York State Energy Research and Development Authority (2023). Solar guidebook for local governments, 148-152. https://www.nyserda.ny.gov/solarguidebook, accessed 23 January 2026.
[16] Thunder Said Energy (2023). Solar power: decommissioning costs? https://thundersaidenergy.com/downloads/solar-power-decommissioning-costs, accessed 26 January 2023.
[17] Weaver, F. (2020, March 20). What to do with old solar panels: cost estimates for decommissioning solar power plants. https://commercial-solarguy.com/what-to-do-with-old-solar-panels, accessed 26 January 2023.
[18] Caunt, G., Simon, B. and Roosa, S. (2018). Renewable energy, chapter 16, in *Energy Management Handbook*, 9th edition, Roosa, S. (ed), 477–478. The Fairmont Press, LLC.: Lilburn, Georgia.
[19] U.S. Government Accountability Office (2018, March 6). Coal mine reclamation: federal and state agencies face challenges in managing billions in financial assurances. https://www.gao.gov/products/gao-18-305, accessed 27 January 2023.
[20] Radmaker, D. (2022, March 3). Blackjewel's catastrophic bankruptcy and the collapse of the mine cleanup system. https://appvoices.org/2022/03/03/bankruptcy-mine-cleanup-collapse, accessed 28 January 2023.
[21] Okon Recycling (2025, July 5). Solar panel decommissioning: costs, regulations, and recycling options in 2025. https://www.okonrecycling.com/renewables-recycling/solar-panel-recycling/solar-farm-decommissioning, accessed 3 August 2025.

19

Sustainable Adaptive Reuse of Renewable Energy Components and Materials

Ahmed Gazel and Stephen Roosa

19.1 Introduction

Renewable energy technologies have seen exponential growth over recent decades and are poised to dominate new power installations through 2050. This transition is crucial for mitigating climate change, yet it brings new sustainability challenges. Chief among these is the management of end-of-life waste from solar panels, wind turbines, batteries, geothermal, and hydroelectric systems. Large-scale deployments imply that by mid-century, tens of millions of tonnes of solar photovoltaic modules, composite turbine blades, spent chemical batteries, and other system components will retire. These components will require reuse, recycling, or disposal. If not properly handled, this waste could offset environmental gains and strain waste management infrastructure. To avoid such unintended consequences, it is essential to understand the types of waste and their volumes under realistic growth scenarios. Strategies for extending the lifespan of renewable infrastructure and closing the material and supply loops are needed.

This chapter presents a comprehensive analysis of global waste generation from key renewables for 2025 to 2050 using linear growth models and current policy scenarios. It proposes multidisciplinary engineering solutions—in design, maintenance, materials, chemistry, and economics—to enhance sustainability. We show how integrative approaches can accelerate progress toward reducing greenhouse gas emissions. The objective of this chapter is to chart a path for renewables to sustainably support the world's circular economy.

DOI: 10.1201/9788743811312-19

19.2 Methodology

We synthesize data from recent projections and technical reports to estimate waste volumes while employing engineering equations with real-world parameters to obtain order-of-magnitude estimates. For each technology, we review engineering practices that prolong service life and reliability. We explore advanced innovations in physics, materials science, nanotechnology, and manufacturing that improve component durability and recyclability. We then discuss green chemistry advances to replace hazardous materials with environmentally benign ones that can be recycled. Finally, actuarial and economic modelling concepts are used to weigh the benefits of longevity and circularity against the impacts on industry business models, proposing policy, and economic mechanisms to align incentives. All are supported by current scientific studies and datasets. We emphasize findings from the Massachusetts Institute of Technology (MIT) and other leading institutions to ensure an up-to-date and scholarly perspective. The analysis is further supported by calculations and references.

19.3 Waste Projections (2025–2050)

Rapid deployment of renewables will lead to a surge of end-of-life (EOL) disposals from 2025 through 2050 as installations of renewable energy equipment and components reach the end of design lifetimes. Our assessment focuses on solar photovoltaics (PV), wind turbines, and battery energy storage systems (BESS)—which are expected to contribute the largest waste streams—plus geothermal and hydroelectric systems. A linear growth scenario under current policies is assumed. Capacity additions are assumed to increase at a roughly constant rate; no extraordinary early retirement or breakthrough extension is assumed beyond standard lifetimes. Table 19.1 summarizes the projected waste quantities by 2050 for each renewable technology using these assumptions. These projections are then detailed in the subsections, along with their underlying assumptions and models.

19.3.1 Solar photovoltaic panel waste

Most solar photovoltaic waste is from modules that use monocrystalline silicone, polycrystalline silicone, or thin-film technologies. Silicone-based solar panels have an operational component lifecycle of about 25 to 30 years, after which their electrical output drops below 80% of its initial rating, and decommissioning is expected. Given the scale of solar PV deployment, components

Table 19.1 Projected global waste from renewable energy technologies by 2050 (current policy per a linear growth scenario).

Technology	Expected EOL waste by 2050	Notes (key components)
Solar PV	~78 million tonnes (cumulative)	Mostly glass, aluminum, silicon [1].
Wind turbines	~43 million tonnes (blades)	Fiber-reinforced polymer blades [2].
Batteries (Lithium-ion)	1.2-8.5 million tonnes per year by 2050	Retired EV and storage batteries (metals) scenarios used in the material flow analysis.
Geothermal	Minimal (<1% of total)	Steel, piping, drill cuttings (mostly reusable) [3].
Hydroelectric	Minimal (few large units)	Concrete, steel from decommissioned dams (rare by 2050)

Sources: IRENA projections for solar PV (University of Cambridge study for wind). Electric vehicle battery retirement scenarios (end-of-life assessment for geothermal). Hydroelectric estimates are based on a 50 to100 year infrastructure lifecycle.

are anticipated to become a major source of waste by 2030 and increase thereafter.[1] Using a linear growth assumption, the approximate annual installations and EOL disposals can be estimated.

- Mass per unit: A standard 1 m² silicon PV module (200W) weighs 10 to 15 kg (glass, aluminum frame, silicon cells, and polymers). This mass (m) equals roughly 50 to 75 tonnes per MW of installed solar PV capacity. For estimation, $m = 60$ t/MW = 60,000 kg/MW as an average mass-to-power ratio.
- Installed capacity growth: If capacity (C) increases from C_0 in 2025 to C_1 in 2050 linearly, the average annual addition is $\Delta C = (C_1 - C_0)/25$. For example, assuming C_0 is roughly 1 TW in 2025 and C_1 is 5 TW in 2050, the ΔC will average 160 GW annually.
- Lifetime distribution: With an average lifetime $L \approx 25$ years, panels installed in year (t – 25) will retire in year t, assuming no premature failures. In a linear growth scenario, EOL waste in year t is roughly the installation rate 25 years prior, times the mass per capacity.

Using these assumptions, the annual PV waste (WPV) generated in 2050 can be estimated using Equation 19.1.

$$WPV_{2050} \approx \Delta C\ 2025 \times mW \qquad (19.1)$$

[1] Global PV capacity grew from roughly 227 GW in 2015 to over 1,000 GW by the early 2020s. It is projected to reach 4-5 TW by 2050.

Inserting ΔC of 160 GW/year and m = 60,000,000 kg/GW (which is 60 kg/kW), we obtain:

$$WPV_{2050} \approx 160 \text{ GW/year} \times 60{,}000{,}000 \text{ kg/GW}$$
$$\approx 9.6 \times 1{,}012 \text{ kg/year}$$

This yields about 9.6 billion kilograms (9.6 million tonnes) of solar PV waste anticipated in 2050. This estimate is consistent with more detailed analyses that account for the age distribution of panels. A landmark study by the International Renewable Energy Agency (IRENA) and the International Energy Agency (IEA) projected that cumulative solar panel waste could reach roughly 78 million tonnes by 2050 under a business-as-usual scenario [4]. This corresponds to tens of millions of panels (on the order of 2 billion panels cumulatively) in need of handling by mid-century. The quantities of waste by the 2040s will surge as the installations in the 2010s and 2020s reach the end of their economic lifecycles.

The bulk of solar PV module waste includes common materials—glass (~75% of module mass), aluminum (frame), and silicon. Trace amounts of hazardous elements (e.g., lead solder or cadmium in certain thin-film panels), if landfilled, could leach toxic substances into soil or groundwater. The sheer volume of waste materials means that without robust recycling programs, solar waste could become a significant part of global electronic waste. One projection warns that by the mid-2030s, the volume of discarded solar panels could outpace new installations (in weight) by a factor of 2.5. When disposal costs are accounted for, the levelized cost of energy (LCOE) for solar increases. However, expenses that are projected to be credited in the distant future are heavily discounted.

The solar industry believes that there is a substantial residual value for solar array components, but cannot ensure that there will be a market for the salvaged components when decommissioned [5]. The feasibility of realizing salvage value may depend on who removes the equipment—the developer, operator, the tenant, or the landowner—and how each estimates the value [6]. However, total decommissioning costs for solar farms have been estimated to offset as much as 40% of the residual salvage value [5]. Given the wide range of costs, it is imperative that asset managers develop a specific cost estimate for each site [7]. Neglecting EOL management can underestimate the LCOE for solar PV component decommissioning. When waste handling is considered, the total cost of disposal could be up to four times higher than projected. This underscores the urgency of developing cost-effective life-extension and recycling strategies for solar PV components.

19.3.2 Wind turbine blade waste

Early 21st-century wind farms are now beginning to retire some wind turbine generators (WTGs); this will accelerate during the 2030s. The turbines have a design life of about 20 to 25 years. The massive turbine towers and large nacelles are mostly steel, and wiring is copper, all of which are largely recyclable. It is possible to refurbish the towers to support new WTGs.

The blades, particularly those of the ubiquitous horizontal axis wind turbines (HAWTs), are made of thermoset composites that are difficult to recycle. The disposal of fiberglass or carbon-fiber reinforced plastic blades is a primary concern. Global wind turbine blade waste, estimated to be about 43 million tonnes of material, would accumulate by 2050 as turbines reach the end of their lifecycle [2]. The largest shares of this blade waste will be generated by China (40%), Europe (25%), and the U.S. (16%) [2]. A scaling calculation can be used to verify these estimates. Modern onshore wind turbines keep getting larger—a typical 2–3 MW turbine might have 50–70 m blades [8]. For example, a 1.5 MW turbine blade set (for all 3 blades) weighs about 36 tonnes or 24 tonnes per MW. Many new 4+ MW onshore or 10+ MW offshore turbines have even larger blades, but also might use materials more efficiently (15–20 t/MW) [5]. Assuming an average of 20 tons of blades per MW installed, a world wind capacity of about 2,000 GW (2 TW) by 2050 would triple present capacity; this implies about 40 million tonnes of blades in service. Over a 25-year period, many WTGs will require replacement. Our estimate roughly aligns with the 43 Mt figure found in the literature.

The quantities of blade waste will continue to increase by the 2040s. If we assume a roughly linear capacity growth and 20-year life, then blades installed in the 2020s (when wind additions are tens of GW/year globally) will become waste in the 2040s. For instance, if 80 GW are installed in 2025 globally (plausible given the current trajectory) with ~20 t/MW of blades, that adds 1.6 million tonnes of blades in 2025, which would retire starting in 2045. By 2050, annual blade waste will be a minimum of a few million tonnes per year worldwide.

While less than the solar PV waste volumes, wind blade waste pose unique challenges because of the difficulties in recycling composite materials. The size of the blades complicates transportation and landfill. Regardless, at 43 Mt cumulative, the scale is manageable relative to municipal waste—one analysis noted this is only 0.015% of all municipal solid waste landfilled in 2015. The key issue is not the volume of material but the hazard and resource value: landfilling composite materials squanders the embodied energy and material content. The inert bulk of the materials can strain landfills located near wind farms.

19.3.3 Battery waste

The rise of renewable energy is coupled with recent increases in the use of BESS—both for battery electric vehicles (BEVs) and grid energy storage—to buffer intermittencies of solar and wind generation. Lithium-ion (L-I) batteries dominate current applications. They have lifetimes of 8 to 15 years in electric vehicles (EVs) and potentially have a second life of 5 to 10 years in stationary storage. Spent L-I batteries are considered hazardous waste because they contain flammable electrolytes and toxic metals like cobalt and nickel. However, they are a rich source of valuable metals (lithium, cobalt, nickel, copper, and aluminum) when recycled.

Projections of global battery waste vary widely based on EV adoption scenarios. By the 2030s and 2040s, we will see increasing battery retirements. A detailed material flow study (considering various EV uptake and recycling scenarios) found that annual end-of-life EV battery retirements could reach between 1.2 million and 8.5 million tonnes in the year 2050.[2] The conservative scenario exceeds 1 million tonnes of waste batteries annually by 2050.

To illustrate, suppose there are N electric cars in 2050 and each has an average battery size of E kWh. If f fraction of those batteries reaches their EOL in that year, the waste mass W is approximately $N \cdot E \cdot mE \cdot f$, wherein mE is the mass per kWh. For instance, if there are 800 million EVs in 2050 (roughly consistent with aggressive electrification scenarios), each with a 50-kWh battery pack, that is 40 billion kWh in use. If 10% retire that year (using a 10-year battery lifecycle), that is 4 billion kWh retiring. With 0.2 kWh/kg energy density (5 kg per kWh), $W \approx 4^9$ kWh × (5 kg/kWh) = 20^9 kg = 20 million tonnes in that year. This is an extreme scenario; more moderate assumptions (fewer EVs or longer life) lower this to single-digit millions of tonnes, which aligns with the cited range. Notably, these figures include primarily EV batteries; grid storage batteries are an additional source, though likely smaller than EVs because EVs are expected to dominate L-I battery volume.

Battery waste management is pivotal for both environmental safety and resource security. Retired L-I batteries present an environmental liability and economic opportunity. If not properly recycled, they can pose fire risks in landfills and leach toxic chemicals (the electrolyte can form corrosive hydrogen fluoride, and metals like cobalt can contaminate soil). A single EV

[2] The 1.2 million tons figure corresponds to slower EV growth and/or extensive second-life usage; the 8.5 million tons figure corresponds to rapid EV deployment with relatively short lifetimes and limited second use.

battery can pollute water or land if improperly disposed of. However, these batteries contain strategic metals. A typical 50 kWh EV battery might contain about 8 kg of lithium, 14 kg of cobalt, and over 20 kg of nickel. Depending on the growth and technology scenario, the projected 2040 demand for lithium and cobalt exceeds today's production by up to 8 times [9]. By 2050, the stock of EOL lithium-ion batteries represents a potentially large resource that could help meet the future demand for lithium, cobalt, and nickel in a high-adoption scenario.

Effective recycling and reuse of these strategic materials could help offset the amount needed from new sources. There are two primary methodologies for EV battery recycling:

- Pyrometallurgy involves melting the batteries at high temperatures to extract metals. While effective, it is energy-intensive and can produce harmful emissions [10].
- Hydrometallurgy uses chemical processes that dissolve battery materials in acids to recover metals. This method is more precise and environmentally friendly, but can be more expensive [10].

If recycling is effectively scaled, it could reduce the 2050 demand for new lithium and nickel by as much as 25%, and cobalt by 40%. The reduced need for mining new minerals provides additional environmental and economic benefits. However, the collection and transport of EV batteries to a recycling facility requires infrastructure. In some cases, recycling is more expensive than mining new materials [10]. It is challenging to massively scale up resource production and focus on the recycling of battery metals to meet the increasing consumption of the electromobility industry [9].

19.3.4 Waste from geothermal plants

Geothermal electrical generation plants are generally low-waste and long-lived. Geothermal plants consist of wells drilled into the Earth's crust, pipelines, heat exchangers, turbines, and generators. Many components (steel casings, pipes, turbines) can last for decades and are often refurbished rather than replaced frequently.

The primary waste outputs from geothermal plants are drill cuttings and muds from well drilling, mineral scales or brines brought up from the reservoir, and worn-out equipment (e.g., pipes corroded by hot fluids). These waste are relatively small in volume and often handled as industrial waste—for instance, drill cuttings can be landfilled or sometimes used in construction fill.

Over its life cycle, waste from geothermal projects are minimal when compared to those produced by other renewable or fossil-fuel sources. At a large scale, the IEA net zero scenario envisions up to 800 GW of geothermal generation by 2050 [11]. The waste would mainly consist of replaced plant equipment and spent fluids, which are minor compared to the massive waste streams of solar PV, wind, and chemical batteries. Geothermal plants typically reinject fluids back underground (closing the loop on liquids) to replenish resources. Geothermal plants have comparatively small site footprints. Decommissioning typically involves plugging wells and removing surface facilities, leaving no hazardous and little persistent waste except concrete and steel, which can be recycled.

19.3.5 Waste from hydroelectric systems

Hydroelectric systems (dams and turbines) are characterized by their longevity—many of the larger high-head dams operate for 50 to 100 years. By 2050, most existing hydroelectric facilities will still be in service, and the waste they create will be negligible. There have been cases of dam decommissioning, especially of smaller or older dams, for safety or environmental restoration reasons. This involves handling large volumes of concrete and steel from demolition.

Globally, only a small number of hydropower plants are expected to be removed by 2050; most scenarios call for increasing hydro capacity modestly, not retiring it. An environmental issue related to decommissioning is the management of accumulated sediment in reservoirs which can contain heavy metals or organic matter. When a dam is removed after many decades, sediment is released downstream or dredged—effectively moving a long-sequestered waste from behind the dam. Studies have shown that the sediment release can cause a pulse of greenhouse gas (GHG) emissions and water quality impacts. This is a localized impact rather than a major solid waste management problem. An older dam being decommissioned might have tens of thousands of tonnes of concrete and steel rebar; these can often be left in place if breached and broken up, or reused as fill if removed. Hydro turbines and generators, when upgraded, yield scrap metal that is readily recyclable.

While hydroelectric power requires attention to ecological impacts, the waste footprint remains is minimal. In fact, hydro facilities sometimes reduce waste by trapping trash and debris flowing in rivers. For example, dams in Bavaria remove tens of thousands of tonnes of plastic and driftwood from rivers annually, acting as giant trash rakes and aiding the waste management process.

Geothermal and hydropower technologies present relatively minor waste challenges through 2050, especially when compared to the looming volumes of solar panels, wind blades, and batteries. An important contribution to the renewable waste discussion is their longevity—these systems show that renewable energy infrastructure can operate for many decades when properly maintained.

19.4 Redesigning for Sustainability

An effective and sustainable strategy for waste reduction is to extend the useful life of renewable energy components and hardware. Every additional year a solar panel or wind turbine operates delays its entry into the waste stream and extracts more value from the resources invested. Extending component lifespan is an engineering challenge in maintenance, reliability, and design. We next review how reliability engineering approaches—including modular design, ease of repair, robust diagnostics, and low-environmental-impact maintenance—can increase longevity and reduce the need for premature replacements.

19.4.1 Design for modularity and repairability

Traditionally, many renewable energy components have been designed to emphasize performance, reliability, and cost rather than reparability. For example, most solar PV modules are sealed—if one cell or interconnect fails, the entire panel is replaced rather than repaired. This leads to unnecessary waste. Recent work is challenging this paradigm by introducing modular components that are easily replaced. In one study, engineers developed a solar panel design that allows sections of the panel to be disconnected and replaced if damaged instead of replacing and discarding the entire panel [12]. After an impact or failure in one section, it can be replaced and the panel reassembled. The modular panel design achieved similar performance to a conventional panel, but with dramatically improved reparability. This concept addresses common damage to modules (e.g., a corner of a panel breaking, thermal damage, or cells cracking due to hail damage) by confining the failure to a module that can be readily replaced. Economic analysis in that study showed that despite slightly higher upfront costs, the modular panels could reduce lifecycle costs for consumers.

Wind turbines are much larger, but modularity can be applied at the component level. For instance, turbine blades are sometimes built in segments (to aid transportation). This allows the possibility of replacing a damaged blade section rather than the entire blade. In practice, larger blades are usually replaced whole, but the concept of sectional repair is emerging. Modular drive-train components (gearboxes, converters, and blade tip modules) are being designed to be replaced with minimal crane time. By using standardized modular components, wind farm operators can store spare modules onsite, quickly replace faulty parts, and ship them for refurbishment. This minimizes downtime and prevents minor issues from causing cascading damage. Ease of disassembly is crucial for end-of-life: designing connections (bolts, adhesives, etc.) such that components can be separated facilitates both repair and recycling.

19.4.2 Example: Power electronics

Inverters and power converters used in solar and wind systems often have shorter lifespans (10 to 15 years) than the solar panels or turbines. By using a modular inverter design, the operators of a large solar PV installation might replace only the failed sub-module of an inverter rack rather than the entire unit, which reduces electronic waste. Similarly, EV battery packs are moving toward modular architectures. When needed, individual cells or modules are replaceable; Tesla builds battery packs using many small cylindrical L-I cells [13]. The cells are grouped into modules, which create the full battery pack. This modular design improves efficiency and capacity [13]. If some L-I modules fail, the pack can potentially be reconditioned by replacing only those modules. Such approaches are aligned with circular economy principles and are increasingly supported by manufacturers as they see value in offering long-term maintenance services.

19.5 Maintenance

Extending the lifespan of equipment is about physical designs and operational strategies. Reliability-centered maintenance (RCM) uses condition monitoring and predictive analytics to service equipment prior to failures. This avoids catastrophic losses and premature component replacements.

19.5.1 Predictive maintenance and diagnostics

Wind turbine operators commonly employ supervisory control and data acquisition (SCADA)-based health monitoring and dedicated sensors (strain gauges, vibration sensors, acoustic emission monitors) on critical components like blades and gearboxes. These feed data into artificial intelligence (AI) or physics-based models that predict the remaining useful life. For example, vibration analysis can detect early bearing wear in a gearbox; by addressing it (replacing a bearing or correcting imbalance), the gearbox's life can be prolonged before a major breakdown forces replacement. Actuarial methods (statistical life analysis) are used to model failure distributions of components—a life table can be made for turbine components, enabling a proactive replacement schedule optimized to minimize unexpected failures. By avoiding the scenario of a blade snapping or a transformer exploding, predictive maintenance ensures components reach or exceed their design life. Regardless, the maintenance frequency of WTGs increases as the turbines age.

One illustrative result from Electric Power Research Institute (EPRI) research showed that overly aggressive wind turbine operation (uprating power output by software optimization) increases fatigue loads and reduces lifetime; operations that respect design limits keep turbines healthy and improve their operational longevity. Solar farms can use infrared imaging drones and current-voltage (I–V) curve tracing to pinpoint solar PV panels starting to degrade by visualizing hotspots and encapsulant browning. These panels can be targeted for refurbishment (e.g., replacing a junction box or bypass diode) prior to complete failures. Modern inverters also perform maximum power point tracking diagnostics; deviations can hint at deteriorating panels or connections, prompting early intervention.

By 2050, widespread use of Internet of Things (IoT) sensors and digital twins for renewable assets could become the norm. This allows each component's operational characteristics to be monitored and recorded. This capability reduces the potential of sudden failures that would necessitate component replacement.

19.5.2 Low environmental impact maintenance practices

Maintenance practices can improve the environmental sustainability of renewable energy systems. When performing maintenance on WTGs and solar PV components, it is important to use methods that do not introduce new environmental problems. Choosing maintenance materials carefully—for instance, using biodegradable lubricants in wind turbine gearboxes and biobased hydraulic fluids in blade pitch systems—eliminates the risk of soil contamination in case of leaks. Similarly, coolant, or refrigerant choices in some power electronics cooling systems can be shifted to eco-friendly options (e.g., using water-glycol instead of oil for transformer cooling where feasible). These initiatives reduce the operational environmental footprint and simplify EOL handling by reducing or eliminating hazardous residues.

Reducing the need for auxiliary resource maintenance offers environmental benefits. Robotic and remote inspection can avoid the need for heavy infrastructure: drones inspecting the blades of WTGs for erosion and cracks means fewer site inspections using cranes, which saves fuel and avoids damage to terrain. Cleaning solar panels is an important maintenance task, especially in dry, dusty regions—traditionally, this uses water (which can be scarce) and sometimes chemical detergents. Robotic dry cleaning and nano-coating-based self-cleaning can maintain panel performance without the expense of water or the use of detergents. By 2050, we envision many solar farms employing autonomous cleaners that brush or vibrate dust off, powered by a small fraction of the array's energy. This not only preserves efficiency (yielding more energy over the panel's life) but also reduces wear—abrasive dust can cause micro-scratches that accelerate degradation; keeping panels clean helps them last longer.

Calibration and monitoring are also part of maintenance: for example, wind farms can obtain calibrated load assessments after 20 years to determine if the life of the turbines can be extended beyond their original design life. This process involves inspecting the structure for fatigue and monitoring actual loads. It is recognized that modern WTGs will generally last 20 years, but the operating life can be extended to 25 years or longer if proper maintenance and minor reinforcements are implemented [14]. The trade-off is that maintenance costs may rise with age, but the overall lifecycle energy output increases substantially. Solar PV panels from the 1990s have been found operating at 80% output after 30 or more years. With careful handling (e.g., avoiding physical stresses, maintaining electrical connections, replacing inverters when needed), many PV modules could also exceed their 25-year design life.

In summary, maintenance and reliability engineering can mitigate waste generation by extending component lifetimes. A combination of modular repairable design, continuous monitoring, and environmentally conscious maintenance practices helps to ensure that renewable energy systems reach and possibly extend their full operating lifespan. Each additional year of operation means fewer new materials are consumed, and there is less waste to manage, which contributes to more sustainable operations.

19.6 Engineering Advancements

Advanced materials and technologies have an objective of providing durable, eco-friendly innovations for renewable energy systems. Innovations in physics and material science are providing new ways to make renewable energy hardware more durable, efficient, and environmentally friendly. By harnessing nanotechnology, novel materials, and advanced manufacturing processes like three-dimensional (3D) printing, engineers can create components that last longer and have lower environmental impact, thereby addressing the root causes of waste. This section highlights numerous innovations across solar, wind, and battery technologies.

19.6.1 High-durability materials and coatings

One straightforward way to reduce waste is to make components that do not fail or degrade as quickly. Materials scientists are actively developing stronger and more weather-resistant materials for renewable systems.

Improved composites for wind turbine blades
Blade damage is a primary cause of wind turbine failures. Traditional fiberglass/epoxy blades suffer fatigue and can crack or delaminate. New composites incorporating carbon fiber and novel resins can increase the strength-to-weight ratio and provide resistance to fatigue. For instance, adding carbon fiber spars in blades reduces flexing and thus lowers fatigue damage. Researchers are exploring nano-reinforcements (like carbon nanotubes or graphene mixed into resins) to improve toughness and reduce crack propagation in blades. Advanced gelcoat coatings on blade surfaces can reduce ultraviolet (UV) degradation and erosion from rain or sand impact. For offshore wind turbines, coatings that resist saltwater corrosion and biofouling help protect blades and towers. These improvements in material durability help extend the life of wind turbine components.

Anti-degradation coatings for solar modules
Solar panels are exposed to UV radiation, moisture ingress, temperature cycling, and abrasion from dust, sand, and hail. Physics-based understanding of degradation (e.g., how UV breaks down polymer back sheets or how thermal stress causes cell microcracks) has led to the development of protective materials.

Nano-coatings are ultra-thin (typically 1 to 100 nm in thickness) layers or chemical structures that are applied to surfaces by a variety of methods [15]. Nano-coatings placed on solar glass can provide multiple benefits: they can be anti-reflective (boosting efficiency) and hydrophobic, which causes water to bead and roll off, thus removing dirt. This keeps panels cleaner with less manual intervention. Some nano-coatings have anti-UV and anti-corrosion properties, protecting the encapsulant and cell interconnects. Field tests have shown that such coatings can slightly improve panel efficiency and maintain performance longer by reducing soiling and weathering.

Researchers are also working on self-healing coatings—materials that can repair minor scratches or cracks when heated by the sun. Additionally, replacing the polymer back sheet with glass (making double-glass modules) has become popular for durability because glass is impermeable to moisture and does not degrade in sunlight. This can reduce module degradation rates.

Thermal management and stress reduction

Managing thermal stresses can prolong the life of solar panels. For example, thermal expansion mismatch in solar cells causes cracks; new cell designs use kerf-less wafers or flexible substrates to alleviate stress.

In wind turbines, loads during extreme wind gusts often determine lifespan—hence smart control systems (aerodynamic or blade pitch control) that reduce loads in extreme winds can prevent damage. This is an application of aeroelastic modelling to actively preserve the structure. By 2050, many WTGs will have advanced control algorithms that lower power capture during storms to avoid heavy loads that shorten the turbine's life. Similarly, advanced physics-based modelling of turbine blade vibrations has yielded tuned mass dampers or other devices installed in blades to reduce oscillations, thereby reducing fatigue. Because of the consistently strong winds found at offshore wind farms, turbines can produce energy more reliably and efficiently [16]. Some typhoon-proof horizontal axis wind turbines with superior blade designs are now capable of withstanding winds up to 220 kph (137 mph) [8].

Durable materials

One exciting initiative is the U.S. Department of Energy's Durable Module Materials (DuraMAT) Consortium, which aims to extend solar PV module lifetime using new materials and designs. By studying how modules fail via accelerated testing and forensics, researchers identify failure modes and develop fixes such as improved encapsulants or corrosion-resistant cells. There is optimism that future solar modules could be certified for 40– to 50-year lifetimes. Each additional decade of life means proportionally less waste and resource use for the same energy output. If modules lasted 50 years instead of 30, fewer new panels would need to be produced to achieve local solar capacity targets—saving raw materials and reducing the strain on recycling systems when those panels finally retire.

19.6.2 Nano-engineering applications

Nano-engineering and nanotechnologies offer tools to improve renewable energy devices at the microscopic scale, often yielding macro-scale durability or efficiency benefits.

- Nano-coatings and surface engineering: Nano-coatings on solar glass (typically silica or titania-based coatings) can impart self-cleaning and anti-reflective properties. Another nanotech application is anti-soiling coatings for mirrors in concentrating solar power (CSP) and anti-corrosion coatings for geothermal plant pipes. In wind turbines, researchers have considered nano-structured surfaces for blades to reduce drag and perhaps mitigate icing. While primarily focused on improving efficiency, these can also reduce the abrasive wear on blades by reducing turbulent buffeting.
- Nano-additives in materials: Adding nanoparticles to polymers or paints can enhance component performance. For wind turbine blades, adding nano-silica or nano-clay to the epoxy resin can improve its toughness (impact resistance) and reduce microfractures under stress. In battery technology, nanoparticles in electrodes (like nano-silicon in L-I batteries) can accommodate expansion and avoid cracking, thus extending cycle life. A nano-engineered battery separator might include a ceramic nanoparticle coating to improve thermal stability and safety (preventing meltdown and fires, which cause premature failures). These changes at the nanoscale improve system safety and longevity.
- Nanotechnology for diagnostics: Nano-sensors and thin-film electronics can be embedded in structures to monitor their health. For example, a nano-strain gauge network printed onto a wind turbine blade could continuously detect strain distribution and crack initiation much earlier than conventional sensors. These data can support predictive maintenance. Carbon nanotubes (CNTs) have been explored as self-sensing materials; a composite material doped with CNTs can change its electrical resistance when deformed, effectively acting as a sensor.
- High-efficiency nanomaterials (to reduce material needs): Although efficiency is not directly a waste topic, higher efficiency means fewer panels or turbines are needed for the same energy, indirectly reducing total material throughput. Nanotechnology is enabling new high-efficiency solar cells, such as tandem cells with perovskite (a calcium titanium oxide mineral) nanocrystals or quantum-dot solar cells. If a solar farm can produce the same energy with panel efficiencies of 30% instead of 20%, it needs 1/3 fewer panels; hence, offering space efficiency and 1/3 less waste at EOL. Likewise, nanostructured anti-reflective coatings on solar cells or plasmonic nanoparticles that help trap light can make thinner cells viable (less semiconductor material per cell).

Nano-engineering allows materials to be manipulated at the smallest scales to enhance macro performance and longevity. Many of these technologies are currently in research or early commercialization. Sometime between 2030 and 2050, we expect them to mature and be implemented widely in new renewable energy installations, thereby yielding more robust, long-lasting equipment. For example, by 2050, tandem silicon-perovskite cells or other advanced cells will likely be mainstream technologies, significantly boosting the efficiency of solar PV modules.

19.7 Advanced Manufacturing

Additive manufacturing techniques, such as 3D printing, are emerging as powerful tools in producing renewable energy components. They will likely revolutionize maintenance and spare parts manufacture and management by 2050. The freedom of shape and the ability to produce parts on demand offer several sustainability advantages.

19.7.1 Optimized designs with less material

3D printing enables complex geometries that traditional casting or machining cannot achieve. For example, engineers can print lightweight lattice structures for wind turbine components (like sections of the nacelle or blade internal supports) that maintain strength while using less materials. Siemens has applied 3D printing to optimize the molds cast for turbine blades and printed metal-gas turbine parts with internal cooling channels that improve efficiency. Similar approaches can print parts of WTGs or hydro turbine runners with shapes that maximize performance and minimize stress concentrations. By reducing excess material and tailoring the structures, 3D printing reduces the raw material consumption and can also eliminate joints or bolts to remove points of potential failure.

19.7.2 Rapid production of replacement parts

One of the challenges in maintaining decades-old equipment is that original spare parts may no longer be produced. In the future, rather than scrapping an old wind turbine due to a single unavailable part, we can print a replacement part from its digital design. There are already cases of obsolete small wind turbine parts being 3D printed in metal as one-offs. By 2050, digital inventories and metal additive manufacturing could allow local fabrication of everything from a custom bushing to a complex gear. This reduces downtime and prevents the decommissioning of an otherwise functional machine due to the lack of a specific component. It also reduces the need to stockpile large inventories of spare parts.

19.7.3 Manufacturing novel components

3D printing can produce entirely new types of components that improve renewable energy systems. For example, researchers are exploring printing ceramic solid-state electrolytes for batteries layer by layer, which could lead to safer, long-lasting batteries (solid-state batteries are non-flammable and potentially have longer cycle life). There is research on printing solar cells with inkjet or roll-to-roll printers—essentially, additive manufacturing of solar PV, which allows coating solar material on flexible substrates and integrating solar PV materials into building components. One vision is 3D-printed perovskite or organic solar cells embedded in devices or building façades; if non-toxic and easily replaceable, these could supplement traditional panels and be more easily recycled.

19.7.4 Large-scale printing for infrastructure

Construction of renewable energy facilities can benefit from additive manufacturing. 3D-printed concrete is being tested for wind turbine towers. A taller tower can greatly increase the energy output of a turbine because the nacelle is located higher into the wind stream. Companies have considered printing concrete tower sections onsite, allowing for wider bases and taller structures without transport limitations. These printed towers can be optimized for material use, potentially reducing the volume of concrete needed (and thus cement-related CO_2 emissions). For some developing countries, this could help meet local content requirements. For hydropower applications, one could imagine printing complex shapes for optimal water flow in small hydro turbines, or using printed polymers for certain low-head turbines.

From a waste perspective, additive manufacturing minimizes scrap materials. Biodegradable or recyclable printing materials (like some bio-plastics or recycled metal powders) are being developed. While challenges remain, the overall impact of advanced manufacturing by 2050 is expected to be more agile, less wasteful, and to reduce maintenance requirements for renewable energy technologies.

19.8 Making Components Greener

One of the core concerns about renewable energy hardware is the use of hazardous or difficult-to-recycle materials (e.g., lead in solar panels, epoxy resins in blades, and toxic metals in batteries). Green chemistry focuses on redesigning materials and chemical processes to be inherently safer and more sustainable. In the context of renewables, researchers are developing non-toxic, biodegradable, or fully recyclable alternatives to the current materials. We next discuss key developments in replacing hazardous components of solar panels, wind turbine blades, and batteries with greener solutions. The objective is to consider ways to reduce environmental risks and improve end-of-life recyclability.

19.8.1 Solar panels: Eliminating toxic elements and designing for recyclability

Modern silicon PV panels are mostly non-toxic (glass, aluminum, silicon). However, they often contain trace amounts of lead in the solder that connects solar cells (amounting to typically ~0.1% by weight); sometimes, fluorinated polymers that can be problematic if incinerated. Certain thin-film solar technologies use toxic metals like cadmium (in CdTe solar panels) or selenium (in copper indium gallium selenide thin-film panels). Green chemistry technologies for solar PV panels offer potential improvements.

Lead-free solder and interconnects: Several solar PV panel manufacturers are transitioning to lead-free solder (e.g., using tin-silver-copper alloys) to comply with European Union (EU) regulations. Research is ongoing to ensure these lead-free solders hold up over decades without cracking. By 2050 it is likely that new solar panels will be fully lead-free, removing the risk of lead leaching from broken panels in landfills. Another approach is using conductive adhesives or novel conductive tapes instead of solder. This eliminates the use of lead and allows easier disassembly because adhesives can be unglued by using heat or solvents.

Non-toxic thin films: Alternatives to cadmium-based solar cells are being developed. For example, perovskite solar cells have achieved high efficiencies, but most use lead in the perovskite layer. Scientists are searching for lead-free perovskites (using tin, germanium, or other elements) that maintain performance. Progress has been made, though efficiencies with tin-perovskites are lower. If a stable, efficient lead-free perovskite is found, it could allow tandem solar cells with no toxic elements. For cadmium telluride (CdTe) solar panels, which offer low cost and good performance, the manufacturer First Solar has a panel return program that ensures that the cadmium is recovered safely. By 2050, one goal is that all solar PV technologies either contain no toxic elements or enable closed-loop recycling. This will ensure the recapture of the most harmful elements so that nothing damaging enters the environment.

Biodegradable or recyclable module components: One intriguing line of research is developing biopolymer encapsulants and back sheets for solar modules. Instead of using ethylene vinyl acetate (EVA) encapsulants, which are difficult to separate, a recyclable thermoplastic or biodegradable polymer could be used to laminate cells to glass. This would make delamination (a key step in recycling) much easier. For example, a thermoplastic encapsulant can be reheated and melted to release the cells and glass, whereas EVA requires chemical or mechanical treatment to break down. Likewise, back sheets made of non-fluorinated, easily recyclable plastics, or dual glass sheets, mean fewer mixed materials. Companies and labs are experimenting with EVA alternatives like thermoplastic polyolefin elastomers, which are easier to process and soluble adhesives. Imagine a glue that holds solar cells but can be dissolved in a specific solvent at module EOL to free the cells and glass—this would greatly enhance recycling requirements.

Recycling-friendly designs: Frames for panels can be made to snap together, using screws instead of permanent bonds where possible. The chemistry ensures that any adhesives or potting compounds used in junction boxes are debondable. Some research has focused on reversible adhesives that lose their bonding when a certain voltage or stimuli is applied. Using smart materials, it is possible that a solar panel could, at its EOL, be disassembled into glass, cells, and metal—all ready for recycling. By adopting such design changes, the industry aims to move from today's situation—where most solar panels end up in landfills due to costly recycling—to a future where recycling is more efficient and economically viable. MIT researchers have noted that boosting solar recycling rates will require a combination of new panel designs, better recycling technologies, and supportive policies [17].

A creative example of green chemistry is an MIT study that proposed recycling old lead-acid car batteries into perovskite solar panels [18]. This concept takes toxic waste (lead from millions of discarded car batteries) and uses a chemical process to convert it into high-purity lead iodide for making perovskite solar cells [18]. The resulting solar cells produce clean electricity; the lead is safely encapsulated in the panel during use and can be recycled. Using waste from one stream as input to another exemplifies integrative solutions to material recycling challenges.

19.8.2 WTG blades: Developing solutions for recycling

Wind turbine blades have been a vexing problem because their thermoset composite materials (typically epoxy resin reinforced with fiberglass, sometimes with balsa wood or carbon fiber components) are very difficult to recycle. Traditional composites cannot be melted down (thermosets are permanently crosslinked), and chemical recycling is expensive. Green chemistry is tackling this via two approaches: new resin chemistry for blades and innovative recycling processes for existing blades.

Recyclable thermoplastic or covalent-adaptable resins: A breakthrough was achieved by researchers who developed a bio-derived thermoset resin that is chemically recyclable. The U.S. National Renewable Energy Laboratory's (NREL) polyester covalently adaptable network (PECAN) resin is made from bio-based feedstocks (like sorbitol, a sugar) and forms a covalently adaptable network—meaning under certain conditions, its crosslinks can break and reform [19]. WTG blades made with PECAN resins can be manufactured using the same fiberglass infusion processes, and they perform on par with conventional epoxy blades. The NREL team demonstrated fully dissolving a 9 m prototype blade in 6 hours with a mild chemical catalyst. This technology means future blades could be 100% recyclable—the fiber and resin loop can be closed, avoiding landfills entirely [19]. The PECAN resin is bio-based (from plant waste) and emits about 40% less GHGs in production than epoxy [20]. The resin is also free of toxic components; it is possible that it can be used for drop-in replacements for wind turbine materials [20]. By 2050, one can envision most new wind turbines using either thermoplastic resins or this new class of covalent-adaptable thermosets. Vestas and other WTG manufacturers are investing in such technologies.

Green additives and natural fibers: There is interest in using bio-based fibers (i.e., flax or hemp fiber) or nano-cellulose in WTG blades, either in hybrid construction with glass fiber or in blades for smaller wind turbines. Natural fibers are renewable and are usually easier to dispose of (i.e., they can be incinerated without releasing toxic gases). Some companies have tested WTG blades using bamboo or wood laminates as structural elements. A contemporary approach is to use engineered wood (glued laminated timber) or bio-composites, which are designed to be biodegradable or easier to recycle. While it is unlikely that large commercial turbines will move entirely to bio-fibers due to strength limitations, niche applications and portions of blades (like the core or sandwich materials) could use them.

Recycling methods for existing wind turbine blades: For the millions of tonnes of blades that are comprised of traditional composites, chemists are developing ways to break them down. The objective is circularity—turning old blades into new blades—which green chemistry is on track to enable. A promising approach is using solvolysis or pyrolysis with special catalysts to recover fibers and potentially some resin constituents. Vestas in 2021 announced a new chemical process that can break down epoxy-based blades by separating fiber from resin and then further breaking the resin into its base chemicals [21]. This process essentially uncooks the thermoset. The output materials are like virgin glass fiber and resin precursors that can be reused. Such processes would allow older blades in landfills or those about to be disposed of to be retrieved and recycled. Scaling these processes economically is the next step—collaborations between turbine makers and chemical companies are an indicator that by 2050 large-scale WTG blade recycling plants could be operating.

Mechanical recycling: Grinding blades into filler material for cement kilns has been implemented. In Europe, some blade waste is cut and used as feedstock in cement manufacturing, which recovers energy and some material value. While not exactly turning wind turbine blades into new ones, this avoids landfilling and uses the energy content to reduce cement fuel use.

Adaptive reuse: At the Almere Poort train station in the Netherlands, shelters for the thousands of daily commuters at the bus stops use parts from 30 m WTG blades. Dutch architects have repurposed blades to make playground components, urban furniture, and bicycle storage enclosures. Other uses include repurposing as utility poles, pedestrian bridges, playground structures, and roofing and girders for small structures.

These innovations mean future wind turbines will be far more environmentally benign at their end-of-life. Instead of giant composite blades directed to nearby landfills, blades can be reused for new purposes, fully melted down or dissolved and remade, or safely burned for energy with no toxic emissions. Coupled with modular construction to ease disassembly, the wind industry is heading toward a circular economy model for turbine materials.

19.8.3 Batteries: Advancing non-toxic, abundant, and recyclable chemistries

Batteries are a type of chemical energy storage. They are an area of active research in materials chemistry. There is a continuous drive to improve their performance in commercial applications. This is often accomplished by reducing weight, upgrading energy density, and improving turnaround efficiencies. Fortunately, many of the performance-driven changes also improve their environmental impacts.

Cobalt-free cathodes: Cobalt in L-I batteries is toxic, expensive, and may not be ethically sourced. A major trend is moving to chemistries like lithium iron phosphate (LFP) and high-nickel cathodes that eliminate or greatly reduce cobalt content. LFP contains no cobalt or nickel. It is composed of lithium, iron, phosphates, and carbon ($LiFePO_4$)—all of which are relatively benign. Iron and phosphates are non-toxic and abundant. LFP batteries are used in China for EVs and stationary energy storage and have a long cycle life.

From an environmental perspective, LFP batteries are non-hazardous and environmentally benign [22]. They are free of toxic heavy metals and do not contain rare-earth elements [22]. They can be recycled, the lithium and phosphates can be recovered, and the materials reconstituted. If a battery cell ruptures, it is not as dangerous as one with soluble cobalt.

By 2050, a large fraction of batteries manufactured may be LFP or other cobalt-free types (i.e., those containing lithium, manganese, and iron phosphate). Nickel is used in many high-energy cells; research is directed toward trying to reduce nickel content to lower cost and to improve stability. Another promising chemistry is sodium-ion batteries, which use sodium and often cathodes like $NaFePO_4$ or other metal oxides that contain no cobalt. Sodium-ion battery (NIB) cells are inherently free of lithium and cobalt. NIB batteries do not rely on poisonous or uncommon materials, which makes them attractive for large-scale stationary storage from a resource and toxicity perspective [23]. Companies are commercializing sodium-ion batteries; by 2050, these could complement L-I batteries, especially if lithium resources become too costly.

Aqueous and solid electrolytes: The typical lithium-ion electrolyte is a flammable organic solvent (with $LiPF_6$ salt) that can release toxic hydrogen fluoride (HF) if exposed to moisture. Green chemistry aims to replace these with safer alternatives. Solid-state batteries use a solid electrolyte (ceramic or polymer) that is non-flammable and prevents leakage. No solvent means no fire risk and no toxic liquid—making these batteries safer during use and disposal. If a solid electrolyte is also made of non-toxic elements, then a failed cell does not leach chemicals. Sodium-sulfur (NaS) batteries are a feasible solution for large-scale energy storage needs and electric grid demand [24]. They provide high energy density, are lightweight, require less water during production, withstand high-temperature conditions, and potentially have longer lifespans than conventional batteries [24]. All components of NaS batteries can be recycled.

The use of water-based (aqueous) batteries is another approach; some researchers are developing water-in-salt electrolytes, or aqueous organic flow batteries such as iron ferricyanide/ferrocyanide batteries. These would be non-toxic and if spilled, cause minimal environmental harm compared to L-I batteries.

By 2050, we may have a portfolio of battery technologies for grid storage, many of which use abundant, non-toxic elements like iron, manganese, or organic molecules (some organic redox flow batteries use molecules like viologens or quinones that can be made from biomass).

Biodegradable batteries: A radical innovation is the concept of batteries that could biodegrade when disposed. Researchers have created prototypes of metal-free, all-polymer batteries with electrodes made from redox-active polymers that can break down into benign molecules. A team at Texas A&M University developed an all-polypeptide battery—the electrodes are made from amino acid-based polymers that on demand can degrade into amino acids [25]. These batteries are in early stages of research and have lower performance than those that are lithium-based; they point to a future where small-scale batteries (for wearables or sensors) could dissolve harmlessly in the environment. For larger batteries, partially bio-based designs could aid recycling (e.g., a polymer binder in electrodes that dissolves in a specific enzyme, freeing active materials gently).

Recycling-friendly cell designs: In addition to chemistry, battery engineers are considering how to make cells easier to recycle. For instance, cell disassembly robots and processes can be supported if cells have removable caps or non-permanent bonding. Some concepts include using screws instead of welds for battery pack assembly, or labelling components for automated sorting. Chemistry has an important role. For example, if different metals are present, they can be separated either by physical means (density, magnetism) or by selective leaching without producing toxic sludge. Green chemistry in recycling leans on solvent extraction and mild leaching agents (like organic acids instead of strong acids) to recover metals with less waste.

By 2050, one can imagine most batteries being made with a defined EOL plan: using a chemistry that has a known recycling process where >90% of materials (metals, graphite, etc.) are recovered. Perhaps batteries will carry a chemical chip that informs recycling facilities as to their composition, so that the optimal recycling methodology (i.e., hydrometallurgical, pyrometallurgical, or direct cathode recycling) can be chosen.

In summary, the batteries of the future might be cobalt-free, largely non-toxic, and designed for circularity. They could be an iron, sodium, or organic. They would pose minimal or no environmental hazards. Their components will either be safely renewed or easily reconstituted into new batteries. This is a shift away from the early 21st-century L-I battery paradigm. It addresses one of the criticisms of battery energy storage systems—that they create new hazardous waste problems. This can be largely mitigated through advances in green chemistry and enhanced recycling.

19.9 Economic Impacts

Implementing technical solutions—those that offer longer equipment lifespans, more durable and resilient materials, and full component recyclability—has broad economic implications. Longevity must be balanced with industry growth. Greater longevity and repairability of renewable energy components will lower costs and create new employment in maintenance services and recycling industries. However, it challenges existing industry business models that rely on continuous sales of new equipment. Next, we analyze the economic dimensions using actuarial and mathematical modelling and propose resilient economic strategies that ensure stakeholders remain incentivized to pursue sustainability. The objective is to reconcile conflicts between building the idea of making long-lasting products and maintaining economic growth.

19.9.1 Cost and benefits of longevity and circularity

From the energy asset perspective, extended lifetimes and greater reliability provide economic benefits for the owner. The levelized cost of energy (*LCOE*), a metric for a system's cost over its lifecycle, improves when system life is extended or when the EOL is extended through adaptive reuse or improved maintenance. This is estimated using Equation 19.2.

$$LCOE = (\text{Capital cost x } CRF + O\&M \text{ expenses}) \div \text{Annual energy output} \tag{19.2}$$

Wherein,

CRF = Capital recovery factor, depending on the lifecycle, typically in years

O&M = Operations and maintenance costs less any related savings, in currency.

If in Equation 19.2 the lifetime increases, the *CRF* decreases for a given discount rate. For example, increasing the lifecycle from 20 to 30 years, assuming a 5% discount rate, the *CRF* might drop from about 0.08 to 0.07. This means that the capital cost is spread over more years. If a wind farm's upfront cost is \$1.5 million per MW, at 20 years' life, the annual capital charge is about \$120k; at 30 years, it is about \$105k. That is roughly a 12% reduction in capital cost contribution to LCOE. Combined with more kWh generated over 30 years, the LCOE reduction can be about 20% to 30%. In practical terms, enabling wind turbines to operate 5 to 10 years beyond their original design lifecycle can increase the return on investment (ROI).

Conversely, if assets fail prematurely or require replacement more often, costs rise. An analysis found in the *Harvard Business Review* pointed out that if widespread early solar panel replacements occurred (due to failure or owners swapping in newer models), the waste disposal costs and lost value could make solar electricity up to four times more expensive than projections assume [27]. Avoiding early retirements by improving durability and proper maintenance lowers the lifecycle costs.

19.9.2 Recycling value and secondary markets

When components are recycled or repurposed, they retain some economic value at end-of-life, which can be a residual value in economic models. For instance, if a solar panel has a residual value (through recovered glass, aluminum, etc.) of \$10 at EOL, that offsets a portion of the initial cost. IRENA has estimated that by 2050, the recovered material from solar PV waste could be worth \$15 billion, enough to produce 2 billion new panels [4]. This implies a circular economy wherein old panels feed into new panel manufacturing and reduce material costs. Actuarial models that utilities use to plan lifecycle costs are beginning to incorporate such residual values, especially in regions like the EU, where manufacturers are obligated to take back and recycle panels—effectively internalizing that future value.

Retired EV batteries still retain 70% to 80% of their capacity; if used in second-life stationary storage, their economic life is extended and the cost is amortized further. A simple economic model might compute an effective cost per cycle for a battery. If new, it costs \$100/kWh and lasts 1,000 cycles, that is \$0.10/cycle. If it gets a second life with 500 more cycles, maybe at a lower efficiency, the effective cost might drop to \$0.07/cycle. This improved value proposition may justify higher upfront costs for better batteries that enable a second life.

In short, the economic benefit of durability is positive and offers more energy output per unit of investment and less expenditure on raw materials over time. A more durable and recyclable renewable energy infrastructure also means reduced strain on mining and material supply. This tends to stabilize commodity prices and improve energy security due to less dependence on imported raw materials.

19.9.3 Resolving threats to traditional manufacturing models

If solar panels lasted twice as long and wind turbines did not require as many replacement blades, manufacturers would likely sell fewer units in the long run. This is analogous to how light-emitting diode (LED) lamps, lasting 10+ years, mean manufacturers sell fewer units overall compared to incandescent lamps. These types of manufacturing shifts can be disruptive. Industries built on a replace-and-grow model could view the weakening demand for replacements as a threat.

For example, consider a hypothetical wind turbine company that anticipates a replacement market as turbines age. If all turbines lasted 30 years or more instead of 20, the replacement cycle would slow. Similarly, if solar PV panels degrade only 0.2%/year instead of 0.8%, solar farm owners might keep them in service 40 years rather than replacing them at 25. To assess this, one could use actuarial models of installed base—a longer mean time to failure flattens the sales curve after the saturation point. Manufacturers might be concerned about new sales declining once capacity is installed and lasts longer. However, there are ways to adapt business models to this new reality and create business opportunities.

Service-based revenue models: Instead of one-and-done sales, companies can adopt a product-as-a-service model. For the renewable energy industry, manufacturers could lease equipment or market an energy service rather than the product. For instance, a solar company could install panels on rooftops and charge for the electricity generated. In this model, longevity is beneficial to the provider because it lowers maintenance costs and increases the duration of revenue from each installation. Today, many residential solar companies lease panels or have power purchase agreements instead of customer-owned systems. If a solar PV panel lasts 40 years, the provider can collect revenue for 40 years—a strong incentive to provide the most durable panels. Similarly, wind turbine makers often have long-term service contracts with wind farm operators. Under those, they might guarantee performance for 20 years; if their turbines can operate reliably for 30 years, they could extend those contracts or have lower warranty costs, thus improving profit margins.

Extended warranties and insurance: Manufacturers could offer fee-based extended warranties that guarantee replacement or repair for 30 years or more. If their products are built to last, they can collect fees with relatively low risk of payout. This creates a revenue stream linked to longevity and effectively monetizes reliability. Some companies already market their long-term service agreements that include EOL refurbishment. Insurers are also interested—insurance models for renewable assets (covering failure, under-performance) will price premiums based on reliability data. If technology improves, premiums drop, which encourages more deployment. In an actuarial sense, making systems more robust lowers the risk (and thus cost) of insuring them, which can save project developers money.

Diversification into recycling and repowering: Manufacturing companies can pivot operations to recycling and repowering at EOL. For example, a turbine manufacturer in 2050 might create a business unit that accepts used WTG blades, recycles the materials, and uses them to produce new blades or other products. This creates a circular supply chain under the same company. Instead of just selling new turbines, they also profit from decommissioning and material recovery services. The skills and facilities needed (material processing, logistics) can be developed gradually. Solar manufacturers could similarly integrate recycling plants into their operations, recovering silicon and metals to supply panel production. A repowering trend in the wind industry is to reduce the total number of installed machines by using larger, more efficient, and higher-capacity WTGs. This increases total electrical production, reduces the geographic footprint needed for the windfarm, and reduces the total number of machines required [8]. The greater height and expansive swept areas of newer WTGs can harvest more consistent winds that are unobstructed by variable landscape conditions or nearby structures. One 15 MW horizontal axis wind turbine, when properly sited, replaces a least ten older 1.5 MW machines [8].

Continuous innovation and component upgrades: Improvements in component performance are possible when hardware is more durable and lasts longer. An innovative business model is to upgrade upon replacement. This way, the manufacturer has ongoing sales without full device replacement. For example, a company could sell new and more efficient solar cells that can be swapped into an old panel's frame. One could imagine in 2040 a solar farm upgrading their 2025 panels by adding a perovskite laminated layer to make them tandem. This might boost module efficiency at a lower cost and reduce waste compared to installing entirely new panels. The original manufacturer could offer this service, keeping their workforce and factories busy. In the wind power industry, retrofitting existing turbines with slightly larger blades or new gearboxes can increase output. Selling upgraded components is a viable revenue stream that coexists with durable initial products. New control systems and gearboxes can be installed to boost a wind turbine's output without installing an entirely new turbine.

19.10 Policy Interventions and Incentives

To ensure an economic transition that aligns with sustainability goals, policy interventions will be critical. The following are key policy and market mechanisms intended to encourage longevity and recycling while keeping industries economically viable.

Extended producer responsibility (EPR): Already required in the EU for solar PV modules, manufacturers must accept responsibility for the end-of-life processing of their products, which includes financing collection and recycling. In this way, EPR effectively internalizes the waste management costs into the product's economics. It creates incentives for companies to design products that are easier and less expensive to recycle and to improve longevity. It supports the recycling industry by assuring a supply of EOL products and funding.

Depreciation and tax incentives: Governments can adjust tax depreciation schedules for renewable assets to reflect longer component lifecycles. If solar farms are depreciated over 30 years instead of 20 years, the annual tax deduction is smaller. However, the asset retains book value longer, which provides an incentive to prevent early system retirements. Incentives for refurbishment (e.g., investment tax credits for upgrading existing plants) help shift owner behavior from replace-to-upgrade.

Carbon pricing and resource taxes: If there is a carbon price or tax on virgin resource extraction, it makes long-lived, recycled-material products more competitive. For example, a carbon tax on steel and concrete would encourage wind farm developers to extend existing turbine life (which avoids new steel/concrete emissions) rather than build new ones as quickly. Similarly, if mining lithium or cobalt carries environmental taxes, recycled materials become cost-advantaged. These economic signals align profit with sustainability.

Support for the recycling industry: Direct support will be provided in the form of research and development (R&D) funding for advanced recycling technologies, offering subsidized loans for recycling plants, and creating quality standards for recycled materials to help support industry growth. For example, if recycled glass and silicon from solar PV modules are certified to meet the same standards as virgin materials, manufacturers will be more willing to use them. Governments can mandate minimum recycled content in new products (e.g., requiring that new EV batteries use a specified percentage of recycled cobalt or lithium by year 20xx). Such requirements create a guaranteed demand for recycled materials and an economic inducement to invest and expand recycling capacity.

Employment transition programs: When manufacturing new units slows, workers can transition to roles in maintenance, refurbishment, and recycling. This transition creates local, skilled jobs that are not easily offshored. Encouraging component and material recycling mitigates job losses from manufacturing and increases employment. Studies suggest that recycling, repair, and reuse create more jobs per ton of material than landfilling or new production. Recycling creates an average of nine times more jobs than landfilling trash materials; adaptive reuse creates as many as 30 times more jobs than landfilling [26]. Governments can provide retraining programs to move workers into solar panel refurbishment or battery remanufacturing.

Anti-obsolescence regulations: These types of policies require that critical components and software remain supported for a number of years. In practice, they help ensure open protocols and the availability of spare parts. Anti-obsolescence standards could make it easier for third parties to extend the life of equipment.

Component longevity and material circularity will reduce costs for operators and society. Though they change manufacturing profiles and require new infrastructure, they create new business opportunities in services and resource recovery. With thoughtful policy and innovation in business models, the renewable energy industry can remain robust and generate greater value while aligning with net-zero imperatives.

19.11 Conclusions

Sustainable development and energy production are linked [28]. Renewable energy is seen as a key solution to meeting environmental and energy supply initiatives. This chapter charts pathways for renewables to support sustainability and the world's growing circular economy. By 2050, renewable energy technologies—solar photovoltaics, wind turbines, battery energy storage systems, geothermal, and hydropower—will form the backbone of a more sustainable energy supply system.

This study examined the end-of-life challenges of this transition: the waste and sustainability issues that accompany the deployment of these technologies at scale. Through detailed projections, we quantified the expected waste streams (on the order of 10^7 to 10^8 tonnes of material) and highlighted the urgency of proactive waste management. We then delved into a multi-faceted strategies that ensure this transition remains sustainable: extending device lifetimes through smarter engineering, employing advanced materials and nanotechnology for durability and recyclability, reinventing chemical compositions to eliminate toxins, and restructuring economic incentives and policies to enhance longevity and circularity.

Our analysis shows that with innovation, each of the challenges can be met. The notion that renewable hardware is destined for landfills can be dispelled by the emergence of fully recyclable WTG blades, longer-lasting solar panels, and BESS designed for adaptive reuse. Engineering equations and models illustrated how improvements in lifespan directly translate to reduced waste generation rates—for instance, increasing a turbine's life from 20 to 30 years reduces its annual waste contribution by one-third. Our research provided evidence that addressing these issues is not only possible but is already underway. Key findings include:

- Massive waste volumes are expected under business-as-usual scenarios, but these are predictable and thus manageable. About 78 million tonnes of solar panels and 43 million tonnes of wind turbine blades could accumulate by 2050. Early action on recycling infrastructure and materials redesign can drastically reduce waste. The recovered material from solar PV waste could power a portion of new installations.
- Maintenance and reliability engineering can extend the operational life of components. We discussed examples like modular solar panels that allow easy repairs and condition-monitoring in wind turbines that prevent catastrophic failures. A consistent theme is that investing in maintenance offers economic and environmental benefits—it is less costly to replace a $50 component than a $50,000 machine, and avoids adding bulk materials to the waste heap. Real-world experiences already show many renewables outlast their design lives; this trend will accelerate with better design and operation, and maintenance practices.
- Advanced materials and nanotech innovations are making renewable equipment more durable and resilient. Self-cleaning nano-coatings for solar panels maintain efficiency and reduce wear on new blade materials. Examples discussed included using bio-based PECAN resin to provide recyclability without performance loss, 3D printing to optimize designs, enabling on-demand spare parts, and reducing waste materials. The integration of these advanced technologies means that renewable

energy components in the future will be superior to those of 2025—not just more efficient, but initially built for a circular lifecycle.

- Green chemistry solutions are eliminating toxic ingredients. The future will have solar modules free of lead and hazardous e-waste, chemical batteries free of cobalt using widely abundant materials, and WTG blades that can be chemically broken down into benign constituents [25].
- Lead-free perovskites are on the horizon, which may fundamentally change how we think about energy storage and generation—making them as environmentally friendly as they are clean in energy terms [25]. Furthermore, by designing materials for recyclability (e.g., reversible bonds in polymers), we ensure that end-of-life processing is not an afterthought but a built-in feature.
- Economic and policy frameworks can align industry incentives with sustainability. We discussed how extended producer responsibility, if widely adopted, will make manufacturers more responsible for the afterlife of their products, thus encouraging durable design and recycling. We also showed that fears of negative economic impacts can be mitigated by shifting to service-oriented models and new revenue streams through adaptive reuse and recycling of components. The overall transition to renewables is expected to create millions of jobs. Many will be in refurbishing, recycling, and maintaining this infrastructure, thus offsetting reductions in manufacturing employment. In essence, a circular economy in renewables creates a jobs-rich economy.

Far from being a fatal flaw, the issues of waste, resource use, and generation intermittency are catalysts driving innovation in renewable energy. Fragility and toxicity concerns are being answered by better engineering and materials. The dependence on mining is a strong motivator for recycling technologies. Intermittencies in generation are spurring advancements in energy storage and grid management. We are witnessing a remarkable convergence of disciplines—electrical engineering, materials science, chemistry, economics, and environmental science—all contributing to a robust renewable energy ecosystem.

A legacy of environmental improvements in the renewable energy industry is upon us. Achieving net zero emissions by 2050 using renewable energy is a plausible reality grounded in ongoing research and pilot projects. However, it will require integrative planning and continued commitment to addressing second-order effects. The journey begins with deploying gigawatts of solar and wind; it extends to crafting the full life cycle and ecosystem around those technologies. If the recommendations and innovations outlined in this research are pursued, we can ensure that the renewable revolution is sustainable on its own terms—yielding clean energy solutions with minimal waste and maximum resource circularity. Renewables will not only combat climate change and reduce pollution in their use phase, but also embody the principles of sustainability in their manufacturing and end-of-life phases. This truly supports a green and thriving global economy by mid-century.

References

[1] International Renewable Energy Agency (2016, June 20). Solar PV waste offers significant business opportunity. https://www.irena.org/news/articles/2016/Jun/Solar-PV-Waste-Offers-Significant-Business-Opportunity#:~:text=A%20new%20report%20released%20today,sold%20into%20global%20commodity%20markets, accessed 20 April 2025.

[2] Liu, P. and Barlow, C. (2017, February 16). Wind turbine blade waste in 2050. *Waste Management.* https://pubmed.ncbi.nlm.nih.gov/28215972/#:~:text=factors%20in%20the%20full%2lifecycle,rest%20of%20the%20world%2019, accessed 23 January 2026.

[3] Goff, S., Brophy, P. and Goff, F. Environmental effects of geothermal power. https://geocom.geonardo.com/assets/elearning/8.12.e4-23-04-20from%20geothermal%20projects, when%20compared%20with%20those%20produced, accessed 23 January 2026.

[4] International Renewable Energy Agency (2016, June). *End-of-Life Management: Solar Photovoltaic Panels.* https://www.irena.org/publications/2016/Jun/End-of-life-management-Solar-Photovoltaic-Panels. ISBN: 978-92-95111-99-8.

[5] Virginia Smart Solar. Decommissioning costs. https://vasmartsolar.com/decommissioning-costs/#:~:text=Decommissioning%20can%20cost%20millions%20in%20today%E2%80%99s%20dollars.%20The,a%20market%20to%20salvage%20the%20equipment%20when%20removed, accessed 22 August 2022.

[6] Maamari, L. Decommissioning of solar sites: a key consideration of the project. https://solunesco.com/2018/09/10/decommissioning-of-solar-sites-a-key-consideration-of-the-project, accessed 25 August 2022.

[7] Roosa, S. (2023, January). The asset manager's dilemma: decommissioning solar farms. *International Journal of Strategic Energy and Environmental Planning*, 5(1), 5–8.

[8] Roosa, S. (2025, September). Wind turbine generators—they just keep getting bigger. *Alternative Energy and Distributed Generation Journal*, 7(3), 15–32.

[9] Maisel, F., Neff, C., Marscheider-Weidemann, F. and Nissen, N. (2023, May). A forecast on future raw material demand and recycling potential of lithium-ion batteries in electric vehicles. *Resources, Conservation and Recycling*, 192, 106,920.

[10] Arab Turbo Team. Are the batteries in electric cars recyclable? https://www.msn.com/en-us/news/technology/are-the-batteries-in-electric-cars-recyclable/ar-AA1CPAj0?ocid=msedgdhp&pc=U531&cvid=513de0a08f86466abfbb41ae027982c6&ei=56, accessed 27 April 2025.

[11] International Energy Agency (2021, May). Net zero by 2050. https://www.iea.org/reports/net-zero-by-2050, accessed 21 April 2025.

[12] Majdi, A., Alqahtani, M., Almakytah, A. and Saleem, M. (2021, May 18). Fundamental study related to the development of modular solar panel for improved durability and repairability. *IET Renewable Power Generation*, 15(7), 1,382–1,396.

[13] Gibson, E. (2024, December 31). Tesla battery pack: how it is made, types, and the process explained. https://poweringautos.com/how-is-the-battery-pack-made-for-tesla-cars/#google_vignette, accessed 22 April 2025.

[14] Phillips, A. (2025, February 6). How long do wind turbines last? Average lifespan explained. Energy Follower. https://energyfollower.com/how-long-do-wind-turbines-last, accessed 22 April 2025.

[15] Nanoslic. Nanocoatings. https://nanoslic.com/about-nanocoatings.

[16] Memija, A. (2024, October 14). World's first 26 MW offshore wind turbine rolls off production line. https://www.offshorewind.biz/2024/10/14/worlds-first-26-mw-offshore-wind-turbine-rolls-off-production-line, accessed 6 December 2024.

[17] Moseman, A. (2025). Can solar panels be recycled? MIT Climate Portal. https://climate.mit.edu/ask-mit/can-solar-panels-be- recycled#:~:text=,designs%2C%20recycling%20technologies%2C%20and%20policy, accessed 24 April 2025.

[18] Stauffer, N. (2015, December 22). Solar energy from discarded car batteries. *MIT News*. https://news.mit.edu/2015/solar-energy-discarded-car-batteries-1222, accessed 24 April 2025.

[19] National Renewable Energy Laboratory (2024). Development of bio-based PECAN resin for fully recyclable wind turbine blades. *Science*, 376, 6,597.

[20] Mercer, E. (2023, December 11). Recyclable, plant-based material could take a spin on next generation of wind turbines. https://www.nrel.gov/news/program/2023/a-recyclable-plant-based-material.html, accessed 25 April 2025.

[21] Aarhus (2023, February 8). Vestas unveils circularity solution to end landfill for turbine blades. https://www.vestas.com/en/media/company-news/2023/vestas-unveils-circularity-solution-to-end-landfill-for-c3710818, accessed 25 April 2025.

[22] Murden, D. (2023, April 14). What is the environmental impact of LiFePO4 batteries? https://ecotreelithium.co.uk/news/lifepo4-battery-environmentally- friendly/#:~:text=LiFePO4% 20batteries%20are%20non-hazardous%20in%20nature.%20They%20are,of%20these%20batteries%20do%20not%20contaminate%20the%20environment, accessed 25 April 2025.

[23] Juqu, T., Willenberg, S., Pokpas, K. and Ross, N. (2022, December). Advances in paper-based battery research for biodegradable energy storage. *Advanced Sensor and Energy Materials*, 1(4), 100,037.

[24] Brahambhatt, R. (2025, January 17). Sodium-sulfur battery breakthrough retains 81% capacity after 200 cycles. *Interesting Engineering*. https://interestingengineering.com/energy/new-electrolyte-for-sodium-sulfur-batteries, accessed 22 April 2025.

[25] Thompson, D. (2021, May 5). Researchers develop new metal-free, recyclable polypeptide battery that degrades on demand. Texas A&M University Engineering. https://engineering.tamu.edu/news/2021/05/CHEN-researchers-develop-new-metal-free-recyclable-polypeptide-battery-that-degrades-on-demand.html#:~:text=batteries%2C%20said%20Wooley%2C%20recently%20honored,%E2%80%9D, accessed 22 April 2025.

[26] Eco-cycle. Jobs and economic benefits of zero waste. https://ecocycle.org/learn-about-zero-waste/jobs-and-economic-benefits, accessed 22 April 2025.

[27] Atasu, A., Duran, S. and Wassenhove, L. (2021, June 18). The dark side of solar power. *Harvard Business Review*. https://hbr.org/2021/06/the-dark-side-of-solar-power?ab=hero-main-text, accessed 25 April 2025.

[28] Roosa, S. (2010). *The Sustainable Development Handbook*, 2nd edition. ISBN: 978-4398-5048-0. Taylor and Francis.

20

Tariffs and the U.S.-Canada Energy Interdependence

Stephen A. Roosa

In the world's developed countries, access to low-cost energy is a prime mover that enables productivity and economic expansion. This chapter considers the energy interdependencies of two major North American countries, namely the United States and Canada. Together, these countries jointly produce more oil than anywhere else on Earth. They are complementary and interdependent regarding their primary supplies of energy. This chapter details the resources and trade relationships of these two allies regarding key sources of energy, particularly oil, natural gas, uranium, and electricity. The infrastructure of these two counties is designed to support bilateral energy distribution and trading. Mapping the energy relationships between countries is a complex process. Canada exports almost all its oil and natural gas production to the regions of the U.S., such as the upper Midwest states and the Northeast. These regions are highly dependent on Canadian crude oil, natural gas, and electricity. The U.S. also exports energy to Canada, albeit in lesser volumes. This chapter delves into the issue of tariffs applied to the importation and exportation of energy resources such as petroleum and electricity. The application of tariffs across international borders changes the cost structures of fuels, materials, components, and energy systems, which hinder and delay the energy transition. The author concludes that the new tariffs being applied to energy resources by the U.S. on Canadian energy imports are counterproductive and create uncertainties for businesses and industries on both sides of the border.

DOI: 10.1201/9788743811312-20

20.1 Introduction

The U.S. and Canada are each other's largest energy trading partners as measured by the value of energy commodities. They share the world's longest common unfortified border, which stretches 8,891 kilometers (5,525 miles), much of it along the 49th parallel. The population of the U.S. is 345.2 million. Canada's population is 41.5 million, slightly more than the U.S. state of California, which is 38.9 million. About 90% of the Canadian population lives within 160 km (100 miles) of the U.S. border. Though Canada is slightly larger geographically, both countries have just under 10 million km^2 (3.9 million square miles) of land area.

Each of these countries is blessed with vast natural resources that include energy. The U.S. is the world's leading oil producer; Canada is the world's fourth-largest producer. In the U.S., Texas produces over 40% of the total volume of oil; the province of Alberta in Canada produces over 80% of the oil in Canada. Proven Canadian oil reserves are 171.0 billion barrels (bbl), far more than the U.S. total of 44.5 billion barrels, including its Strategic Petroleum Reserve (SPR) which contains only 0.4 billion barrels. Neither country is a member of the Organization of the Petroleum Exporting Countries (OPEC). Petroleum encompasses both crude oil and refined petroleum products. Petroleum products result after crude oil is refined into fuels like gasoline, jet fuel, or propane.

Canada is a major and dependable energy supplier to the U.S., providing 58% of the volume of hydrocarbons imported by the U.S. in 2023. Hydrocarbons are organic compounds consisting entirely of hydrogen and carbon, which serve as the foundation for crude oil, natural gas, and coal. In 2021, Canada supplied 61% of U.S. crude oil imports, 98% of natural gas imports, 93% of electricity imports, and 28% of its uranium purchases [1]. These are major contributions to the energy needs of the U.S. [1]. The energy transition for the U.S. has been directed toward greater use of natural gas, reducing the use of coal, lowering energy imports from the middle eastern countries, increasing the use of renewables, and relying on energy efficiency improvements. Canada has been an important partner in this evolving energy transition.

The oil production of these countries is complementary. Canada's oil exports to the U.S. are primarily heavy oil from oil sands [2]. U.S. production is primarily light oil from the Permian Basin in Texas and New Mexico [2]. Many U.S. refineries, primarily in the Midwest and Gulf Coast regions, are configured to process heavy oil. Interestingly, the amount of oil that Canada exports to the U.S. is about equal to total U.S. oil exports. This partnership is crucial for both economies. It creates employment and ensures energy security. A network of pipelines, along with fleets of rail, marine, and freight vehicles, transport crude oil, natural gas, natural gas liquids (NGLs), and refined petroleum products (RPPs) between Canada and the U.S.

20.2 Mapping the Interdependence

It is physically beneficial and economically efficient for the U.S. and Canada to partner in the distribution and sale of energy. This is particularly true for oil, natural gas, and electricity. Though the U.S. is now the world's largest oil producer, this was not always the case. The U.S. in the 1970s through the 2000s was not energy independent and was heavily dependent on costly imported oil. Canada was and remains an important crude oil supplier. As of February 2025, total U.S. crude oil production reached 13.5 million barrels per day (bbl/day). Canada is the world's fourth-largest oil producer with 4.8 million bbl/day. New pipeline access and crude-by-rail have allowed more Canadian oil to reach refineries on the U.S. Gulf Coast, increasing from about 140,000 bbl/day in 2010 to about 450,000 bbl/day in 2024 [2]. Because of a vast network of pipelines and refineries, Canadian and U.S. oil production are complementary. Infrastructure is in place and was designed to support this interdependency, especially regarding supplies of crude oil and electricity. Despite increased domestic oil production, the U.S. reliance on oil imports from Canada has never been greater [2].

20.2.1 Mapping energy flows

Mapping the energy flows between counties is a complex process because of the different types of energy resources involved. Challenges include benchmarking and obtaining relevant databases without missing data. There have been past attempts to map the energy flows across the countries in North America. Surprisingly, the trade of energy between the countries impacts all U.S. states and all Canadian provinces. While the province of Alberta is the center of Canadian oil and gas production, oil is also sourced from Saskatchewan, Newfoundland, and Labrador. Most of the oil and natural gas production from Alberta flows into the U.S. Electricity, mostly generated using nuclear energy and hydropower, flows from the provinces of Ontario and Quebec to the U.S. New England states and Michigan. In addition to vast amounts of electricity, Ontario exports uranium to the U.S. [1].

Mapping energy flows provides a snapshot of the movement of energy across international borders. Perhaps one of the better past efforts to map the relationship was performed by the Center for Strategic and International Studies in 2018, which focused on the energy trade between the U.S. and Canada. It highlighted most aspects of the energy interdependence between these two allies [1].

20.3 Energy Trading

The cross-border trade of energy between the U.S. and Canada is substantial. The extent of the trade activities of key energy products between these major energy producers is an indicator of the interdependence between these countries.

Despite the massive scale of oil and natural gas production in the U.S., the country imports vast volumes of hydrocarbons from Canada. In 2023, Canada provided 58% of the hydrocarbons imported by the U.S. More specifically, Canada provided 60% of the crude oil and close to 100% of the natural gas imported by the U.S. Canada also provided 21% of the refined petroleum products (RPPs), and 95% of the NGLs imported by the U.S. [3]. Despite being a hydrocarbon-exporting nation, Canada imported 1.1 million barrels (bbl) of oil equivalent (BOE) per day in 2023, 83% of which came from the U.S.

20.3.1 Crude oil

Crude oil and petroleum products are key examples of the interdependence. In 2023, Canada exported 3.9 million bbl/day of crude oil. Of this, 97% or 3.8 million bbl/day went to the U.S. [3]. The value of U.S. imports of crude oil from Canada was $93 billion in U.S. dollars (or $130 billion Canadian), a decrease of 18% from 2022 because of the lower average price for crude oil [3].

Unlike the trade of crude oil, the exchange of refined petroleum products between the two countries is relatively balanced (see Figure 20.1). RPPs include a wide range of products refined from crude oil, such as motor gasoline, distillate fuel oil (diesel), heating oil, kerosene, and jet fuel [3]. In 2023, Canada exported 0.37 million bbl/day of RPPs, with 87% going to the U.S.; it imported 0.47 million bbl/day, with about 80% sourced from the U.S. [3]. Many U.S. refineries, particularly those located in the Midwest, are designed to process the types of heavy crude oil that are extracted from the Alberta oil sands.

20.3.2 Natural gas

The natural gas industry is one of the oldest utilities in North America and has provided a key resource for a growing U.S. population [4]. Natural gas

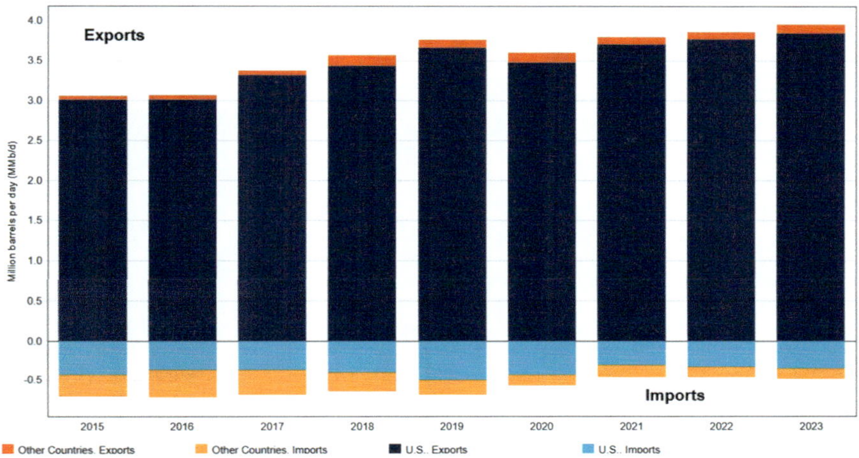

Figure 20.1 Annual crude oil exports from Canada to the U.S. and imports from the U.S. to Canada in million barrels/day, 2015–2023 [3].

processing requires multiple steps that begin at the wellhead and end at the burner tips [4]. Each production stage increases its value. Today, natural gas is primarily used to produce heat and electricity.

The U.S. is the world's leading producer of natural gas. In 2023, three regions—Appalachia, Permian, and Haynesville—accounted for 59% of the total natural gas production in the U.S. [5]. Associated natural gas accounted for about 37% of U.S. natural gas production and most of this production came from the Permian region. As of year-end 2024, U.S. total natural gas proved reserves—estimated as wet gas—totaled about 625 trillion cubic feet (Tft^3) down from 691 Tft^3 in 2022 [6].

Canada is the world's fifth largest producer of natural gas. Canadian natural gas production in 2023 averaged 17.9 billion cubic feet (Bft^3) per day [7]. Total Canadian natural gas proved reserves total about 77 Tft^3.

The volume of natural gas imports and exports between the U.S. and Canada has remained relatively stable since 2015. The U.S. imports about four times as much natural gas from Canada as it exports to Canada (see Figure 20.2). Canada exports 7.8 Bft^3/day of natural gas, all of which goes to the U.S. Canada imported 2.0 Bft^3/day of natural gas; almost all was imported from the U.S.

Liquefied natural gas (LNG) is a form of natural gas that has been cooled to about –162°C (–260°F) to enable shipping and storage. Once liquified, the volume of natural gas is about 610 times smaller than its volume in

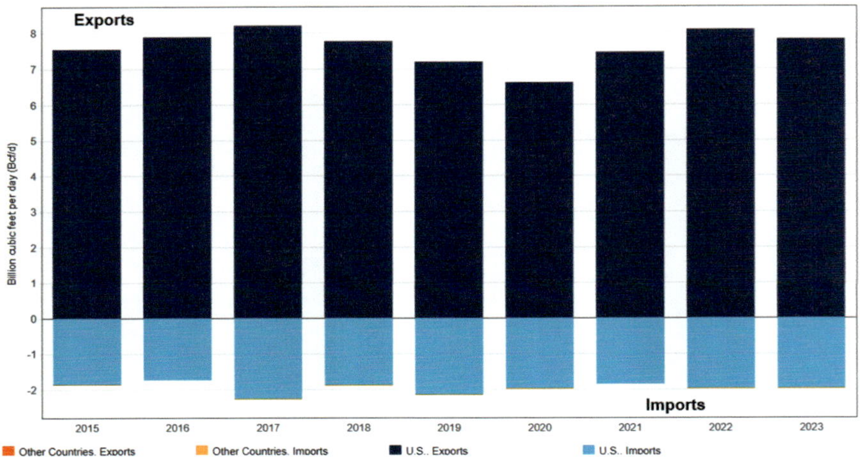

Figure 20.2 Annual natural gas exports from Canada to the U.S. and imports from the U.S. to Canada in billion ft^3/day, 2015–2023 [3].

a gaseous state. The U.S. became the world's largest exporter of LNG beginning in 2023, overtaking Qatar. In 2024, the U.S. exported approximately 123 Bft^3 (12 Bm^3)/day of LNG. Though Canada is not yet a major exporter of LNG, the country has seven export projects in various stages of development.

20.3.3 Natural gas liquids

Natural gas liquids (NGLs) are recovered from natural gas or condensate as liquids when they are under slightly higher pressures and lower temperatures [3]. These liquids include ethane, propane, butane, and pentane, among others. Canada exported 245.0 thousand bbl/day of NGLs, with 66% or 159.0 Mb/day going to the U.S. Canada imported 91.8 thousand bbl/day of NGLs—almost 100% from the U.S. [3].

Figure 20.3 presents yearly stacked bar charts, one each for butane, propane, and ethane, from the Canadian import/export perspective. Each chart displays imports and exports in thousand barrels per day (Mb/day) from 2015 to 2023, categorized by source and destination (U.S. versus other countries) [8]. Exports are displayed as positive values, with dark blue representing exports to the U.S. and dark orange for exports to other countries. Imports are shown as negative values, with light blue representing imports from the U.S. and negligible quantities of imports from other countries shown in orange [8].

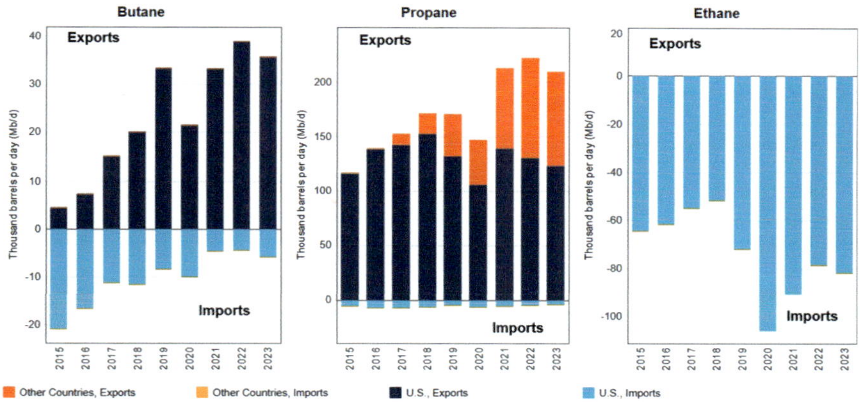

Figure 20.3 Annual NGLs exports from Canada to the U.S. and imports from the U.S. to Canada in 1,000 barrels/day, 2015–2023 [3].

20.3.4 Electricity

The U.S. and Canada share one of the world's most integrated electricity grids. Resources are pooled to improve reliability and economic dispatch. There are 86 international power lines that move electricity through various provinces and states [3]. Electricity is more costly south of the border. U.S. consumers in 2024 paid an average retail price of $0.184/kWh for electricity. Consumers in Canada paid an average of only $0.126/kWh. This is the result of the greater use of hydropower in Canada.

Canada is the world's second largest producer of hydroelectricity after China. Just over 60% of Canada's electricity is generated by hydropower. The country's ability to provide renewable electricity is highly valued in New England and elsewhere in the U.S. It provides baseload generation and enables some regions to reduce their greenhouse emissions. The U.S. needs Canadian electricity. Despite energy efficiency efforts that stabilized usage from 2006 to 2019, the U.S. demand for electricity has since steadily increased (see Figure 20.4). Electricity consumption was 4,527,969 gigawatt-hours in 2025, up 3.0% from 2024. This growth in the demand for electricity is due to the added electrical needs of regional data processing centers, air conditioning, electric vehicles, and other uses.

Together, electricity imports and exports have a crucial role in ensuring the reliability of the electricity systems in several Canadian provinces and U.S. states [3]. Because of interconnected transmission infrastructure, all of Canada's electricity trade is with the U.S. Despite increases in electricity use,

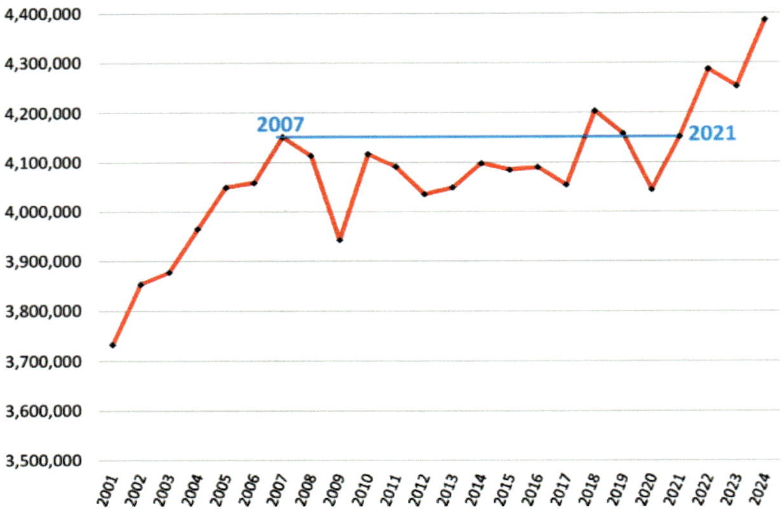

Figure 20.4 U.S. electricity generation, 2001–2024 in GW hours/year (sources: EIA, wolf-street.com).

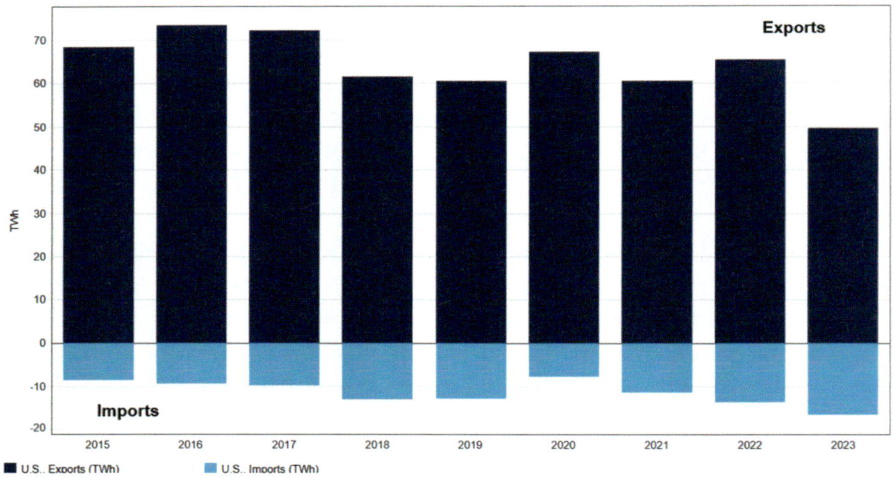

Figure 20.5 Annual electricity exports from Canada to the U.S. and imports from the U.S. to Canada in TWh, 2015–2023 [3].

exports of electricity to the U.S. from Canada declined from 2015 through 2023 (see Figure 20.5). In 2025, Canada exported 49.4 terawatt-hours (TWh) of electricity; alternatively, the U.S. exported an estimated 18.1 TWh of electricity to Canada [3].

20.4 Impact of Tariffs on Energy Supplies

In early March 2025, the U.S. announced tariffs on goods, including energy, from Canada and Mexico, some of which were delayed to be implemented later in the year. The term "energy" or "energy resources" as applied to the tariffs is defined as crude oil, natural gas, lease condensates, natural gas liquids, refined petroleum products, uranium, coal, biofuels, geothermal heat, the kinetic movement of flowing water, and critical minerals [9]. The tariffs chiefly targeted oil, natural gas, uranium, and electricity imports to the U.S. from Canada. Why were the tariffs being threatened and implemented? Donald Trump, the U.S. President, said that a primary reason for his new tariffs against the country's three largest trading partners (Canada, Mexico, and China) was the "national emergency" brought on by fentanyl flowing across the country's borders [10]. However, nearly all the fentanyl entering the U.S. was via its southern border with Mexico; overdose deaths had declined in the previous two years [10]. He also pointed to concerns regarding illegal immigration, most of which is not via Canada.

Justin Trudeau, then Canada's Prime Minister, responded that "While less than 1% of the fentanyl intercepted at the U.S. border comes from Canada, we have worked relentlessly to address this scourge that affects Canadians and Americans alike. We implemented a $1.3 billion border plan with new choppers, boots on the ground, more coordination, and increased resources to stop the flow of fentanyl" [11]. Referring to the imposition of tariffs, he stated that "Canada will not let this unjustified decision go unanswered" [11].

Regardless of Canada's efforts, the U.S. placed a 10% import tariff on Canadian oil. "Virtually all economists think that the impact of the tariffs will be very bad for America and for the world… they will almost surely be inflationary" said Nobel Prize winner Joseph Stiglitz, an economics professor at Columbia University [12]. The damage of the tariffs will be compounded by the retaliatory actions that Canada and Mexico vowed to take [13]. "It is going to be an absolute disaster for both countries. We are your largest trading partner," said Doug Ford, the premier of Ontario, Canada's most populous province. "I do not want to respond, but we will respond like they have never seen before. You [referring to the U.S.] need our uranium. If he [referring to U.S. president] wants to destroy our economy and our families, I will shut down the electricity going to the U.S." [14]. He also said that, "we keep the lights on for 1.5 million homes and manufacturing in New York, Michigan and Minnesota." [14]. Ford later backed off his threat but said he will not be afraid to retaliate by matching tariffs, noting Canada's importance regarding

U.S. usage of its energy and electricity [15]. Counter or retaliatory tariffs by Canada would likely adversely impact electricity sales to the U.S.

20.4.1 Crude oil

The Canadian oil drilling sector has endured a prolonged downturn dating back to the oil price collapse of 2014. After peaking at 626 rigs in February 2014, by March 2025, there were only 188 active rigs in operation; this dropped to 153 rigs in February 2026 [16]. After reaching a peak of 1,609 active oil rigs in October 2014, the U.S. had 416 operating rigs in February 2026. In 2024, total employment in the drilling sector stood at approximately half of what it was a decade before, underscoring the fragile nature of the post-COVID recovery [16]. Yet Canada remains the dominant supplier of imported oil to the U.S. In 2024, Canada supplied 62% of U.S. oil imports, with Mexico, Saudi Arabia, and Venezuela together providing just 14% [2].

In Canada, there are two common benchmarks for oil. The variables that typically drive price differences between the benchmark crudes are quality, marketability, and logistics. Western Canadian Select (WCS) is a blend of diluted bitumen and conventional heavy oil priced in Hardisty, Alberta. Canadian light sweet (CLS) crude oil, once known as Edmonton Par, is a basket of light sweet crudes priced in Edmonton, Alberta [14]. Most Canadian crude sells at a lower price than U.S. crude. In Canada, light and heavy benchmarks largely trade at a discount to West Texas Intermediate (WTI). Both Central Alberta oil prices and WTI prices have been generally declining in price since April 2024.

About two-thirds of Canada's exports are shipped to the Midwest via Enbridge's Mainline [14]. The Midwestern U.S. states are Canada's ideal customer. The region is closer and lacks major producers and nearby import terminals. Refineries in the Midwest invested billions to retool their facilities to process the heavy, sour, and acidic crudes from the oil sands [14]. Canada's largest crude pipeline operator, Enbridge, constructed most of its infrastructure to serve the region. Unfortunately, having a single large buyer of Canadian crude oil reduces Alberta's ability to compete for higher prices [14].

In addition to tariffs announced on goods, there was a 10% tariff on Canadian energy imports. The impacts of tariffs are far-reaching. For example, on April 9, 2025, the price of Canadian Central Alberta crude oil was \$51.43 (U.S.) per barrel. An import duty of 10% on Canadian oil increases the price to \$56.57 per barrel, which is slightly less than the WTI price of \$57.35 per barrel. This means that tariffed Canadian crude, despite having a less challenging and costly supply chain to regions of the U.S., has little price advantage. When the export pipelines are at capacity, the incremental barrels of oil must be shipped by rail, which has a higher transportation cost and increases pricing discounts [14].

Canadian suppliers could ship their oil to other countries and sell at international prices, or lower their prices to continue selling to the U.S., which would reduce their margins. As most of this Canadian crude flows to the upper Midwestern U.S. states, consumers there will ultimately pay the increased costs; the added funds raised by the tariff will flow to the U.S. treasury. If Canada is unable to sell its oil to the U.S., it could, within the boundaries of its vast lands, construct pipelines to coastal ports to ship oil to other countries. This would be costly. Reinventing the proposed Keystone XL pipeline to the Gulf Coast offers an opportunity for beneficial marketability because it opens a large market for heavy crude, which has a notable shortage of stable suppliers [14]. However, the oil would flow through U.S. territory [14]. With the added tariffs on Canadian crudes, the economics of the project are less favorable. If reconsidered, this alternative would once again face strong opposition from landowners and environmentalists in the U.S. In the meantime, Canadian oil has nowhere else to go.

Tariffs will eventually impact industry supply chains. For example, one major concern is the availability and price of sand, a resource included in Canada's proposed list of counter-tariffs. Sand is a key component in hydraulic fracturing (fracking) processes [16]. It acts as a proppant, holding open the fractures created in rock layers during the fracking process. Fracking sand (also called frac sand) must meet stringent specifications in terms of grain shape, size, and durability to be used for hydraulic fracturing. A price increase for this essential resource would raise operating costs for oil and natural gas companies, further straining the industry [16]. The price for fracking sand varies from \$25/ton to \$200/ton depending on the type of sand, its availability, and delivery costs.

20.4.2 Natural gas

The 10% energy tariff would likely apply to the importation of natural gas into the U.S. from Canada. If so, how it would be applied in the pricing formula for natural gas is undefined and difficult to sort out. It is uncertain whether the import tariff basis for natural gas will be applied to the posted index price, the contract's index price at or near the U.S.-Canadian border, or the price at a trading hub [17]. Because of the nature of the tariff, a legal argument can be proposed that the added tariff costs are defined by contract

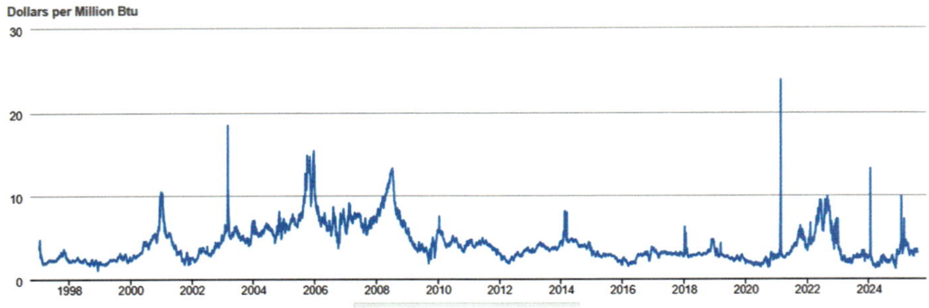

Figure 20.6 Henry Hub natural gas prices from 1998 to mid-2025 ($/million Btu) [19].

as government-imposed taxes [17]. There are two options: 1) the buyer pays taxes at and after the delivery point; or 2) the seller pays taxes before and at the delivery point [17]. To further confound the issue, there remains the question of whether the tariffs would apply to the costs of delivering the product to the Canadian border. While the contractual responsibility for payment is uncertain, the costs must be paid to guarantee carriage.

In North America, prices for natural gas tend to vary seasonally as demand increases in the winter. However, market-specific demand shocks are the main source of natural gas price fluctuations. Figure 20.6 shows the price of natural gas from 1998 through to mid-2024 [19]. The price of natural gas in the U.S. more than doubled from April 2024 to December 2025 to $5.06/ MMBtu. While this increase may seem substantial, the price is much less than was posted in August 2022 when the market price for natural gas spiked at $9.40/ MMBtu. It fell by 27 February 2022 to $2.85/MMBtu. While the price elasticity of natural gas demand is relatively high, natural gas supply prices are relatively inelastic.

For tariffs on natural gas imports into the U.S., the pricing formula could be based on the posted market index price. For example, if the pricing formula were based on market pricing, a 10% tariff on natural gas priced at $2.91 (U.S.) per MMBtu would be about $0.29, which would increase the price to $3.20/MMBtu. However, AECO-C (a virtual trading point), Canadian natural gas is priced as low as $1.48/MMBtu [18]. If the tariff is levied at this point and price, it would increase the cost by only $0.14/ MMBtu, yielding a total price of $1.62/MMBtu.

Natural gas prices have become increasingly volatile due in part to the exports of LNG. An increase in price would increase the tariff applied. Ultimately, the tariff would vary, add to price volatility, and be paid by U.S. consumers.

20.4.3 Uranium

Canada, the world's second-largest producer and exporter of uranium, has reserves about 10 times larger than those held by the U.S. The U.S. has an estimated 59,400 tons of uranium, accounting for 1% of the global supply—Canada possesses 588,500 tons [20]. Canada boasts the world's largest deposits of high-grade uranium with concentrations reaching up to 20% uranium—100 times higher than the global average [20]. A 25% tariff on Canadian mineral imports that include uranium could cost U.S. off-takers an additional $11.75 billion—a figure that would likely increase as uranium prices recover [20].

The U.S. relies on foreign sources for 70% of its enriched uranium. Though Canada supplies 27% of U.S. uranium, it has not threatened to disrupt supplies. Though an unlikely event, it would be difficult for the U.S. to quickly develop high-grade domestic supplies. The alternative would be to obtain more uranium from Kazakhstan or Russia. Uranium is a commodity whose prices are also highly volatile (see Figure 20.7).

The application of tariffs clearly counters the intent of Executive Order 14156, the declaration of a national energy emergency, signed on 29 January 2025. It explicitly prioritized uranium and other energy resources and directed that adding uranium to its critical minerals list be considered [9].

20.4.4 Electricity

The impact of tariffs on electricity imported to the U.S. is regional. After the U.S. tariffs on Canadian oil imports were announced, Canada immediately hit back by imposing 25% tariffs on $155 billion worth of U.S. exports to Canada [22]. The province of Ontario quickly threatened a 25% surcharge to be applied to electricity sales to the U.S. states of Michigan, New York, and Minnesota [23,24]. The electricity surcharge would immediately affect consumers in these states because they rely on imports of electricity from Canada to supply their businesses and homes [24]. Regional wholesale electricity markets would be disrupted. Adding about $277,000 per day to the cost of imported electricity, the 1.5 million consumers in the affected U.S. states would see their average costs for electricity increase by about $830 per year [25]. Within a few days, Canada backed off this threat and instead retaliated by increasing tariffs on other goods.

If a more broadly applied 10% energy tariff were placed on wholesale electricity, the impact would affect additional U.S. states. For example, electricity exported to the U.S. from Canada is also consumed in California, Nevada, and parts of Arizona.

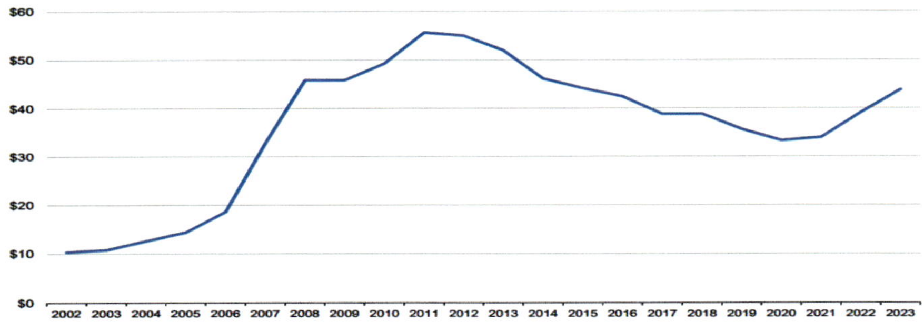

Figure 20.7 Weighted-average price of uranium purchased by owners and operators of U.S. civilian nuclear power reactors, 2002–2023, $/lb. [21].

20.5 Conclusion

The availability of low-cost energy is a prime mover that enables productivity and economic expansion. The gross domestic product (GDP) of the U.S. in 2025 was $30.62 trillion, by far the world's largest. Canada's GDP was $2.28 trillion, ranked 14th worldwide, slightly larger than many of the world's other fossil-fuel exporters, including the Russian Federation[1] [26]. The U.S. and Canada are increasingly interdependent in terms of their importation and exportation of energy resources. It is physically beneficial, strategically important, and economically efficient for the U.S. and Canada to partner in the distribution of energy resources, particularly for the sale of oil, natural gas, uranium, and electricity. Their energy infrastructures are interconnected and optimized to support energy trading between the two allies. Though Canada's exports of hydroelectricity to the U.S. are only a fraction of total U.S. electricity generation, the shared grid between the two countries is critical to maintaining system reliability [23].

Canada generates large amounts of electricity from hydro and nuclear power stations, which do not create air pollution nor emit greenhouse gases. The U.S. is dependent on Canadian uranium and its comparatively less costly hydroelectricity. In the U.S., electricity users have reduced their costs by investing in more efficient equipment such as new lighting, appliances, electronic equipment, heating and air-conditioning systems, and controls [27]. This drive for greater efficiency has, until recently, kept the demand for electricity stable despite the growing economy and population [27]. Due to the

[1] Data is indicated in U.S. dollars.

market penetration of electric vehicles and the demand from new data centers, electricity consumption is now increasing [27].

International energy markets tend to focus on oil and petroleum derivatives. According to the Energy Information Administration (EIA), in the next 25 years, the U.S. will need to import somewhat less oil than it does today—7.0 million bbl/day in 2050 compared to 8.42 million bbl/day in 2025 [2]. With no anticipated increase in demand and the U.S. and Canada involved in trade disputes... who will control the oil markets? OPEC has indicated it will increase oil production, flooding world markets with excess production, which would increase downward pressure on petroleum prices. Alternatively, world oil prices could increase, especially if a major event were to occur in the Middle East. In any case, the U.S. will likely miss its newfound leverage to control pricing. Given most any scenario, the new tariffs being applied to energy resources by the U.S. are counter-productive.

20.5.1 The consequences of tariffs

Given the size of its economy, the U.S. is certainly in a stronger position to sustain the negative consequences of an impending trade war that may result from the institution of import duties. Canada appears to be more vulnerable because about 19.5% of Canada's GDP is due to exports to the U.S. [28]. However, certain U.S. regions and states are highly dependent on their exports to Canada. North Dakota derives about 10% of its GDP from exports to Canada alone [28]. Many U.S. states have worked for decades to foster close trading partnerships with Canada; these relationships are threatened and could be permanently damaged. Tariffs are an escalation of the trade tensions between the two countries [23]. Already, certain U.S. products are being boycotted by Canadians, and their leisure travel to some U.S. states has declined precipitously.

Attempts to solve the problems brought on by fentanyl flowing into the U.S. and illegal immigration by creating tariffs on energy flowing from Canada into the U.S. are mistargeted. Little illegal drug trafficking has been proven to cross into the U.S. from Canada, and most of the immigration issues happen elsewhere. If tariffs are the solution, then why are there no commitments to direct the funds raised toward resolving them? These tariffs create distractions that stymie more effective solutions. In context, they seem burdensome, unjustified, and untimely. U.S. consumers will incur the burden of the added taxes on energy. This situation will not be improved by redistribution of benefits.

The conflicting rationales for tariffs are not only troublesome but also create uncertainties and chaos for North American consumers and businesses.

These tit-for-tat, on-again/off-again, threats of tariffs are counterproductive. U.S. consumers will pay more than Canadians due to the sheer volume of energy the country imports. The application of tariffs significantly increases the price of energy—particularly oil, natural gas, and uranium—paid by U.S. consumers and industries. It directly opposes stated policies, found in Executive Order 14156, that intend to reduce the "high energy costs that devastate American consumers by driving up the cost of transportation, heating, utilities, farming, and manufacturing, while weakening our national security" [9]. The U.S. president's past pledges to reduce both taxes and the cost of energy are undercut by actions to institute tariffs on energy supplies.

If the tariffs on certain goods are fully imposed, the U.S. will violate the United States-Mexico-Canada Agreement (USMCA), a multilateral compact that replaced the North American Free Trade Agreement. It was brokered during Donald Trump's first term in office and went into effect in July of 2020. The stated policy goal of the USMCA was to support competitiveness and help lower the reliance on energy imports from outside the region [29]. It provided for the free flow of energy across borders in North America by extending the continued zero-tariff treatment of U.S. energy exports to Mexico and Canada [30]. The treaty facilitated the movement of oil and natural gas by pipeline, revised rules of origin certification for oil and gas movement, and streamlined the regulatory process of U.S. LNG exports [30]. Many of the manufactured goods, equipment, electronics, spare parts, and chemical products covered by the treaty are value chain inputs to key U.S. energy industries. The application of U.S. tariffs jeopardizes the intent of the USMCA and may directly violate the terms of the treaty [30]. With the initial announcements and threats of spiraling tariffs, oil prices fell sharply, U.S. stock market volatility increased, and the dollar weakened as investors looked for safer places for their assets [22].

Simultaneous rounds of increased and retaliatory tariffs with its other two leading trade partners, China and Mexico, will likely stall U.S. economic growth.[2] The counteroffensive responses by these countries, plus Canada, will inevitably create inflation, disrupt employment, stress supply chains, reduce economic growth, and eventually lower corporate earnings. The U.S. could enter an extended period of stagflation, one with sluggish growth, higher-than-anticipated inflation and unemployment. Inflation in the U.S. could increase by an additional 0.6% and economic growth could slow by 0.7% to

---

[2] The 2025 GDP of other countries impacted by U.S. tariffs: China ranks second with a GDP of $19.4 trillion, Mexico ranks 19th with a GDP of $1.86 trillion. These data are not adjusted for purchasing power parity.

1.1% lower over 9 to 12 months as an add-on result of the tariffs [31]. Small business owners who operate on low margins and rely on low-cost energy will be threatened by the increased risks, and some will shutter their companies. The uncertainties and added risks will impact investments in energy. If the U.S. proves to be an unreliable partner for Canadian oil, the country will likely consider other markets. Though capital-intensive, these options might include constructing new pipelines into the Northwest Territories, Manitoba, Hudson Bay, or Eastern Canada, and selling its oil to other countries. Canada could also retaliate by halting or preventing the purchase of other U.S. goods, including durables and military hardware.

20.5.2 Will tariffs on energy achieve the intended policy goals?

Using tariffs as an economic tool to protect national economies usually backfires and often fails to achieve the stated policy goals. This is particularly true in the case of U.S. tariffs on energy, as there is no proven correlation between the increase in taxes and the resolution of the targeted objectives. An example of an undesirable national security consequence of tariffs on Canadian uranium is that they will make imports from Russia and elsewhere more competitive. Tariffs will increase the costs of imported oil rigging equipment, pumps, motors, energy storage components, solar panels, and wind turbine generators. As a result, they will ultimately increase U.S. dependence on foreign energy and negatively impact domestic energy resilience and security.

Some advocates argue that the cost of tariffs can be offset by the increased production of some industries. For example, it is argued that the 25% tariff on imported metals could create an incentive to expand U.S. aluminum production. The U.S. uses about 5 million metric tons annually. Because only 678,000 metric tons are domestically produced, large amounts of aluminum are imported from Canada [32]. Producing aluminum is energy-intensive. Canada has comparatively low-cost electricity and has been a reliable supplier of aluminum [32]. Creating a metric ton of this versatile material takes nearly 15,000 kWh of electricity—enough to power a typical U.S. household for 16 months [32]. Increasing U.S. production is not probable for the near-term, as it takes years to construct smelters, and markets may shrink in real terms. If the U.S. were to onshore all 2.7 million metric tons of Canadian aluminum it imported in 2024, the increased energy requirement would be over 40 million MWh of electricity—nearly four and a half times the annual electricity production of the Hoover Dam [32]. A more cost-effective solution for the U.S. is to enhance and mandate aluminum recycling programs.

20.5.3 Manufacturing a national energy emergency

It is difficult to sort out the real reasons a U.S. national energy emergency (per Executive Order 14156) was declared by the U.S. President, or how tariffs on energy will meet its goal of strengthening "relations with allies and partners, and support international peace and security" [9]. This newly fabricated energy crisis is defined more by its lack of evidence than its essence. There are no major natural gas shortages, no oil shortages or price shocks, no long lines at the pumps, no rationing of petroleum fuels, and no widespread electrical outages nor extended brownouts. While the increased need for electricity to supply new data centers and electric vehicles will be challenging, U.S. electric generating capacity is presently adequate and 41% is free of carbon emissions [33]. The energy transition for the U.S. has been directed toward greater use of natural gas, reducing the use of coal, lowering energy imports from the Middle Eastern countries, increasing the use of renewables, and relying on energy efficiency improvements. Canada has been an important partner in this evolving energy transition.

The U.S. is the world's largest energy producer, maintains a surplus in total primary energy production, and is the dominant exporter of natural gas [33]. It is ironic that the fossil-fuel industry has argued strongly against the application of modest carbon taxes but has readily accepted the larger cost of tariffs on their products. After all, a tax is a tax. Based mostly on their lobbying efforts and media-driven claims, the substance of Executive Order 14156 withers when faced with actual data [33]. The tariffs will not resolve a national emergency if none exists, and that is perhaps the intention. In early 2026, the U.S. Supreme Court ruled the initial set of 10% tariffs were illegal. The President immediately imposed a new 150-day, 15% tariff using Sections 232 and 301. Further extensions will require congressional action. At any time, the U.S. administration can declare a victory and reset the policy structure.

The tariffs as proposed appear random, unjustified, untimely, and entropic. The use of the added taxes being collected is unclear and untargeted. Most importantly, they are misdirected and will alienate a long-standing and key U.S. ally and trading partner. The U.S. administration's policies that affect or implement tariffs on energy imports from Canada are a form of inverse protectionism that is aggressive and counter-productive.

The economic interdependence between the U.S. and Canada means that tariffs will create ripple effects in energy markets. The imposition of new tariff structures on energy supplies will lead to higher consumer costs, investment uncertainties, and unintended consequences. Some businesses will fail under the weight of the tariffs.

Perhaps the most effective way to resolve these interrelated problems is to eliminate the tariffs that are increasing the cost of energy. The trade and supply chains associated with delivering sustainable energy supplies are challenging enough without incurring the added costs of tariffs and surcharges. When energy prices increase, demand will ultimately decline, and consumers will reduce consumption or find substitutes. Another way is to direct the funds collected to supporting clean energy generation and improving energy infrastructure.

One result of the tariffs will be to improve the comparative economic competitiveness of certain domestic energy resources and technologies. This may ultimately and inadvertently align with the recent Presidential Action intended to "unleash America's affordable and reliable energy and natural resources" [34]. However, because the tariffs increase the cost of imported fossil fuels, in some regions domestic renewable energy resources and technologies will become more economically competitive.

References

[1] Center for Strategic and International Studies (2018). Mapping the U.S.-Canada energy relationship. https://www.csis.org/analysis/mapping-us-canada-energy-relationship, accessed 4 March 2025.

[2] Jaremko, D. (2025, January 30). Explainer: why Canadian oil is so important to the United States. Canadian Energy Centre. https://www.canadianenergycentre.ca/explainer-why-canadian-oil-is-so-important-to-the-united-states, accessed 4 March 2025.

[3] Canada Energy Regulator (2025, February 12). Market snapshot: overview of Canada-U.S. energy trade. https://www.cer-rec.gc.ca/en/data-analysis/energy-markets/market-snapshots/2025/market-snapshot-overview-of-canada-us-energy-trade.html, accessed 4 March 2025.

[4] Roosa, S. and Fredenthal, C. (2018). Natural gas processing, chapter 24 in *Energy Management Handbook*, 9th edition (ed. by Roosa, S.). Taylor and Francis Group, LLC. 978-1-138-66697-9.

[5] U.S. Energy Information Administration (2024, March 27). U.S. Natural gas production grew by 4% in 2023, similar to 2022. https://www.eia.gov/todayinenergy/detail.php?id=61646.

[6] U.S. Energy Information Administration (2024, March 27). Natural gas explained. https://www.eia.gov/energyexplained/natural-gas/how-much-gas-is-left.php.

[7] Canada Energy Regulator. Provincial and territorial energy profiles—Canada. https://www.cer-rec.gc.ca/en/data-analysis/energy-markets/provincial-territorial-energy-profiles/provincial-territorial-energy-profiles-canada.html, accessed 23 March 2025.

[8] Canadian International Merchandise Trade Database (2025, February 5). https://www150.statcan.gc.ca/n1/pub/71-607-x/71-607-x2021004-eng.htm.

[9] Trump, D. (2025, January 19). Declaring a national energy emergency. Federal Register, Executive Order 14156. https://www.federalregister.gov/documents/2025/01/29/2025-02003/declaring-a-national-energy-emergency, accessed 14 March 2025.

[10] Ortiz, E. (2025, March 4). Trump says tariffs are to curb fentanyl, but U.S. overdose deaths are already declining. https://www.msn.com/en-us/news/world/trump-says-tariffs-are-to-curb-fentanyl-but-u-s-overdose-deaths-are-already-declining/ar-AA1Agcue?ocid=msedgdhp&pc=U531 &cvid=1812fba178214c3598d3ea993c062e07&ei=85, accessed 5 March 2025.

[11] Trudeau, J. (2025, March 3). Statement by the Prime Minister on unjustified U.S. tariffs against Canada. https://www.pm.gc.ca/en/news/statements/2025/03/03/statement-prime-minister-trudeau-on-unjustified-us-tariffs-against-canada, accessed 5 March 2025.

[12] Hamid, J. (2025, March 4). Trump's global trade war is now officially live and markets are crashing. https://www.msn.com/en-us/money/markets/trump-s-global-trade-war-is-now-officially-live-and-markets-are-crashing/ar-AA1AcdS1?ocid=msedgdhp&pc=U531&cvid=11bb983a42894955abc250d7e15d9c84&ei=102, accessed 4 March 2025.

[13] Bickerton, J. (2025, March 4). Canada to cut off electricity to U.S. states: 'Need to feel the pain'. https://www.newsweek.com/canada-cut-off-electricity-us-states-need-feel-pain-2039125, accessed 4 March 2025.

[14] *Oil Sands Magazine* (2018, December 13). Differentials explained: Why Alberta crude sells at a deep discount. The National Law Review. https://www.oilsandsmagazine.com/market-insights/crude-oil-pricing-differentials-why-alberta-crude-sells-at-deep-discount-to-wti, accessed 5 March 2025.

[15] Villarroel, M. (2025, March 3). Canadian Premier apologizes to Americans for considering cutting energy exports: 'It's Not You, It's Your President'. https://www.msn.com/en-us/money/companies/canadian-premier-apologizes-to-americans-for-considering-cutting-energy-exports-it-s-not-you-it-s-your-president/ar-A1AfvFS?ocid=msedgdhp&pc=U531&cvid=7ec4b5e0b74a4b53ab1f164b68898fd0&ei=63, accessed 5 March 2025.

[16] Mahfouf, L. (2025, March 4). Canada's oil industry braces for impact as U.S. tariff threat looms. https://www.msn.com/en-us/money/other/

canada-s-oil-industry-braces-for-impact-as-us-tariff-threat-looms/ ar-A1AdZsX?ocid=msedgdhp&pc=U531&cvid=1812f-ba178214c3598d3 ea993c062 e07&ei=47, accessed 5 March 2025.
[17] Rich, R., Tracy, S. and Donovan, J. (2025, February 5). Tariffs loom on natural gas imports: be prepared. https://natlawreview.com/article/tariffs-loom-natural-gas-imports-be-prepared, accessed
[18] Oilprice.com (2025, April 25). Oil and natural gas price charts. https://oilprice.com/oil-price-charts/4.
[19] U.S. Energy Information Administration (23 March 2025). Henry Hub natural gas spot price. https://www.eia.gov/dnav/ng/hist/rngwhhdd.htm, accessed 5 August 2025.
[20] Baskaran, G. (2025, January 29). Canadian tariffs will undermine U.S. minerals security. https://www.csis.org/analysis/canadian-tariffs-will-undermine-us-minerals-security, accessed 5 March 2025.
[21] U.S. Energy Information Administration (2023, June). 2023 Uranium marketing annual report. https://www.eia.gov/uranium/marketing/pdf/2023%20UMAR.pdf, accessed 5 August 2025.
[22] Marre, K. (2025, March 3). Trump's border boast reveals true motivation behind tariffs. https://www.msn.com/en-us/news/world/trump-s-border-boast-reveals-true-motivation-behind-tariffs/ar-AA1AbA-j2?ocid=msedgdhp&pc=U531&cvid=193f00feaaa0492cbf31106a1d-c181ab&ei=8, accessed 4 March 2025.
[23] Morehouse, C. (2025, March 6). Ontario premier sticks to electricity tariffs despite Trump trade reversal. https://www.eenews.net/articles/ontario-premier-sticks-to-electricity-tariffs-despite-trump-trade-reversal, accessed 14 March 2025.
[24] Puelma, C. (2024, March 8). Canada hikes electricity prices for the U.S.: the impact of Ontario's energy surcharge. https://www.msn.com/en-us/money/companies/canada-hikes-electricity-prices-for-the-u-s-the-impact-of-ontario-s-energy-surcharge/ar-AA1AwuVe?ocid=msedg-dhp&pc=U531&cvid=a3e48cb6931946ca9b15a62eacd743df&ei=13, accessed 9 March 2025.
[25] Hamdeh, Y. (2025, March 10). How much states will pay in electricity bills after Canada tariffs revealed. https://www.msn.com/en-us/money/other/how-much-states-will-pay-in-electricity-bills-after-canada-tariffs-revealed/ar-AA1AEKwb?ocid=msedgdhp&pc=U531&cvid=a9172a0c3b6140 b1bae0b131c68dd9bc&ei=14, accessed 10 March 2025.
[26] World statistics (2025). GDP, countries and continents. https://worldstatistics.net/global-gdp-rankings-2024-countries-continents.

[27] Richter, W. (2025, February 27). Demand for electricity takes off. U.S. power generation by source in 2024: natural gas, coal, nuclear, wind, hydro, solar, geothermal, biomass, petroleum. Wolf Street.

[28] Kliment, A. (2025, March 6). Graphic truth: Who's most vulnerable in the North American trade war? https://www.gzeromedia.com/gzero-north/graphic-truth-whos-most-vulnerable-in-the-north-american-trade-war, accessed 14 March 2025.

[29] International Trade Administration. United States-Mexico-Canada Agreement. https://www.trade.gov/usmca.

[30] United States-Mexico-Canada Agreement. Fact sheet. https://www.mediaite.com/wp-content/uploads/2025/03/USMCA-Energy.pdf, accessed 13 March 2025.

[31] Smith, A. (April 2025). The high stakes of a trade war. Kiplinger, 16.

[32] Kennedy, A. (2025, March 8). Visualized: what cutting Canadian aluminum would cost the U.S. https://www.visualcapitalist.com/sp/aac01-canadian-aluminum-energy, accessed, 18 March 2025.

[33] Mazria, E. (2025, March 10). Don't buy the fossil fuel hype: There is no "national energy emergency". https://commonedge.org/dont-buy-the-fossil-fuel-hype-there-is-no-national-energy-emergency/, accessed 24 January 2026.

[34] The White House (2025, January 20). Unleashing American energy. https://www.whitehouse.gov/ presidential-actions/2025/01/unleashing-american-energy, accessed 14 March 2025.

21

Sharing Best Renewable Energy Practices in the Developing World

Nancy L. Najarian

Increasing the use of renewable energy and energy efficiency technologies in developing countries is crucial to combating climate change, improving people's lives, and supporting the energy transition. An array of tools, online and in print, is available to developing countries to assist with project preparation and management. The ultimate objective is to help countries and their citizens achieve carbon reduction goals set on a national or community level. *Are these tools and technical assistance being accessed as fully as intended?* Asking and answering targeted questions will help determine if and how these tools are being applied by countries, leadership, businesses, and residents to share lessons learned and best practices. The answers to these questions help to ascertain if sharing information among these entities is enabling measurable increases in the use of renewable energy and energy efficiency practices. Are lessons learned and best practices being shared effectively? Can those in remote regions find out what worked thousands of kilometers away? What tools are available to a developer interested in privately funding a project? Can data and lessons learned be freely shared in the private sector? What can energy professionals do to provide needed expertise and best practices to the developing world? What lessons can the U.S. learn from activities in the developing world? This chapter explores the answers to these questions.

21.1 Introduction

With increasing urgency, developing countries are responding to the need to reduce carbon emissions (CO_2e) by developing renewable energy projects,

DOI: 10.1201/9788743811312-21

practicing energy conservation, and implementing energy efficiency measures. They provide electricity using renewables to regions that previously lacked reliable sources of energy.

By the end of 2024, renewables accounted for 46% of global installed power capacity. An unprecedented 15.1% increase in renewable energy capacity compared to 2023, this was the biggest annual increase the world experienced since 2000. This was largely due to growth in solar and wind power [1].

In 2019, the International Renewable Energy Agency (IRENA) reported that nearly two-thirds of all new power generation capacity added in 2018 was from renewables, *led by emerging and developing economies* [2]. IRENA's Renewable Energy Statistics 2025 Report states that despite record renewable additions in 2024, growth rates indicate that the world is not on track to triple the installed renewable power capacity by 2030—the goal adopted at the Conference of Parties (COP) 28. To meet this objective, the world needs to achieve a much faster pace of growth in the development of new renewable power plants and distributed electricity generation [1]. Developing countries have an important role in meeting this collective goal. From their successes, important lessons may be learned by developing and developed nations alike.

21.1.1 What has impacted the growth in renewable energy in the developing world?

Development financial institutions (DFIs) such as the International Finance Corporation of the World Bank Group, the Asian Development Bank (ADB), or the U.S. Development Finance Corporation provide financial resources for energy projects [3]. Often simultaneously, government-to-government technical assistance is provided to build the capacities of organizations and individuals who operate and expand their use of renewable energy. Enabling tools and measures to increase renewable energy use include policy changes, public utility and national outreach coordination, capacity building, organized civil society, financial institutions, and investors.

Many tools are available to assist these countries with project preparation and implementation. These tools include:

- national, regional, or global studies, and maps locating sites favorable to develop wind or solar farms;
- software tools that help analyze the economics of a particular type of renewable energy; and

- workshops that connect policy makers and implementers, providing them with forums to share their experiences and best practices.

A study funded by the National Renewable Energy Laboratory (NREL) underscored the value of countries sharing their experiences [4]. Any country's ability to successfully integrate variable renewable energy depends on many factors, including technical requirements, resource options, planning processes, market rules, policies and regulations, institutional and human capacity, and events in neighboring countries. The more diverse and robust the experiences from which a country can draw, the more likely it will be to implement appropriate, optimized, and system-wide approaches. This is as true for countries in the early stages of renewable energy integration as it is for countries that have already had substantial success.

Going forward, successful renewable energy integration will depend upon the ability to maintain broad perspectives, to organize and make available the wealth of experiences, and to ensure that there are pathways from analysis to enactment [4]. The goal is to help countries and their citizens achieve carbon reduction objectives set by nations and their communities. Are countries, leaders, and stakeholders utilizing the lessons learned by others? How are these tools and technical assistance being accessed? Is their impact sufficient to make the changes our world needs to combat climate change? Are best practices being used to deploy and/or improve their use of renewable energy? Are the lessons learned and best practices being effectively shared? These questions beg for answers.

International, regional, and national organizations provide data, expertise, and support to guide governments as they connect with private sectors, individuals, and communities, and make commitments to reduce greenhouse gas emissions (GHG) by using renewable energy and implementing energy efficiency.

21.1.2 It begins with vision and leadership

An enabling environment is necessary to move a country toward increased renewable energy usage. A multitude of organizations assist governments with creating this environment. As has been demonstrated across the developing world and industrialized nations, a strong leader who provides a vision for moving the country toward carbon reduction is important. Leadership sets the goals and builds excitement within the populace and stakeholders who form coalitions that commit to low CO_2e energy sources. The coalitions determine what is needed and how to create an enabling environment. Leadership's goal is to facilitate policies, create programs, and foster an environment that offers inviting opportunities for investment in new renewable energy projects. The countries of Costa Rica, Uruguay, and India provide examples of how leadership and policy support transformed individual developing countries.

21.2 Costa Rica

Costa Rica, a country of 5.3 million, has produced many leaders who promoted and acted upon commitments to combat climate change with progressive environmental policies at home and internationally [5]. Figure 21.1 shows electricity production by source.[1]

- 2007: President Oscar Arias Sanchez announced that his country would mark the bicentenary of its independence from Spain in 2021 by becoming carbon neutral.
- 2012: The first Country Carbon Program Neutrality 1.0 was launched with a national strategy and action plan. The plan encouraged private companies to adopt practices to reduce their emissions and compensate for their carbon footprints.
- 2015: Costa Rica generated 99% of its electricity from renewable sources—80% from hydropower and 11.5% from wind generation. Costa Rica and its rapid transition to renewable energy were helped by substantial amounts of rainfall in 2014/15 and the subsequent increase in the potential of hydropower in the country.
- 2017: For 300 days the country's grid operated on 100% renewable energy.
- 2018: Costa Rica passed Law 9518 to promote electric vehicles (EVs) through improved access to credit and economic incentives, setting the stage to decarbonize the transportation sector. It obligates many government institutions to take specific actions to promote the use of EVs. For instance, the Ministry of Environment and Energy (MINAE) must make policies and publicize information that promote the use of EVs, and coordinate institutional cooperation among government agencies, municipalities, and public institutions to promote the acceptance and use of EVs [6].

[1] In Figures 21.1, 21.2, and 21.4, the category other renewables include geothermal, wave and tidal power.

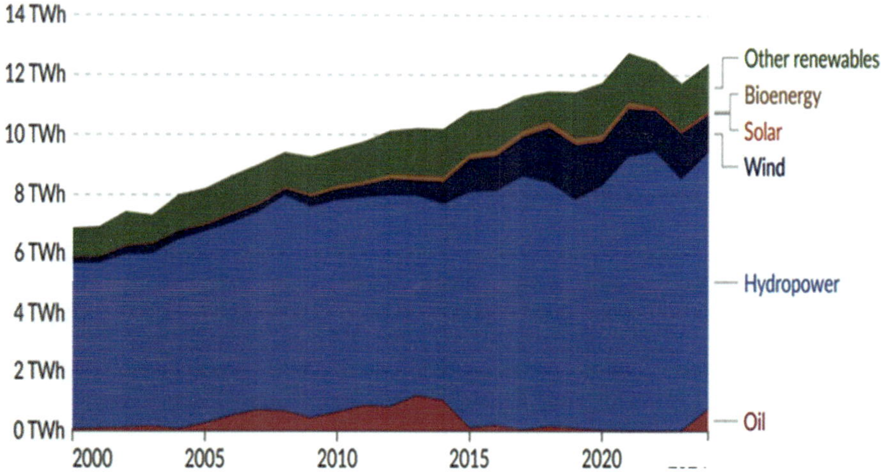

Figure 21.1 Costa Rica, electricity production by source (TWh).[2]

- 2019: President Carlos Alvarado unveiled the detailed 2050 National Decarbonization Plan. This initiative was a roadmap for a transition away from fossil fuels and more polluting ways of producing food and managing waste. The Environment Minister Carlos Manuel Rodríguez noted that if the plan is achieved, his grandchildren in 2035 will have the same carbon footprint as his grandparents in the 1940s [5].

- 2020 to 2025: For several years, Costa Rica generated nearly 100% of its electricity from diverse renewable energy sources—hydroelectric power, wind, and geothermal energy that taps into volcanic activity.

- Reduced rainfall and climate change are testing the resilience of Costa Rica's energy model, particularly hydroelectric power. Reductions in rainfall and prolonged El Niño events have disrupted water flows, impacting electricity production from hydropower. In 2022 technicians from the Costa Rican Electricity Institute observed a sharp decline in reservoir levels that diminished the efficiency of hydroelectric plants. They noted that this trend was expected to continue; their predictions projected a drop in renewable energy contributions to the country's grid from 99% in 2021 to 93% by 2024 [7].

[2] Data sources: Ember (2025) and Energy Institute, Statistical Review of World Energy (2025).

- 2025: The Costa Rican government is addressing these challenges with a variety of solutions. As a temporary remedy, they will activate thermal plants to supplement energy production during periods of low rainfall. Despite this short-term reliance on fossil fuels, the country remains committed to its long-term sustainability goals.
- The government is diversifying energy sources and improving infrastructure to ensure a stable energy supply while minimizing environmental impacts. Integrating wind, solar, and geothermal solutions strengthens grid power, reduces reliance on any single energy resource and ensures long-term sustainability. The government is funding research and development to improve energy efficiency, investing in smart grid initiatives, and introducing smart meters. These efforts are expected to boost the country's renewable energy matrix by 2040 [7].

21.2.1 Costa Rica's environmental leadership

Costa Rica's leadership places great importance on the environment, and that includes how government is organized. "When you have the ministries of energy and environment in the same house you can make big leaps forward. Same person, same agency," says Carlos Manuel Rodriguez, who served three terms as Minister of Environment and Energy for Costa Rica. Afterwards, he became CEO and Chairperson of the Global Environment Facility (GEF). GEF is a family of environmental funds, defined as a financial mechanism for six multilateral conventions to address global environmental challenges including biodiversity loss, climate change and pollution [8]. Other examples of environmental leadership include former Costa Rican President José María Figueres who served on United Nations Secretary General Ban Ki-Moon's Advisory Group on Climate Change and Energy. Christiana Figueres, led the UN Framework Convention on Climate Change, the group that convened the 2015 Paris climate agreement [5].

In 2019, Monica Araya, from the non-profit Costa Rica Limpia (Clean Costa Rica), stated, "This is the new Costa Rica… we know that the future is renewable and electric. We have all the conditions—clean electricity, a young president who wants to do right, and technologies on our side." Renewable energy has become a part of the country's identity. People proudly believe that it is a Costa Rican imperative to go green. The president knows that he can set a precedent at a time when the world is trying to determine how to transition to electric mobility. He believes that it is doable, beneficial, and works technologically [9].

21.2.2 Costa Rica's challenges

There is urgency and need to revolutionize Costa Rica's transportation system; half of the country's cars are more than two decades old and very polluting. Demand for automobiles is rising. More than 60% of the country commutes by diesel buses or trains, and transportation accounts for 60% of GHG emissions. Since 2002, there has been a moratorium on oil exploration and exploitation in the nation's continental and marine territories, which is in effect until 2050 [5].

21.2.3 New policies

Costa Rica has become a leader among Latin American countries in electrifying the transportation sector. The nation's approaches provide a model that can be replicated across Central America [5]. The country's policy to decarbonize transportation and its National Decarbonization Plan focus on changing the composition of passenger, light, and heavy vehicles as well as freight trains. The government's transportation goals include: 1) placing new electric passenger and freight trains in service by 2022, 2) nearly a third of all buses will be electric by 2035, 3) building vehicle charging stations, and 4) by 2050, nearly all cars and buses on the roads will be electric [5]. As of 2025, Costa Rica has made important strides toward achieving these goals by:

- Signing the Zero Emission Vehicles Declaration (November 2024), a public commitment to work toward all new sales of cars and vans being zero emission by 2035.
- Actively transitioning its public bus fleet to electric, with several cities already piloting electric buses. Law 9518 mandates a gradual replacement of the bus fleet with electric models at a rate of at least 5% biennially.
- As of 2025, 15% of cars in Costa Rica are now EVs. Because almost all the electricity in Costa Rica comes from renewable energy sources, charging EV batteries adds little to carbon emissions. The diversity of EVs available for purchase also adds to the adoption. Unlike some other Latin American countries, the market is not dominated by a single foreign manufacturer. No company has more than a 25% market share. Costa Rica has 11 electric vehicle models priced under $25,000 with 24 brands and 45 models available.
- In 2025, there are over 260 EV chargers in Costa Rica. The goal is to have a charger every 80 km (49.7 miles) on highways and every

120 km (74.6 miles) in county areas. The government knows that to keep pace with the growing EV purchases, it must develop EV charging infrastructure.

- Tax incentives and other initiatives under Law 9518 (passed 2018) are having a marked impact on the increase in EV sales and charging infrastructure. New and used EVs are exempt from the 13% value-added tax. Municipalities can waive parking fees for electric vehicles. EVs used as taxis receive specific tax benefits. National banks are required to offer favorable financing options for consumers interested in purchasing EVs [10].

21.2.4 Sharing best practices

To share best practices from Costa Rica's experience, the World Bank promotes workshops, training programs, and discussions of international best practices [11]. The project:

- Helps key decision makers strengthen their technical capabilities to launch electric public transportation initiatives for Costa Rica's cities.
- Offers valuable takeaways for countries in the region where the demand and support for clean transport systems are growing.

Costa Rica's example holds important lessons for the rest of the world. It took a courageous stance based on the scientific evidence in 2007. The country established a series of commitments to reduce CO_2e with ambitious goals that focused on the long-term avoidance of climate change. Their mission inspired people with a positive vision [11].

21.2.5 Lessons from Costa Rica, for rapid transition to renewable energy

Finding the best pathway for a rapid transition to renewable energy involved Costa Rica working to harness its unique resources and taking a strong leadership stance to attract investment.

- Policy, infrastructure, and cultural development are needed concurrently to make rapid transitions.
- Countries with less historical responsibility for tackling climate change can show leadership and provide successful examples to other countries with greater culpability [5].

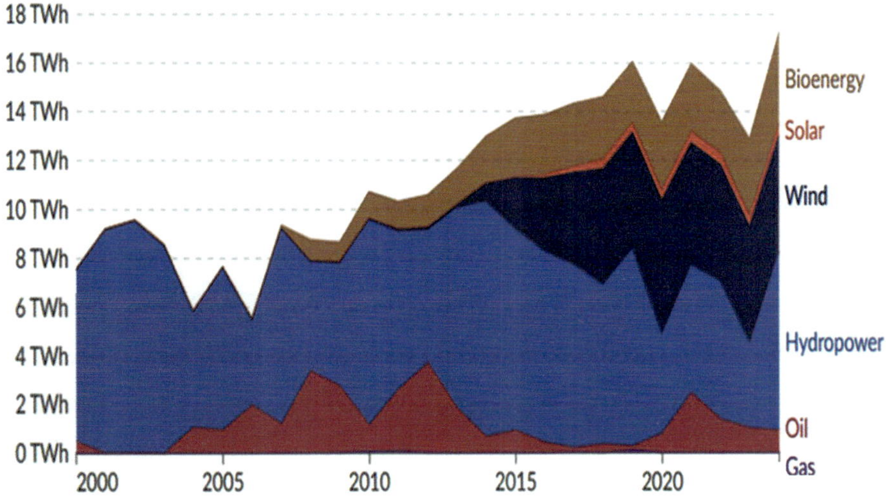

Figure 21.2 Uruguay, electricity production by source (TWh).[3]

21.3 Uruguay

Uruguay's population is about 3.4 million. In 2000, oil accounted for 27% of its energy use. In 2002, a 218 km (135 mile) long pipeline was constructed to supply natural gas from Argentina with an annual capacity of 1.8 billion m^3. At that time, the country relied heavily on fossil fuels. In 2015 at the Paris Climate Talks, Uruguay's head of climate change policy (and former Environmental Minister) Ramón Méndez pledged an 88% cut in CO_2e by 2017, compared with the average for 2009 to 2010 [12]. Today, renewables provide 95% of the country's electricity, and prices are lower than in the past relative to inflation. There are also fewer power outages because the country's diverse energy mix means greater resilience to outages caused by droughts. Figure 21.2 shows electricity production by source.

21.3.1 Attracting investment

There have been no new nuclear or hydroelectric power additions in Uruguay in over 20 years. According to Méndez, the keys to success are encouragingly

[3] Data sources: Ember (2025) and Energy Institute, Statistical Review of World Energy (2025).

replicable: "clear decision-making, a supportive regulatory environment, and a strong partnership between the public and private sector. What we have learned is that renewables are a financial business. The construction and maintenance costs are low, as long as you give investors a secure environment, it is very attractive" [12].

21.3.2 Highlights of renewable energy adoption

An important aspect of Uruguay's successful adoption of renewable energy has been the public and private sectors' ability to work together—with remarkable results. The government of Uruguay maintained a stable democracy and provided reliable partnerships with private companies. The highlights of Uruguay's adoption of renewable energy include:

- From 2010 to 2015, energy investment in Uruguay, mostly for renewables, surged to $7 billion, or to 15% of the country's total annual gross domestic product; this was five times the average of Latin American countries.
- Given the country's reliable wind speeds (averaging 3.6 m/s or 8 mph), the state utility guarantees a fixed electric price for 20 years. Maintenance costs are low and stable and investors are guaranteed a profit.
- A mix of renewables creates greater resiliency to climate change.
- A lower risk of electrical power disruptions exists because windfarms supply electricity to hydropower plants, thus reducing vulnerability to drought conditions by 70%. As a result, Uruguay no longer imports electricity. In 2014, it exported a third of its electricity generation to Argentina [12].
- In March 2018, wind power became the principal source of electrical generation, surpassing hydropower in contributing electricity to the national grid. According to the state electric company (UTE), wind power contributed 40.5% of total generation compared to hydropower's 38.4% [13]. Biomass (8.9%), thermoelectric (7.6%), solar power (4.6%), and imports contributed the remaining electricity [14].

21.3.3 Uruguay's challenges

The country's challenges to meeting its renewable transition goals mainly involve transportation, a sector dependent on oil and derivatives that accounts for 45% of its energy use. Some have labeled this Uruguay's second energy

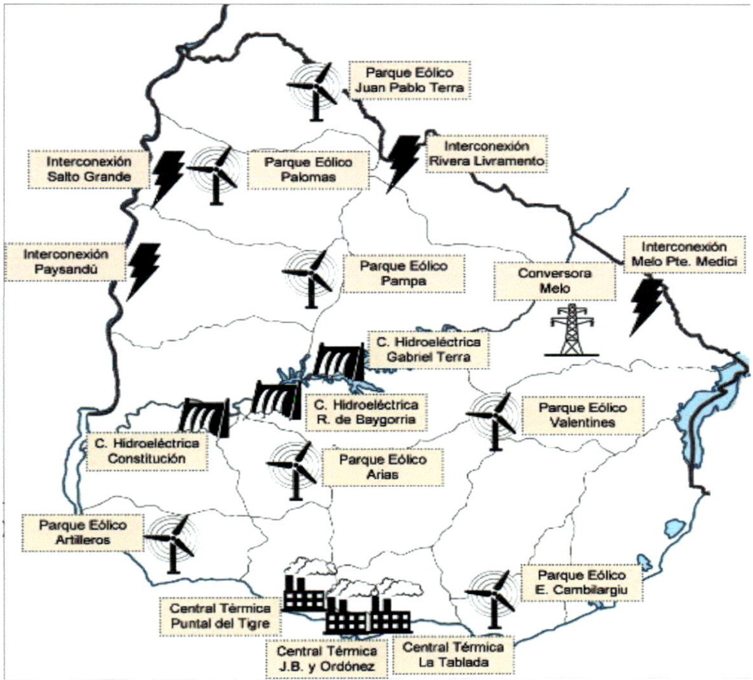

Figure 21.3 Power generation in Uruguay [13].

transition—the first being their successful transition to renewable energy. Power generation resources are mapped in Figure 21.3. The electromobility program was launched in June 2018 in Montevideo. This initiative promotes the transition towards electromobility, a sustainable and efficient urban transportation system. The program objectives included establishing a regulatory framework for a low-carbon transportation system, promoting the adoption of EVs for public transport and utility use, and fostering cultural change [15]. The results of the program include:

- The Joint Sustainable Development Goals (SDG) Fund's program, Renewable Energy Innovation Fund (REIF), supports the achievement of the Sustainable Development Goals by 2030. The fund has made a strategic investment partnering with seven major Uruguayan banks that collectively represent more than 80% of Uruguay's banking sector. The initial investment grew to $22 million USD for renewable energy projects across various sectors, including transportation. Transportation companies are thus able to afford to switch to electric buses.

- The initial phase of the program financed the acquisition of 50 electric buses for Montevideo's public transportation system. REIF incorporated gender considerations throughout the project lifecycle. A result of the specialized study identified barriers to women's participation in the transportation sector and led to the creation of a targeted training program for women drivers and mechanics. This electric bus program demonstrates that innovative financing mechanisms can accelerate the transition to a more sustainable and inclusive urban transportation system. It is a model that other countries throughout Latin America are studying [16].
- In January 2025, the Latin American Energy Organization (Olade) demonstrated Uruguay's commitment to electric mobility. The data showed that Uruguay had approximately 5,950 registered electric vehicles (17.4 EVs/10,000 residents), a larger market penetration than Brazil or Mexico.
- The government's initiatives have a crucial role in the growth of EVs. For example, the Uruguayan Ministry of Industry, Energy, and Mining (MIEM) launched the Subite program. This initiative offers financial incentives, including tax exemptions and rebates for electric vehicle purchases. In addition, the government eliminated the Specific Internal Tax (IMESI) for EVs until August 2025, another encouragement for consumers to transition from traditional combustion engines.
- EV charging infrastructure has expanded with 332 charging stations nationwide in 2025. The state-owned utility company UTE has the goal of establishing charging points every 50 km (31.1 miles) along major routes to ensure accessibility for all EV users. Like Costa Rica, having 98% of electricity supplied by renewable energy makes charging EVs more sustainable and economically viable [17].

21.3.4 Lessons learned from Uruguay's experience

"Uruguay has proved that renewables can reduce electricity costs, meet over 90% of electricity demand without the back-up of coal or nuclear power plants, and that the public and private sectors can work together effectively in this field," says Ramón Méndez, Uruguay's head of climate change policy [12]. He attributed Uruguay's success to three key factors:

Credibility: a stable democracy that has never defaulted on its debts, making long-term investment attractive.

Helpful natural conditions: good wind, ample solar radiation, and lots of biomass from agriculture.

Strong public companies: being a reliable partner for private firms who can work with the government to create an attractive operating environment [8].

21.3.5 Strong decision making

The lesson that Uruguay provides is the importance of strong central decision-making. Uruguay was once paralyzed by a seemingly endless and rancorous debate about energy policy. The government finally agreed to a long-term plan that drew cross-party support, ending the paralysis. "We had to go through a crisis to reach this point," Méndez said … "but in 2008, we launched a long-term energy policy that covered everything … Finally, we had clarity" [12].

21.4 India

India is the world's most populous country. In 2008, 400 million people in India had no access to electricity. In rural locations that were grid-connected, people were without power for 12 to 14 hours each day. By 2025, only 6.4 million people lacked access to electricity. Though most electricity is sourced from coal, India could be described as a regional leader in renewable energy deployment. Figure 21.4 shows the increase in electricity production by energy source.

21.4.1 India's development of renewable energy

In 1988, the United States Agency for International Development (USAID) and the Association of Energy Engineers (AEE) launched a series of workshops on energy efficiency and renewable energy in five of India's largest cities. Led by Dr. Stephen Roosa, the AEE representatives met with India's newly appointed renewable energy secretary and helped outline ways to introduce renewable energy to India's remote locations that lacked electricity. These meetings and workshops were instrumental in seeding a series of events that spawned India's governmental initiatives to support renewable energy development.

The 1992 Kyoto Protocol initiated the Clean Development Mechanism and enabled numerous Joint Implementation projects that supported solar power development in India. With the election of Barack Obama as president in 2008, the U.S. started a new energy partnership with India. By 2009, India and the U.S. launched a multi-pronged initiative called the Partnership to Advance Clean Energy [18], the highlights of which included:

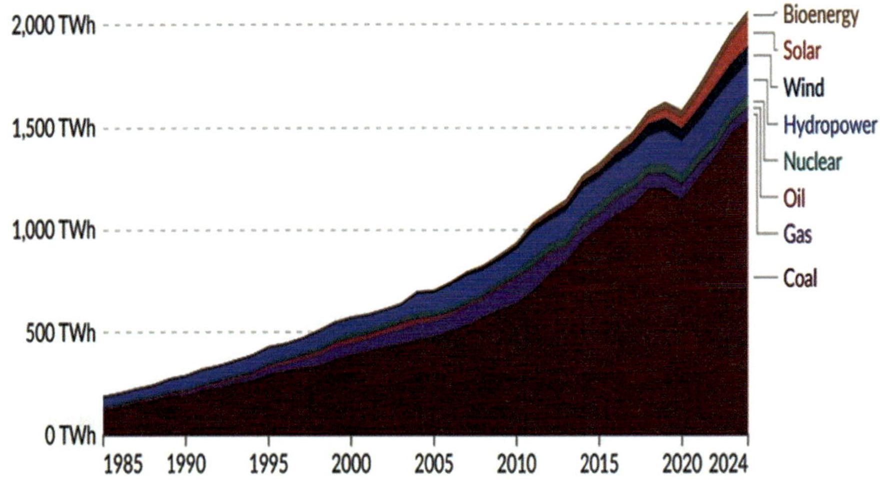

Figure 21.4 India, electricity production by source (TWh).[4]

- A joint research and development (R&D) initiative that brought together scientists and industry from both countries to support renewables (solar and biofuels) and building energy efficiency.
- When Prime Minister Narendra Modi entered office in 2014, the cost of clean energy technologies declined due to technological innovation. Modi previously worked at the state level, experimented with the deployment of new energy technologies, and wrote a book on climate change [18].

21.4.2 What changed in India?

By 2015, during the UN climate negotiations, India, the U.S., other countries, and the European Union pledged to double their investments in clean energy R&D. India's federal government at the time was investing $72 million in clean energy R&D; as part of this coalition, it served as the laboratory for many climate technology innovations [19].

- Policies that once seemed too progressive for India were embraced. India increased its goal for net-zero emissions by 2070, establishing

[4] Data sources: Ember (2025) and Energy Institute, Statistical Review of World Energy (2025).

targets of 500 GW of renewable energy by 2030, and a 45% reduction in GHG emissions intensity. Its progress toward these goals is laudatory.

- As of October 2024, India's renewable energy capacity represented 46.3% of the country's total installed capacity. India, once a coal-dependent economy, in less than a decade became the world's third-largest renewable energy producer. This transition shows how developing nations can successfully balance rapid economic growth with sustainable energy development. It is a testament to the feasibility of large-scale clean energy transitions in emerging economies.
- India's original 2015 goal was to install 20 gigawatts (GW) of solar capacity by 2022. As of January 2025, India's solar capacity rocketed to 97.9 GW, an extraordinary 10-fold increase from the prior 9 years. Creating renewable energy clusters optimizes land use and transmission efficiency.
- Commitments were made to add renewable generation from wind and small hydro, and to boost the nation from one that lacked adequate electricity to one with a surplus. India is committed to issuing bids for at least 10 GW of wind capacity annually from 2024 to 2028.
- Ranked fifth globally in hydroelectric power capacity, both large-scale projects and distributed small hydro installations contribute to India's renewable energy portfolio. The country is exploring technologies to develop India's estimated 40,000 MW wave energy potential [20].
- The country announced the goal of developing a "100% national electric vehicle fleet" by 2030 [18].

21.4.3 India's challenges

Coal is a crucial part of India's energy strategy. Its use ensures grid stability and fuel security as demand for energy rises. Coal use contributes to employment, government revenues, and transportation. In 2023 and 2024, record coal production highlighted its continued importance in maintaining base-load needs. Efforts to make coal plants more flexible and to renovate and modernize the plants address the need for coal to balance the intermittency of renewable energy generation. Upgrading generation stations that use coal stabilizes the electric grid while improving plant life, efficiency, and environmental compliance.

Strengthening India's transmission infrastructure is crucial to expanding renewable energy. Under the National Electricity Plan, India's transmission planning aligns national and state strategies to support renewable energy growth. The government must improve intrastate transmission to better manage renewable energy variability, balance power imports, and reduce energy curtailment [20].

21.4.4 Lessons learned from India

India has made substantial strides in moving toward the use of renewable energy. It is a model of a populous nation enacting policies, making investments, and carving a path toward meeting its Paris Climate Agreement Nationally Determined Contribution (NDC) goals. Taking further actions will help India aim for reductions in CO_2e and exceed government targets [21]. Suggestions for policymakers that go beyond a sole focus on the renewable energy supply include:

Better frameworks for electricity: Production variability is a feature of certain types of renewable energy. This requires flexible grid infrastructure and custom contractual terms. Time-of-day pricing for electricity is an important first step. In 2019, only about 3% of electricity was purchased through exchanges that reflect supply and demand conditions; the balance was purchased through power purchase agreements that treat all power the same, regardless of its variability.

Smart grids: Smart grids are more than just smart meters. The grid of the future will allow demand to react to variable renewable energy supplies.

Energy efficiency: Energy efficiency needs more attention and funding. There is no regulated rate of return for efficiency investments. New buildings will be part of the capital stock for decades, but builders often disregard occupant energy costs. Lack of energy efficiency programs in the housing sector is a major gap.

Benefits: Greater market penetration of clean energy technologies, including EVs, will accelerate because they offer additional benefits, including energy security, employment, and reduction of air pollution. These benefits need to be recognized during planning processes at the central, state, and municipal governmental levels.

Global support: An influx of global capital would increase funding and lower interest rates. India needs more than the transfer of current renewable energy technologies; it also needs innovative regulation and business models [21].

Figure 21.5 Town-sized, solar-powered smart grid in Les Anglais, Haiti, energized in 2012 [16].

21.5 How Can Those in Remote Villages Learn What Worked in Similar Communities?

Costa Rica, Uruguay, and India are examples of countries that benefitted from strong and visionary leadership, which sparked substantial progress toward reducing CO_2e. The lessons learned and best practices from each country's experience are being shared globally and impacting other developing countries' leadership gains and policies.

The challenge of sharing information across the world, be it from the most populated urban centers to the most remote of areas, is being met by organizations that choose a specific focus and offer lessons learned to benefit others. EarthSpark International is an example of a not-for-profit entity that works to bring electricity to remote areas and shares best practices, including innovative technologies [22]. Their mission includes researching and developing business models that can be scaled to address energy poverty.

21.5.1 Developing innovative technology and transferring business models

Partnerships are key to successful renewable energy solutions. Important innovations in decarbonizing the global energy supply may come from remote villages that lack electricity. The absence of incumbent infrastructure presents opportunities to build energy systems with today's best technologies and business models—leapfrogging the developmental solutions. These models leverage clean energy, storage technologies, smart grids, and customer participation. They can be adapted elsewhere and inform the evolving utility business models in established markets [22].

Les Angles, Haiti, (Figure 21.5), was a town that lacked grid-supplied electricity. EarthSpark developed a new, privately operated, microgrid, expanding the grid in 2015 to 430 connections. Over 2,000 people were directly served with 24-hour electricity powered primarily by solar energy and battery storage. Energy costs were reduced by as much as 80% [23].

In December 2019, EarthSpark launched its second solar-powered microgrid in Haiti, the first community-power grid to be approved under Haiti's new national regulatory body. Building on a model of community-scale smart solar infrastructure, the grid will serve 500 homes and businesses with 24/7 electricity in Tiburon, a small fishing town in the southern peninsula. The launch marks a major step in mainstreaming microgrid power in Haiti [24].

On August 14, 2021, Southern Haiti experienced an earthquake, affecting both microgrid locations; Les Anglais was hit harder than Tiburon. The two microgrids powered the towns through the tragedy with a total downtime of less than three hours. The Les Anglais grid was immediately de-energized after the earthquake. After checking for damage, only minimal structural damage was found. No specialized repairs were necessary. EarthSpark was able to continue powering Les Anglais and Tiburon through the earthquake, confirming the value of microgrids and distributed renewable energy. To mitigate the risk of future damage to the two microgrids and others being planned, the organization began to work with municipalities, clinics, and government officials to prepare for future disasters [24].

Fast forward to the end of 2024. By then, six mayors from across the country's southern peninsula signed contracts to bring electricity to their towns. Together, the six new grids will bring reliable electricity to over 28,000 people [25].

21.5.2 Smart metering

EarthSpark built an innovative new infrastructure, taking advantage of the opportunity to develop a technology and business model that provided a necessary low-cost, high-functionality smart meter to meet its needs when it installed the inaugural grid in 2012. It is now enabling grid operators to increase energy access and improve operations in 22 countries. After developing a prototype, EarthSpark spun off the smart metering company [26]. The company is developing a scalable, project-based model for microgrid development and operation by bundling technical innovation, community engagement, diverse partnerships, and novel financing [23].

21.5.3 Challenges faced and overcome in Haiti—applying lessons learned

Since 2020, EarthSpark, working in Haiti, has endured and thrived during hurricanes, earthquakes, COVID-19, and acute political and security crises, among others. Exacerbating everything was the national fuel crisis that stopped the importation of most liquid fuels for many weeks.

Applying lessons learned from working for over 10 years in Haiti, EarthSpark undertook a solar project in a small rural market town of Wanyjok, South Sudan. While working in a different climate, culture, and context, a similar need and opportunity existed for innovation to solve energy poverty. Like Haiti, South Sudan is one of the least electrified countries in the world, with only ~7% of people having access to electricity, leaving 10.4 million people living without electric power. This limits South Sudan's potential economic development and the resiliency of communities, households, and businesses.

Adam Eberwein, the technical lead of EarthSpark International, described applying lessons learned. Our "team spent nearly a decade, day in and day out, living and working in the towns where we deliver microgrid services in rural Haiti, making improvements at every step along the way. In spite of all the differences between these two regions, it was a privilege to see how our lessons learned on one side of the world applied so well to the other."

For two years, EarthSpark has worked with a solar installation company in South Sudan. Together they planned and launched the country's first solar-powered community-level microgrid in September 2022. The result was an increase in the resiliency of the community and its economy. Over 500 people received connections. 98% of the customers reported no change or decrease in energy expenses compared to previous kerosene and diesel expenses. 61% of business customers reported improvements in business operations—most of the customers were small businesses within the community. The lives of the entire community of Wanyjok (~20,000 people) have improved with microgrid power, bolstering resilience. Reliable access to electricity means streetlighting for the downtown market area, healthcare improvements, longer operating hours for local entertainment businesses, and the improved security of a well-lit marketplace [27].

21.5.4 A testament to public private partnerships

Over the years, EarthSpark has worked to help clarify the Haitian legal and regulatory landscape for microgrid development and operation. By leveraging public private partnerships (PPPs), the company works directly with local mayors and communities to facilitate public and private cooperation among enterprises in emerging markets. In doing so, the organization creates ownership and buy-in, empowering local stakeholders to participate in their own energy governance while improving the long-term sustainability of the systems [25].

The contractual agreements between EarthSpark and the Haitian mayors (signed in late 2024) to build the six new microgrids show how these partnerships work. The Haitian public servants were composed of diverse teams across agencies who persevered through difficult circumstances, at times risking personal security just to go to work. They navigated complex rules and processes to build a new rural energy program. It required time and effort to get these first six grid contracts signed. The company was honored to have formal PPPs with these six municipalities and the Ministry of Public Works in Haiti to build new microgrids [25].

21.6 What Tools, Data, and Lessons Are Shared with Developers Interested in Funding Projects?

Private sector actors in the renewable energy arena often rely on data and analysis that are available in the public domain. While private data is usually proprietary, organizations have often provided critical data that a country and investors need to structure successful renewable energy projects. This section provides a sampling of organizations focused on helping countries and their private sector partners access and understand the potential for renewable energy, opportunities for projects, and funding that will make renewable energy projects a reality. Many of these organizations' missions are aligned with the UN's 17 sustainability goals [28]. They work to support countries as they implement their NDCs to reduce CO_2 emissions and limit global warming to 2°C or less. These organizations offer accessible information and opportunities to interact with experts from all over the world who can apply best practices to specific projects.

21.6.1 United Nations Framework Convention on Climate Change

With 197 parties, the United Nations Framework Convention on Climate Change (UNFCCC) has nearly universal membership and developed the 2015 Paris Climate Change Agreement. UNFCCC is also the parent treaty of the 1997 Kyoto Protocol. The ultimate objective of all agreements under the UNFCCC is to stabilize atmospheric GHG concentrations at a level that prevents dangerous human interference with the climate system, within a time frame that allows ecosystems to adapt and enable sustainable development [29].

21.6.2 International Renewable Energy Agency (IRENA)

Founded in 2009, IRENA is an intergovernmental organization that supports countries in their transition to a sustainable energy future; 180 countries are actively engaged with IRENA [30]. IRENA provides a wide range of products and services that include [31–41]:

- Annual reviews of renewable energy employment.
- Renewable energy capacity statistics.
- Renewable energy cost studies.
- Renewables readiness assessments, conducted in partnership with governments and regional organizations, to help boost renewable energy development on a country-by-country basis.
- The Global Atlas maps provide resource potentials by source and by location.
- Renewable energy benefits studies.
- REmap, a roadmap to increase renewable energy use and meet the 2050 Paris Agreement goals.
- Renewable energy technology briefs.
- Facilitation of regional renewable energy planning.
- Renewable energy project development tools like Project Navigator (an online platform providing comprehensive, easily accessible, and practical information, tools, and guidance to assist in the development of bankable renewable energy projects), the Sustainable Energy Marketplace (an investment catalyst), and the IRENA/ADFD Project Facility that funds renewable energy projects.

Sharing lessons learned, IRENA holds seminars and workshops. One example happened in June 2019 during which representatives from countries with enterprising renewable energy goals pledged 70% to 100% renewable power targets by mid-century. IRENA publishes the findings from these events [42].

In August 2019, the IRENA and UN Climate Change (UNFCCC) announced that they are jointly increasing efforts to fight climate change by promoting the widespread adoption and sustainable use of renewable energy [43].

21.6.3 The Low Emission Development Strategies Global Partnership (LEDS GP)

Founded in 2011, LEDS GP enhances coordination, rapid information exchange, and cooperation among countries and international programs. LEDS leaders and practitioners from over 350 institutions and 118 countries collaborate through innovative peer learning, forums, and networks to support the formation and implementation of low-emission development strategies. The organization focuses on providing support to developing countries and regions. LEDS GP engages leaders from institutions across government agencies, technical institutes, international agencies, and non-governmental organizations (NGOs). It operates regionally in Africa, Asia, Europe and Eurasia, Latin America and the Caribbean, and has six technical global working groups plus a global secretariat [44].

21.6.4 American Cities Climate Challenge and the Renewables Accelerator

Bloomberg Philanthropies' American Cities Climate Challenge, founded in June 2018, is a two-year acceleration program with powerful resources and access to cutting-edge support to help cities meet or exceed their near-term carbon reduction goals.

The Renewables Accelerator, an initiative of the Climate Challenge, supports U.S. cities in procuring renewable energy as they seek to meet these goals. Jointly led by Rocky Mountain Institute and World Resources Institute and facilitated by the Urban Sustainability Directors Network (USDN), the Renewables Accelerator offers technical support to 25 selected challenge cities and to over 100 USDN members interested in powering their cities with renewable energy. They facilitate peer exchange among cities and share lessons learned from other large renewable energy buyers [45].

21.6.5 Replicating the solarize campaign across U.S. cities

The City of Portland, Oregon, established its Climate Action Plan with a goal to reduce carbon emissions by 40% by 2030 and 80% by 2050. Its city council set a goal to be 100% renewable in all economic sectors by 2050. The first Solarize campaign started as a community-led movement to help stakeholders overcome the financial and logistical barriers to installing solar power at their properties. Three years of Solarize Portland campaigns resulted in over 600 solar installations and 1.7 MW of new residential solar PV capacity. Solarize has been replicated over 170 times in 18 different U.S. states and abroad [46].

21.6.6 National Renewable Energy Laboratory (NREL)

NREL is a national laboratory of the U.S. Department of Energy, Office of Energy Efficiency and Renewable Energy. This U.S. government organization advances the science and engineering of energy efficiency, sustainable transportation, and renewable power technologies, and provides the knowledge to integrate and optimize energy systems. NREL works with organizations through research partnerships, licensing NREL technologies, supporting cleantech stakeholders, and fostering the clean energy economy. Some of the tools and opportunities to receive support from NREL on renewable energy development include [47–50]:

Data visualization and geospatial tools: NREL's geospatial data science researchers have developed tools that allow users to apply its data. The tools help determine how much electricity can be produced from residential solar systems or what renewable resources are available in a specific area.

Maps: NREL offers an array of web-accessible maps to support renewable energy development and electrical generation projects.

Data resources: NREL offers geographic information system (GIS) data resources for a variety of renewable energy technologies. The datasets are designed to be used in GIS software applications.

21.6.7 Case study—NREL tools help determine the economic benefits of rooftop solar in Mexico

In 2016, NREL provided technical assistance to the Government of Mexico (GOM) under the USAID Mexico Clean Energy Program. This assistance contributed to the country's planning efforts as it prepared to meet its clean energy targets by deploying rooftop solar technologies [51].

Background
In 2015, the GOM declared its clean energy participation targets—30% by 2021 and 35% by 2024—under its Energy Transition Law and General Climate Change Law. This regulation established the framework for advancing Mexico's clean energy and climate goals, thus advancing its emerging clean distributed generation (DG) market. To foster that growth in an inclusive and equitable way that serves investors, ratepayers, utilities, and society, the GOM needed hard data to answer tough questions [52].

In 2016, the NREL team used its System Advisor Model (SAM) software tool to explore the economics of distributed solar PV in Mexico. SAM is a free, state-of-the-art platform for simulating the performance and economics of renewable energy projects. Among other renewable energy technologies, SAM offers a user-friendly, detailed treatment of distributed photovoltaic (DPV), including rooftop solar systems. The model considers customer consumption patterns, retail electricity rates, DPV compensation schemes, system pricing, and a variety of other technical and economic inputs. It is used to create hourly generation profiles for rooftop solar projects, perform return-on-investment analyses, and test the impact of various DPV public policies on customer economics [53].

21.6.8 Solar compensation policy with the energy regulatory commission

The first analysis conducted with the Energy Regulatory Commission of Mexico (CRE) in 2016 was designed to inform internal discussions at CRE, focusing on how customers might be impacted by prospective changes to DG metering and billing arrangements in Mexico. Results of the analysis were used to inform and update Mexico's DG regulatory framework in February 2017 [54].

21.6.9 Potential benefits of residential sector DG and energy efficiency

The second analysis was conducted in July 2016 under the leadership of Mexico's Secretariat of Energy, analyzing the potential benefits of residential sector DG and energy efficiency from the perspectives of individual customers, the Mexican Secretariat of Finance and Public Credit, and the environment. The analysis uncovered that there is a disconnect between where the greatest potential benefits exist for the state, versus where it is most financially attractive for customers to invest in DPV systems. Mexico's Secretary of Energy concluded that without a public policy intervention, DPV's "penetration in the market will be very slow, and the environmental and social benefits will not be observed in the short- and medium-term." Well-supported public policy changes can bridge the gaps between benefits for governmental and commercial interests. In January 2017, Mexico's Secretary of Energy Joaquin Coldwell released a report summarizing the findings [53].

21.7 Are Lessons Learned in the Developing World Applicable to the U.S.?

Yes! Leadership and vision are critical; if, at present, it does not come from the top leaders, it can come from the private sector advocating for and working with the state and local governments. In a recent report to its shareholders, a U.S. energy company relies upon the studies and expertise that national and international groups provide, employing the information in much the same way as developing countries. The U.S. also contributes to best practices. A study done in western Texas is now the basis of an international model that can be modified based on each country's unique circumstances. It supports integrated transmission planning and renewable energy generation development [52]. Co-produced by USAID and NREL, *The REZ Process: A Guidebook for Practitioners* [54], is based on the Texas 2008 initiative, Competitive Renewable Energy Zones Transmission Optimization Study.

As we globally accept the challenge of building a low CO_2e energy supply, the free exchange of shared information helps build the political will to free the flow of funds and deploy new technologies.

21.8 Can Energy Professionals Do More?

Organizations that support renewable energy expansion are found in nearly every country in the world. Individual researchers and consortia of organizations across countries, regions, and continents are pooling their resources. Their members conduct research, providing tools and analysis, and promoting transparency through information exchange. The information is shared through print, digitally, via human interaction, or at ministerial meetings. The commitment to sharing information and expertise is ubiquitous. As practitioners, there are a multitude of opportunities to share expertise. The opportunities are as varied as working with a local community group desirous of moving to deploy renewable energy systems, serving on advisory boards, educating elected officials about policy development, or reaching out internationally to help not-for-profit entities. Sharing expertise, best practices, and lessons learned helps us reach the goals of renewable energy application and deployment.

21.9 Conclusions

Costa Rica, Uruguay, and India provide important examples for other developing and developed countries that seek to effectively use renewable energy and allay their dependence on fossil fuels. There are similarities in actions that helped each country to rapidly progress in its adoption of renewable energy: 1) leadership with a vision and the will to develop policies that support the adoption and use of renewable energy, 2) a government that is a stable partner to private investment in renewable energy, and 3) commitment to electrify the country's transportation systems.

Are lessons learned and best practices being shared effectively? The lessons and best practices freely communicated in print and digitally, as well as through worldwide conferences, webinars, and workshops, provide evidence that they are being shared effectively. There is a commitment from both governmental and non-governmental organizations to provide tools and resources for accurate information to those making energy policy, including funding commitments, in their respective countries.

Can those in remote regions learn what worked in places 1,000s of kilometers away? In this era, when most of the world relies on mobile phones to access information, there is irony in the fact that access to electricity dictates when and if personal phones are charged. However, those who work in developing countries are creating their own innovative solutions, new technologies, and business models suited to bringing electricity to the world's remote regions. These models and technologies can be replicated throughout the world.

What tools are available to a developer interested in privately funding a project? Can data and lessons learned be freely shared in the private sector? The private sector can access tools, critical data, and analysis that assist in the planning, funding, and implementation of renewable energy projects. Access to proprietary information can be challenging. There are many examples of successful renewable energy projects of varied sizes and locations. Private entities wanting to find viable opportunities now have access to accurate and timely information. In addition, the most successful countries are those where the policies clearly support partnerships with the private sector, and the lessons learned are not being lost on countries that have yet to develop similar ones.

What more can we as energy professionals do to provide needed expertise and best practices to the developing world? The answer is a clear and resounding, "get involved!" There are needs for every level of expertise. Opportunities abound to help both locally and internationally. The time is now to become part of the information exchange.

Are there lessons we can apply to the U.S. from the developing world? Though the U.S. government has backed away more than once from taking a leadership role in passing policies to support and promote renewable energy, there is much that the U.S. can and does learn from other countries. States and cities, along with private companies, continue to independently exercise strong commitments to increase the use of renewable energy. Those who lag will have the benefit of learning from their neighbors.

The sentiments of Monica Araya of Costa Rica Limpia (Clean Costa Rica) express what we must do. "It is important to communicate that the situation is tough, but it is also important to pivot to resilience and to ideas of what is possible for us to protect ourselves. Even bigger countries like India have told me, 'Maybe India can't move forward the same way that Costa Rica can, but that doesn't mean that a city in India the size of Costa Rica cannot think big and move faster to clean energy.' That was a very empowering idea… In my country, if you want to get people excited, you have to say that this will make us a country that could inspire others" [9].

References

[1] International Renewable Energy Agency, Abu Dhabi (2019). Renewable capacity statistics 2019.

[2] International Renewable Energy Agency, Abu Dhabi (2025). Renewable capacity statistics 2025.

[3] Organization for Economic Co-operation and Development. Development finance institutions and private sector development. http://www.oecd.org/dac/stats/development-finance-institutions-private-sector-development.htm, accessed 15 September 2019.

[4] Cochran, J., Bird, L., Heeter, J. and Arent, D. (2012, April). Integrating variable renewable energy in electric power markets: best practices from international experience. National Renewable Energy Laboratory. https://www.nrel.gov/docs/fy12osti/53732.pdf, accessed 25 January 2020.

[5] Rapid Transition Alliance Staff (2019, August 8). Can Costa Rica's path to carbon neutrality be replicated by other countries? Rapid Transit Alliance. https://www.resilience.org/stories/2019-08-08/can-costa-ricas-path-to-carbon-neutrality-be-replicated-by-other-countries, accessed 14 September 2019.

[6] Santos, L. (2025, August 26). The electric car taking Costa Rica by storm for its low ownership cost. Mobility Portal Europe. https://mobilityportal.eu/electric-car-costa-rica-ownership-cost, accessed 26 August 2025.

[7] Tellier, G. (2025, February 27). Costa Rica renewable energy: a leader in sustainability. Costa Rican Immigration Experts. https://crie.cr/costa-rica-renewable-energy-a-leader-in-sustainability, accessed 21 August 2025.

[8] The Global Environment Facility. https://www.thegef.org/newsroom/publications/gef-glance, accessed 27 August 2025.

[9] Belkhyr, Y. (2018, July 18). In conversation. Electric and empowered: Monica Araya on Costa Rica's clean energy future. Ted Blog. https://blog.ted.com/electric-and-empowered-monica-araya-on-costa-ricas-clean-energy-future, accessed 2 September 2019.

[10] Admin (2024, December 9). Costa Rica's electric vehicle revolution: a quiet but rapid transformation. *Costa Rican Times*. https://www.costaricantimes.com/costa-ricas-electric-vehicle-revolution-a-quiet-but-rapid-transformation/78960, accessed 26 August 2025.

[11] Gonzalez, M. (2018, November). Championing Costa Rica's transition to clean transportation systems. The World Bank. https://www.worldbank.org/en/news/feature/2018/11/01/championing-costa-ricas-transition-to-clean-transportation-systems, accessed 3 September 2019.

[12] Watts, J. (2015, December 5). Uruguay makes dramatic shift to nearly 95% electricity from clean energy. *The Guardian*. https://www.theguardian.com/environment/2015/dec/03/uruguay-makes-dramatic-shift-to-nearly-95-clean-energy, accessed 28 August 2019.

[13] The National Administration of Power Plants and Electrical Transmissions (Spanish: Administración Nacional de Usinasy Trasmisiones Eléctricas, aka UTE), is Uruguay's government-owned power company.

[14] Bnamericas (2018, April 4). Uruguayan wind power takes top spot in March generation figures. https://www.bnamericas.com/en/news/uruguayan-wind-power-takes-top-spot-in-march-generation-figures, accessed 25 August 2019.

[15] Bnamericas (2018, June 1). Uruguay launches electromobility program. https://www.bnamericas.com/en/news/uruguay-launches-electromobility-program, accessed 15 September 2019.

[16] Joint SDG Fund (2025, May 5). Electrifying public transport: REIF powers Uruguay's sustainable mobility. https://jointsdgfund.org/article/electrifying-public-transport-reif-powers-uruguays-sustainable-mobility, accessed 2025, 26 August 2025.

[17] Martinez, J. (2025, January 26). Uruguay leads Latin America in electric vehicle adoption. https://www.riotimesonline.com/uruguay-leads-latin-america-in-electric-vehicle-adoption/, accessed 26 August 2025.

[18] Sing, K. (2018, January 24). India's switch from environmental victim to renewable energy champ focusing on cooperation, not blame has turned the subcontinent into a green power pioneer. Center for Strategic and International Studies. https://www.csis.org/analysis/indias-switch-environmental-victim-renewable-energy-champ, accessed 14 September 2019.

[19] Partnership to Advance Clean Energy-Deployment. Technical assistance program. https://www.usaid.gov/sites/default/files/documents/1861/PACE-D-Brochure-Feb-2016-1.pdf, accessed 16 September 2019.

[20] Kumar, S. (2025, February 13). How India is paving the way for a sustainable energy future. Ernst Young. https://www.ey.com/en_in/insights/energy-resources/how-india-is-paving-the-way-for-a-sustainable-energy-future. Accessed 28 August 2025.

[21] Tongia, R. (2018, September 18). Understanding India, its energy needs and ambitions, and the global implications for carbon emissions. Brookings. https://www.brookings.edu/blog/planetpolicy/2018/09/18/understanding-india-its-energy-needs-and-,ambitions-and-the-global-implications-for-carbon-emissions, accessed 12 September 2019.

[22] EarthSpark. Solidarity in Southern Haiti. http://www.earthsparkinternational.org, accessed 3 September 2025.

[23] EarthSpark. http://www.earthsparkinternational.org/our-model.html, accessed 3 September 2019.

[24] EarthSpark (2019, December 21). Tiburon Turns On! EarthSpark International launches second solar microgrid in Haiti, sets precedent. http://www.earthsparkinternational.org/blog/previous/2, accessed 1 September 1, 2025.

[25] Archambault, A. (2024, November 12). We did it! Sharing determination + sunshine. https://www.earthsparkinternational.org/blog/we-did-it-sharing-determination-sunshine, accessed 3 September 2025.

[26] Spark Meter. https://www.sparkmeter.io, accessed 4 September 2019.

[27] Setting up South Sudan's first solar microgrid (2023, July 25). http://www.earthsparkinternational.org/blog, accessed 1 September, 2025.

[28] United Nations Sustainable Development Goals. About the sustainable development goals. https://www.un.org/sustainabledevelopment/sustainable-development-goals, accessed 16 September 2019.

[29] Archambault, A., McManus, R. and Madison, S. (2018, November 12). Feminist electrification: gender planning for pro-women outcomes in rural

electrification. http://www.earthsparkinternational.org/blog/feminist-electrification, accessed 4 September 2019.
[30] United Nations Climate Change. https://unfccc.int/ accessed, 5 September 2019.
[31] International Renewable Energy Agency. https://www.irena.org, accessed 5 September 2019.
[32] International Renewable Energy Agency (2018). Renewable energy and jobs annual review 2018, https://www.irena.org/publications/2018/May/Renewable-Energy-and-Jobs-Annual-Review-2018, accessed 5 September 2019.
[33] International Renewable Energy Agency (2019). Renewable energy statistics. https://www.irena.org/publications/2019/Jul/Renewable-energy-statistics-2019, accessed 5 September 2019.
[34] International Renewable Energy Agency (2019). Renewable energy assessments. https://www.irena.org/publications/2019/May/Renewable-power-generation-costs-in-2018, accessed 5 September 2019.
[35] International Renewable Energy Agency. Publications. https://www.irena.org/publicationsearch?initiative=7c1c6e32deb6440cb-76c0287a1d5605f, accessed 5 September 2019.
[36] International Renewable Energy Agency. Global atlas. https://www.irena.org/globalatlas, accessed 5 September 2019.
[37] International Renewable Energy Agency. Benefits. https://www.irena.org/benefits, accessed 5 September 2019.
[38] International Renewable Energy Agency. REMap. https://www.irena.org/publications/Our-Collections#technology_briefs, accessed 5 September 2019.
[39] International Renewable Energy Agency. Project navigator. https://www.irena.org/navigator, accessed 5 September 2019.
[40] International Renewable Energy Agency. Sustainable energy marketplace. https://www.irena.org/marketplace accessed, 5 September 2019.
[41] International Renewable Energy Agency. Abu Dhabi fund for development. IRENA/ADFD Project facility fund. https://www.irena.org/ADFD, accessed 5 September 2019.
[42] International Renewable Energy Agency (2019, June 6). Online workshop on identifying challenges for achieving 100% renewable power systems by mid-century. https://irena.org/events/2019/Jun/Online-workshop-on-identifying-challenges-for-achieving-100pc-renewable-power-systems-by-mid-century, accessed 5 September 2019.

[43] United Nations Climate Change (2019, August 1). IRENA and UN Climate Change join forces to accelerate renewables as climate solution. UN Climate Press Release. https://unfccc.int/news/irena-and-un-climate-change-join-forces-to-accelerate-renewables-as-climate-solution accessed, 5 September 2019.
[44] Low Emissions Development Strategy Global Partnership (LEDS). http://www.ledsgp.org/?loclang=en_gb, accessed 5 September 2019.
[45] American Cities Climate Challenge Renewables Accelerator. https://cityrenewables.org, accessed 6 September 2019.
[46] American Cities Climate Challenge. Portland, Oregon. https://cityrenewables.org/story/portland, accessed 6 September 2019.
[47] National Renewable Energy Laboratory. https://www.nrel.gov, accessed 6 September 2019.
[48] National Renewable Energy Laboratory. Geospatial data science data and visualization tools. https://www.nrel.gov/gis/tools.html, accessed 6 September 2019.
[49] National Renewable Energy Laboratory. Geospatial data science maps. https://www.nrel.gov/gis/maps.html, accessed 6 September 2019.
[50] National Renewable Energy Laboratory. Geospatial data science data resources. https://www.nrel.gov/gis/data.html, accessed 6 September 2019.
[51] National Renewable Energy Laboratory. System advisor model. https://sam.nrel.gov, accessed 7 September 2019.
[52] Billo, J. and Lee, N. (2017, September 6). Transmission planning for a high renewable energy future: lessons from the Texas competitive renewable energy zones process, webinar transcript. Clean Energy Solutions Center. https://cleanenergysolutions.org/sites/default/files/documents/2017-09-06-transcript.pdf, accessed 25 January 2026.
[53] Zinaman, O., Aznar, A. and McCall, J. (2018, May). Clean Energy Transition Partners, LLC, and National Renewable Energy Laboratory. Informing Mexico's distributed generation policy with system advisor model (SAM) analysis. https://www.nrel.gov/docs/fy18osti/71038.pdf, accessed 7 September 2019.
[54] Lee, N., Flores-Espino, F. and Hurlbut (2017, September). Renewable energy zone (REZ) transmission and planning process: a guidebook for practitioners. National Renewable Energy Laboratory and United States Agency for International Development. https://docs.nrel.gov/docs/fy17osti/69043.pdf, accessed 25 January 2026.

Index

Continued

About the Editor

Dr. Stephen A. Roosa has over 35 years of experience in commercial energy management, energy engineering, and performance contracting. During his career, he worked for over a decade as the corporate energy manager for a Fortune 100 company and afterwards in various business development capacities with several leading energy services companies. His experience includes renewable energy projects and thousands of energy studies. He has developed over $500 million in energy conservation, efficiency, management, and sustainable development projects for his customers. He is a Certified Energy Manager (CEM), Certified Sustainable Development Professional (CSDP), Certified Building Commissioning Professional (CBCP), Certified Measurement and Verification Professional (CMVP), Certified Demand Side Management (DSM) specialist, Certified Business Energy Professional (BEP), Certified Renewable Energy Professional (REP), Certified Carbon Auditing Professional (CAP), Corps of Engineers' Certified Energy Monitoring and Control Systems Designer, and a LEED Accredited Professional.

Stephen holds a Ph.D. from the University of Louisville, a Master of Business Administration from Webster University, and a Professional Bachelor of Architecture degree from the University of Kentucky. He co-instructed his first graduate-level course at the University of Kentucky at the age of 19, implementing Kentucky's first course in the use of the built environment as an educational medium. He has since taught graduate and undergraduate courses at leading U.S. universities. He has developed and delivered live seminars, workshops, and presentations throughout the U.S and in Norway, Ireland, Italy, Hungary, Egypt, Saudi Arabia, Jordon, Kuwait, the Maldives, the Dominican Republic, India, the Russian Federation, and South Africa.

Stephen is the former Editor-in-Chief of the *International Journal of Strategic Energy and Environmental Planning*. He is the present Editor-in-Chief of the *International Journal of Strategic Energy and Environmental*

Planning and the *Alternate Energy and Distributed Generation Journal.* He serves on the Editorial Board of the *International Journal of Energy Management*. His published books include *International Solutions to Sustainable Energy: Policies and Applications, Microgrid Fundamentals: Development and Implementation, Carbon Reduction: Policies, Strategies and Technologies, The Sustainable Development Handbook*, and the *Energy Management Handbook*, 9th edition. He has published over 60 journal articles relating to energy conservation, energy engineering, energy management, alternative energy, and sustainable development.

Dr. Roosa is a past President of the Association of Energy Engineers (AEE). He currently serves as its Director of Sustainable State and Local Programs, as Chairman of both the AEE's Certified Sustainable Development Professional Board and Renewable Energy Professional Certification Board. He is a member of the World Energy Engineering Congress Advisory Board and served on the Certified Energy Manager certification board for over 10 years.

Stephen has received numerous related awards during his career including the AEE International Energy Manager of the Year Award, the AEE International Energy Project of the Year Award, the U.S. Army Corps of Engineers National Energy Engineers Systems Technology Award, the U.S. Army Energy Conservation Award, the Federal Energy Water & Management Project Award, U.S. Department of the Defense Citation of Achievement, and the U.S. Joint Chiefs of Staff Citation for Energy Management. He is a member of the Energy Manager's Hall of Fame, a Legend in Energy, and an AEE Fellow. He was a two-time recipient of the Rockefeller Family Funds, Energy and Environmental Education Award, and a four-time recipient of the Kentucky Colonel's Medal of Honor. In 2025, he was selected as runner-up for the AEE's Energy Innovator of the Year Award.

For Product Safety Concerns and Information please contact our
EU representative GPSR@taylorandfrancis.com Taylor & Francis
Verlag GmbH, Kaufingerstraße 24, 80331 München, Germany